Mathematics Education

The Singapore Journey

SERIES ON MATHEMATICS EDUCATION

Published

Vol. 1 How Chinese Learn Mathematics
Perspectives from Insiders
Edited by: L. Fan, N.-Y. Wong, J. Cai and S. Li

Vol. 2 Mathematics Education
The Singapore Journey
Edited by: K. Y. Wong, P. Y. Lee, B. Kaur, P. Y. Foong and S. F. Ng

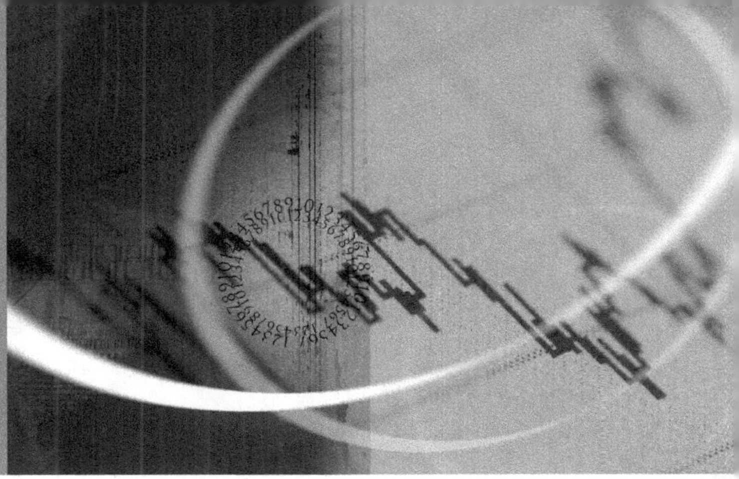

Series on Mathematics Education Vol. 2

MATHEMATICS EDUCATION

The Singapore Journey

Edited by

Wong Khoon Yoong
Lee Peng Yee
Berinderjeet Kaur
Foong Pui Yee
Ng Swee Fong

Nanyang Technological University, Singapore

World Scientific

NEW JERSEY • LONDON • SINGAPORE • BEIJING • SHANGHAI • HONG KONG • TAIPEI • CHENNAI

Published by

World Scientific Publishing Co. Pte. Ltd.
5 Toh Tuck Link, Singapore 596224
USA office: 27 Warren Street, Suite 401-402, Hackensack, NJ 07601
UK office: 57 Shelton Street, Covent Garden, London WC2H 9HE

British Library Cataloguing-in-Publication Data
A catalogue record for this book is available from the British Library.

Series on Mathematics Educaiton — Vol. 2
MATHEMATICS EDUCATION
The Singapore Journey

ISBN-13 978-981-283-375-4
ISBN-10 981-283-375-7

Managing Editor
LEE Peng Yee, Nanyang Technological University, Singapore

Editorial Assistant
JIN Haiyue, Nanyang Technological University, Singapore

Contents

Part III Comparative Studies in Mathematics Education

Introducing the Landscape of the Singapore Mathematics Education Journey

Mathematics education encompasses both practices and research. This book is a state-of-the-art review, though not an encyclopaedia or a handbook, of current practices and research studies in mathematics education and mathematics teacher education in Singapore. It traces the journey from the development of the Singapore mathematics curriculum in the 1950s to the present day and reports on how that curriculum is put into practice and what research can enlighten the readers about the effects of those practices. This review extends the first state-of-the-art review on mathematics education in Singapore published in 1991 (Chong, Khoo, Foong, Kaur & Lim-Teo, 1991). This former review was commissioned by the Southeast Asian Research Review and Advisory Group (SEARRAG), and it included only summaries of research studies but did not describe practices. During the intervening one and half decades, many changes have taken place in the practices of mathematics education in Singapore, including several revisions of the mathematics curriculum with reduction of topic coverage, infusion of Information and Communications Technology (ICT) into mathematics instruction, recruitment and training of trainee teachers and in-service teachers by the National Institute of Education (NIE), and the important roles of mathematics education within the national education system. The education research culture in Singapore has also blossomed with more mathematics educators at NIE participating in international comparative projects, an increasing number of research students in mathematics education at NIE (at the time of writing, there are 16 PhD students in mathematics education), higher expectations on mathematics educators at

NIE to conduct research and to publish, stronger funding of educational research from the Singapore Ministry of Education (MOE), in particular, the establishment of the multi-million dollar Centre for Research in Pedagogy and Practice (CRPP) at NIE, and a growing awareness and acceptance of senior education administrators and policy-makers of the contributory functions of high-quality research studies in the debates to formulate policies about critical education issues to prepare Singapore students for the uncertain future.

Members of the editorial team have duly recognised the above trends and needs, and have invited or commissioned mathematics educators and senior mathematics teachers in Singapore to contribute chapters on different aspects of this education journey. Each chapter has been reviewed by at least two editorial members. Collectively the chapters describe the mathematics education landscapes, relate personal experiences, peep into classroom snippets, offer historical sketches, tell well-worn or novel stories, report heavy statistics, and much more, including the important theme of students' learning in most of the chapters. These are idiosyncratic voices describing some features of the Singapore mathematics education landscape from the authors' perspective; they do not represent any official views. Nevertheless, we hope that the issues these authors so earnestly explore in this book will be of interest not only to local readers, in particular graduate students and researchers, but also to the international mathematics education community, because many local studies are not readily available in international publications. Policy-makers and readers outside of mathematics education may be able to tease out relevant lessons in their own domains beyond mathematics. However, we also wish to reiterate that what has apparently worked for Singapore mathematics education at a particular period in its history may not work elsewhere or at different times, a fundamental belief that has guided many of us in what we have learned and adapted from other countries and domains of knowledge.

There is no one "right" way to begin a journey, nor is there one "best" way to read a huge book like this one about an education journey. Nevertheless, when one embarks on a journey, it is prudent to have on hand a good map that indicates the major signposts of the terrain. These signposts can inform the travellers about the peaks and valleys, forests and deserts, islands and currents along the journey. For many travellers, be they tourists or natives, such a map or guidebook, either printed or mental, also helps them to visualise the relationships among the objects and scenes encountered along the way. In the case of an abstract journey of how Singapore students have become very competent in mathematics or how Singapore has developed a world-class education system in the eyes of some international educators (e.g., Barber & Mourshed, 2007; National Mathematics Advisory Panel, 2008), there are actually many challenges along the timeline and in the curriculum space. The chapters in this book describe these challenges narrated in many cases by authors who have participated in national curriculum development, published mathematics textbooks, and conducted the research studies they are writing about. Furthermore, what the readers finally know about certain aspects of the journey depends to a great extent on who are the narrators; hence, the biographical data given at the back of the book should be read together with the respective chapters. The rest of this introductory chapter describes our mental map for the chapters.

The 23 chapters are divided into three main parts. Part I contains six chapters about the Singapore education system and mathematics teacher education. A good place to begin especially for readers who are unfamiliar with the Singapore education system is Chapter 1 by Wong and Lee N. H. This chapter traces the development of the Singapore education system and its mathematics curriculum over the past five decades. It explains the terminologies commonly used to describe this system and several major initiatives introduced by the Singapore MOE during this period. For example, terms such as *Special, Express, Normal, model drawing method, the Pentagon framework, PSLE,* and *clusters,* have specific local meanings and they will appear frequently throughout the book. Understanding this background information will help the readers follow the discussions in the other chapters of the book.

Singapore teachers have been known to work very hard, and a recent McKinsey report (Barber & Mourshed, 2007) has highlighted the important contributions of high-quality teachers in a world-class education system. Chapter 2 by Lim-Teo describes how mathematics teachers are trained at NIE in programmes that range from pre-service to in-service to graduate levels. The chapter also contributes to international discussion about balancing the roles of mathematics content knowledge, pedagogical content knowledge, and theory-practice link in teacher education, in particular, drawing on some findings of a major study led by the author on mathematics pedagogical content knowledge (MPCK) among novice primary school mathematics teachers.

Helping teachers to develop through a learning community that stresses mutual support has been strongly promoted by many educators. Chapter 3 by Chua P. H. explains several measures undertaken in Singapore along this path, giving details about Learning Circles, action research, and the functions of the Centre of Excellence on Mathematics at the East Zone. An emerging form of such a learning community in Singapore is the adaptation of "lesson study" that has its origin in Japan but has since spread to several countries. In Chapter 4, Fang, Lee C. K. E., and Sharifah document their experiences in trialling this form of professional development in a primary school, working on lessons in long division, area and perimeter, and equivalent fractions at Primary 3 and 4 levels. Another way to bring about teacher change in pedagogy is to engage them as collaborators in research. Yeap and Ho in Chapter 5 explore this approach by describing three case studies and situating it within their 4-I Model of teacher change. Singapore MOE recognises the need to appoint Master Teachers who have shown very strong pedagogical practices in the disciplines, as catalysts to promote teacher professional learning by organising various activities. Currently there are only two Master Teachers in mathematics, and in Chapter 6, Ng J. D. and Foo K. F., as Master Teachers at the primary and secondary level respectively, explain the nature of this scheme and sketch their own personal journeys as teacher leaders and curriculum leaders at the cluster level.

Mathematics teaching and learning covers numerous features and this diverse topography adds richness to the journey. Part II consists of 12 chapters that offer interesting details (as in close-up shots) and more macro descriptions (as in medium shots) of these features.

A unique Singapore mathematics landscape is the model drawing method. It was created in the 1980s to help students solve so-called "challenge problems" or algebra word problems without the use of algebra. This systematic method has now attracted some attention from the international community. However, the experience of teaching this method in primary schools over the past twenty years has raised several issues in curriculum design, assessment, and instruction (see Chapter 1) that have attracted the attention of several researchers. There are two chapters about this model drawing method: Chapter 7 by Ng S. F. and Lee K. takes a pedagogical perspective drawing on a series of studies of students' work in solving algebra word problems, and Chapter 8 by Lee K. and Ng S. F. examines the role of working memory using functional neuro-imaging technologies to conclude that symbolic algebraic method demands more of working memory than the model drawing method. This latter research attempts to link problem solving behaviours to physiological basis, a path of applying neuroscience to study important education problems.

The next two chapters (9 and 10) deal with two specific mathematics topics and the studies were conducted by recent PhD graduates. In Chapter 9, Wu and Wong provide data about the performance of a large sample (n = 907) of secondary school students in graph reading, graph interpretation, graph construction, and graph evaluation. A concern is that many students could not explain their statistical reasoning with statistical graph items. In Chapter 10 about another large-scale survey (n = 1002) of Primary 6 to Secondary 2 students, Jiang presents the problem solving strategies used by these students to solve three word problems on speed and their common errors. Indeed, there are many more studies of problem solving heuristics, learning difficulties, and errors among Singapore students that have not been included in this book. Nevertheless, to provide further glimpses of this part of the terrain, Foong in Chapter 11 provides a narrative review of more than 60 studies about mathematics problem solving, conducted mostly by graduate students. It is not possible to give a one-line summary this complex landscape, and it suffices here to just highlight the need for researchers to work closely with teachers to address their beliefs, practices, and dilemmas to teach mathematics via problem solving or to teach about problem solving.

In line with international trends, Singapore has invested billions of dollars into equipping schools with essential and adds-on ICT resources, developing e-learning materials, training teachers to integrate the technologies into mainstream teaching, and introducing policies of ICT use, for example, allocating 30% of curriculum time to embed ICT. International findings about ICT use in mathematics education have been equivocal, and large-scale studies tend to report lack of impacts of ICT on student achievement (e.g., Dynarski et al., 2007). In Chapter 12, Ng W. L. and Leong describe the Singapore practices and report some findings on how ICT has been harnessed in mathematics teaching and learning. They take note of the many unresolved issues about ICT use down the technological path and stress the need to use ICT not as a technological toy but as a tool that can promote active learning.

Although the learning of mathematics involves mainly cognitive processes, there is convincing evidence and strong theoretical arguments to show that emotions, attitudes, values, and other affective characteristics held by the students can influence the effectiveness of their learning. Two chapters address this issue. Chapter 13 by Yeo reports a weak correlation of 0.39 between mathematics anxiety (as measured by the Fennema-Sherman Mathematics Anxiety Scale) and test anxiety (as measured by the Test Anxiety Inventory) among a group of 621 Secondary 2 students. In Chapter 14, Lui, Toh, and Chung describe a relationship between the perceptions held by about 700 Secondary 2 students about the social climate of their classrooms and their self-concept in mathematics. In a follow-up action research project, two groups of Secondary 2 students engaged in several cooperative learning activities, and their feelings were generally positive. Studies that extend from those reported in these two chapters should examine the relationships between affective variables and mathematics achievement at different grade levels.

Students differ widely in their aptitude for mathematics. Singapore has followed international practices to develop special curriculum for the gifted and for those who need special assistance. Khong describes in Chapter 15 several programmes that cater to the needs of the gifted and the talented in mathematics. The key approach is through curriculum differentiation, and 11 types of differentiation are briefly explained, ranging from using challenging problems to out-of-class activities. On the other hand, in Chapter 16, Cheam and Chua W. L. J. explain the Learning Support for Mathematics (LSM) framework that MOE has developed in the past few years to include the types of experiences to be offered to lower primary pupils who have been identified through normed tests to need special remediation in mathematics. This framework is a 4-pronged intervention approach covering cognition, metacognition, motivation, and environment. These "special needs" students might have missed crucial learning and development experiences in kindergartens and pre-schools, which are not part of the national, formal education system. Sharpe in Chapter 17 describes three studies about numeracy at these early years, showing that some kindergarten teachers were not knowledgeable about how young children think about number work. She proposes several activities that are more developmental in nature. Indeed, there is only sketchy information about the terrain of early numeracy in Singapore.

Ever since Singapore students achieved top performance in mathematics in the TIMSS series of international comparison in 1995, 1999, and 2003, a constant stream of overseas educators and delegations has visited Singapore trying to find out the factors that might contribute to this success. However, within Singapore, the main education stakeholders, namely, students, parents, teachers, schools, and ministry officials, do not place much emphasis on TIMSS findings to set the benchmarks for success of the education system. The local stakeholders are more likely to use as main indicators the performance of students in the public examinations, namely, the Primary School Leaving Examination (PSLE), Singapore-Cambridge General Certificate of Examinations at Ordinary level, Normal level, and Advanced level. This book does not provide any in-depth analysis of these important examinations, but the interested readers can locate the examination results from the *Singapore Education Statistics Digest* at http://www.moe.gov.sg/. The evolvement of the Singapore examination systems is clearly explained in the recently published book by Tan, Chew, and Goh (2008). Set against this national examination system, Quek and Fan argue in Chapter 18 for the need to devise a new examination paradigm to encompass what is now known as "alternative assessment." They then describe a large study that incorporated four types of alternative assessment (journal writing, self-assessment, oral presentations, and projects) in classroom lessons.

The five chapters in Part III cover four international comparative studies that Singapore has participated in. The performance of Singapore students in TIMSS is dealt with by Kaur in Chapter 19. Although Singapore students scored very well in many TIMSS items, Kaur cautions that a sizeable proportion of these students are still weak in solving problems unfamiliar to them in terms of context, language, and format. She proposes no fewer than nine possible factors of success, but TIMSS analyses conducted to-date have not been able to provide convincing direct links or effects between these factors and mathematics achievement. In the case of background variables, Boey in Chapter 20 provides some findings from TIMSS 2003 on seven categories that cover students' attitudes toward mathematics, home resources, and school environments.

Besides TIMSS, mathematics educators in Singapore have also participated in smaller comparative studies. Chapters 21 (Kaur and Yap) and 22 (Kaur, Koay, and Yap) describe two longitudinal studies undertaken in the past decade. The Kassel project, reported in Chapter 21, followed more than 2000 students from Secondary 2 (13+) in 1995 to Secondary 3 (14+) in 1996, whereas the International Project on Mathematical Attainment (IPMA) (see Chapter 22) tracked the mathematics performance of about 800 students from Primary 1 (6+) in 1999 to Primary 5 (10+) in 2003 on a yearly basis. Longitudinal studies are rare in education, so these two chapters should be of particular interest to mathematics educators. Unlike the other studies that concentrate on students' mathematics performance, the last chapter in Part III by Wong, Kaur, Koay, and Jamilah reports on students' perceptions about their "best" mathematics teachers using the relatively new "Draw a teacher" technique, supplemented with written responses.

With this necessarily brief introduction, we invite you to embark on the Singapore mathematics education journey, beginning with the terrain that best captures your interests or needs. Bon voyage. In the final concluding chapter, we will discuss new landscapes that invite local explanation and international collaboration.

Editorial Board
May, 2008

References

Barber, M., & Mourshed, M. (2007). *How the world's best-performing school systems come out on top*. London: McKinsey & Co.

Chong, T. H., Khoo, P. S., Foong, P. Y., Kaur, B., & Lim-Teo, S. K. (1991). *A state-of-the-art review in mathematics education in Singapore*. Singapore: Institute of Education.

Dynarski, M., Agodini, R., Heaviside, S., Novak, T., Carey, N., Campuzano, L., Means, B., Murphy, R., Penuel, W., Javitz, H., Emery, D., & Sussex, W. (2007). *Effectiveness of reading and mathematics software products: Findings from the First Student Cohort. Report to Congress*. Washington, DC: U.S. Department of Education, Institute of Education Sciences. Retrieved August 31, 2007 from http://ies.ed.gov/ncee/pdf/ 20074005.pdf

National Mathematics Advisory Panel (2008). *Foundations for success: The final report of the National Mathematics Advisory Panel*. Washington, DC: Author.

Tan, Y. K., Chow, H. K., & Goh, C. (2008). *Examinations in Singapore: Change and continuity (1891 – 2007)*. Singapore: World Scientific.

Part I

Singapore Education and Mathematics Teacher Education

Chapter 1

Singapore Education and Mathematics Curriculum

WONG Khoon Yoong **LEE Ngan Hoe**

Education is about opening doors for our children, and giving them hope and opportunities. It is more than filling a vessel with knowledge – it is to light a fire in our young people.

Prime Minister Lee Hsien Loong, 12 August 2004[1]

The first part of this chapter describes the key features of the Singapore education system. This description sets the background for the second part of the chapter, which is about the Singapore mathematics curriculum. Mathematics contents and public examinations in mathematics have remained fairly stable for the past four decades, even though emphases have changed in tandem with national priorities. These include greater attention to higher order thinking and reasoning skills, alternative assessment techniques, group work, mathematics communication, and use of Information Technology and calculators. Several lessons about mathematics education reforms are raised throughout the chapter and links to relevant chapters in the book are indicated.

Key words: education system, "Pentagon" framework, model method, mathematics textbooks, Thinking Programme

1 OVERVIEW OF CHAPTER

This chapter has two main sections to give readers who are not familiar with the Singapore education system an overview to understand the

[1] Cited in Ministry of Education (2006a), p. 8.

issues discussed in the other chapters of this book. The first section describes the structure and aims of the Singapore education system. This provides a sketch of the numerous changes and initiatives implemented over the past four decades to realise the twin aims of the national desired outcomes of education, namely to "develop the individual and educate the citizen." A striking message from this first section is that the Singapore Ministry of Education (MOE) will continue to make changes, at times quite drastically, to respond to rapid challenges from the economy, politics, culture, and religious practices of the nation, spurred on by unremitting globalisation and international upheavals such as terrorism, epidemic, fundamentalism, and international benchmarking of educational achievement in key subjects such as mathematics, science, and languages.

The second section explicates the nature of the Singapore intended mathematics curriculum from primary to pre-university levels. It traces the development of this national curriculum from the 1960s based on the British mathematics curriculum to a new conceptualisation in the 1990. Despite several reforms to the mathematics curriculum, the major mathematics topics and the assessment practices have been retained in the past four decades.

For the past ten years, MOE has made almost annual changes, both minor and major ones, to the education system, with the consequence that some of the information given in this chapter may become obsolete pretty soon. Thus, we aim to depict the big picture in this chapter and encourage the interested readers to check the latest official information at the following four comprehensive websites. Unless otherwise stated, specific information given in this chapter is from the following sources:

- Singapore Ministry of Education (http://www.moe.gov.sg/): education statistics (http://www.moe.gov.sg/esd/ESD2007.pdf), curriculum documents, school directory, press releases, details of various programmes, etc.
- Singapore Examinations & Assessment Board (http://www.seab.gov.sg/): assessment objectives, nature of examination papers, examination timetables, etc.

- Ministry of Information, Communications and the Arts (http://www.sg/): general information about Singapore.
- Yearbook of Statistics Singapore 2007 (http://www.singstat.gov.sg/pubn/reference/yos.html).

2 SINGAPORE: A BRIEF PROFILE

Singapore is a small country of area 700 sq km, comprising one main island (88% of total area) and more than 60 surrounding islets. It does not have any natural resources of great commercial value, but its strategic geographical location, coupled with a well educated workforce, substantial foreign investment, and political stability, has made it a nation with high per capita gross national income of S$45 353 (US$32 000). The Changi airport, Keppel harbour, Jurong Bird Park, and Singapore Airlines (SIA) are some of the Singapore icons that have gained international recognition.

Around 1360, there was a settlement on the Singapore island called Temasek (Danmaxi in Chinese), meaning "water town." It gained the name *Singapura*, meaning "Lion City," in the 14th century, through a legend about a lion on the island. Four hundred years later in 1819, Stamford Raffles signed a treaty with the Sultan of Johor to establish a British settlement on Singapore. It became a self-governing state in 1959 under the People's Action Party (PAP). In September 1963, it became part of Malaysia, but shortly after this merger, it separated from Malaysia to become fully independent on 9 August 1965. It has been ruled by PAP for more than forty years.

The population in 2006 was 4.48 millions. Singapore citizens and permanent residents make up 80% of the population, and there is a relatively high proportion of foreign workers at 18%. The profile of local population by races is: Chinese (75.2%); Malays (13.6%); Indians (8.8%); Others (2.4%). The main religions practiced are Buddhism (42%), Islam (15%), Christianity (15%), and Others (28%). Government policies are implemented cautiously in order to maintain racial and religious harmony among its people of diverse backgrounds and to promote social cohesion. A crucial decision to address social cohesion was made in 1956, even before full independence, to adopt the bilingual policy: English as the language of administration because of its international standing as well as its neutral status among the different ethnic groups; the three main Mother Tongue languages, namely Malay (also the national language), Chinese, and Tamil as official languages. In 1966, the bilingual policy was applied to education so that all academic subjects are taught in English from primary to tertiary levels and the Mother Tongue language becomes a compulsory subject. As a consequence, most Singaporeans are now bilingual in English and their Mother Tongue. The general literacy rate of residents aged 15 years and above is 95.4%.

3 SINGAPORE EDUCATION SYSTEM

3.1 *Education Structure*

Singapore has a 6-2-2 school system: 6 years of primary, 2 years of lower secondary, and 2 years of upper secondary education. Beyond this, the students may continue with 2 or 3 years of pre-university education at Junior Colleges and Centralised Institutes, enrol in the polytechnics and other institutes, or join the workforce.

The plan to provide six years of free primary education for Singapore children of all races was first launched in 1947 (Lim, 2007). This phase of education became compulsory only recently in January 2003, where "compulsory school age" is above 6 years and below 15 years of age. The penalty on parents who do not send their children to schools is a fine of up to S$5000 or jail of up to a year, or both. Indeed, most parents will ensure that their children receive ten years of general education because they understand the importance of education. Singapore follows a "meritocratic" yet competitive education system "to provide equality of opportunity to all regardless of their ethnic groups or socioeconomic status" (Kang, 2005, p. 152).

Pre-school education (Nursery at age 3 and Kindergartens 1 and 2 at ages 4 and 5) is not part of the formal education system. It is provided by private organisations. These pre-school centres and their teachers must still register with MOE. The Pre-school Education Unit of MOE (2003) has developed a framework and a package of materials for the kindergarten curriculum. The desired outcomes of this curriculum include healthy habits, desirable social skills, physical coordination, and basic speaking and listening skills. About 5% of each cohort of Primary 1 pupils have not attended pre-schools (Liaw, 2006). Some of these Primary 1 pupils are found to lack basic skills in listening, writing, and counting, and they need special assistance provided by the Learning Support Programmes in English and Mathematics (see Chapter 16).

In Singapore, all public schools from primary to pre-university levels are funded by MOE. These schools are classified into four types: *government* (252), *government-aided* (65), *autonomous* (26), and *independent* (11) schools. The main difference is the degree of autonomy the schools enjoy in managing their budget, staff, admission criteria, and curriculum. The *independent* schools enjoy the highest degree of autonomy, whereas the *government* schools are the least autonomous. The *independent* and *autonomous* schools have outstanding academic records and well-rounded education programmes but they charge much higher school fees. Primary education is free for all. For *government* or *government-aided* schools, the monthly school fee is S$5 (free for Malay students) and miscellaneous fees range from S$8 to S$16. *Autonomous* schools collect monthly fees ranging from S$3 to S$18, with additional miscellaneous fees. *Independent* schools charge a monthly fee from S$125 to S$255. As an example, the monthly school fee for local students at Anglo Chinese School (Independent) is S$235 compared to only S$5 for its *government-aided* "version." Information about all the schools is available from the MOE website and the websites of individual schools. Open sharing of information in the Internet is a strong marketing feature of the "marketisation of education" in Singapore, which, as discussed by Tan (2005), leads to intense competition among schools for promising students, niche areas of excellence, and various kinds of awards offered annually by MOE. Singapore schools are placed into several clusters which in turn are grouped under four zones (see Chapter 3). These clusters and zones are given additional authority and resources to address common problems among the schools within their jurisdiction and to develop niche areas to enhance the quality of education.

In addition to the above types of schools, there are three specialised independent schools: the Singapore Sports School (started in 2004), the

National University of Singapore Mathematics and Science High School (started in 2005), and the Arts School (started in 2008). A fourth one called School of Science and Technology will begin in 2010. These specialised schools aim to widen the education choices open to those Singapore students who have specific talents and good academic abilities.

The evolvement of different types of schools reflects the shift from a centrally controlled system to one that offers greater autonomy at the school level. Until quite recently, MOE makes major decisions about the education structure, aims and objectives of the national system and individual subjects, national examinations, approval of textbooks, entry to different courses of study (formerly called *streams*), employment and training of teachers, and many other areas. For a small country like Singapore, there are certain advantages of a central system: economy of scale; common standards for promotion in the education system and career choice; shared values among pupils in a multiracial and multicultural society; and coherent and well-structured programmes. Teachers, in particular novice ones, will know what to teach and can avoid the frustrations that some teachers experience without such clear guidelines. However, in recent years, MOE has encouraged "bottom-up initiatives" and "top-down support" to allow more schools to develop niche areas within broad guidelines. Many schools look to the National Institute of Education (NIE), the sole teacher education institute in the country, and private consultancy agents to provide school-based training of their teachers to initiate and carry out their own new programmes.

In 2006, expenditure on education was S$6966 millions. The recurrent expenditure on education per student was S$4270 (US$3000, primary), S$6186 (US$4370, secondary), and S$10 000 (US$7070, pre-university). According to the McKinsey report (Barber & Mourshed, 2007), Singapore spends less than what many developed countries do on education, yet it can still provide quality education. As a comparison, the public expenditures on education per student for the US were US$8243 (primary) and US$9644 (secondary) (UNESCO Institute for Statistics, 2007).

The basic data about student enrolment and teachers are shown in Table 1. "Mixed level" schools refer to those that include different levels in the same school, for example, primary and secondary.

Table 1
Information about Students and Teachers

Level	Students	Teachers	Average class size
Primary	277 291	12 268	34.6
Secondary	200 291	10 570	36.4
Mixed level	29 176	2131	na
Pre-university	23 665	1835	23.0

The average class size at Primary 1 and Primary 2 has been reduced in recent years to enable teachers to carry out more engaged learning at these two levels through Project SEED (Strategies for Effective Engagement and Development), piloted in 2003 and fully implemented for Primary 1 in 2005. This program helps pupils to make the transition from Kindergarten to Primary 1, in particular, from informal and incidental learning in Kindergarten to formal, discipline based lessons at Primary 1, and to adjust to longer school hours at Primary 1 compared to Kindergarten.

The education structure has changed over the past four decades and it will continue to evolve in the future. This has resulted in a complex national system. However, the main features can be summarised in Table 2.

The current structure is based on ability-based streaming (called "tracking" in some education systems), implemented in early 1980 as a result of the Goh's Report in 1978 on education. This system was designed to address the unacceptable attrition rates of about 30% in schools at that time (many students could not master two languages), to raise the standard of English in line with the bilingual policy, and to maximise pupils' learning by matching the course contents and instructional pace to their assessed ability.

Primary education begins when the child is 6 years old. Primary 1 to 4 is called the *Foundation* stage, and Primary 5 and 6 the *Orientation* stage. Initially, Primary 4 pupils were assigned into the *Normal* (the more capable ones), *Extended,* and *Monolingual* streams, and a small minority (top 1% who are outstanding in all the four subjects, namely English, Mother Tongue, Mathematics, and Science) were admitted to the Gifted Education Programme (GEP; see Chapter 15). The *Normal* stream was six years long, whereas the *Extended* and *Monolingual* streams were eight years long. Over the years there was concern about the disparity

Table 2
Structure of Singapore Education (simplified)

Typical Age Range (years)	Education Level	Singapore Nomenclature
3+ to 5+	Pre-School (Informal)	Nursery, Kindergartens 1 and 2
6+ to 9+	Primary: Foundation stage	P1 – P4
10+ to 11+	Primary: Orientation stage Subject Banding	P5 – P6; PSLE (Primary School Leaving Exam)
12+ to 13+	Lower Secondary: Special/Express; Normal (Academic); Normal (Technical)	S1 – S2
14+ to 15+	Upper Secondary: Special/Express	S3 – S4 (Arts, Commerce, Science); O Level Exam
	Upper Secondary: Normal (Academic); Normal (Technical)	S3 – S4: N Level Exam N5: O Level Exam
16+ to 17+	Pre-university at Junior Colleges or Centralised Institute (CI)	JC1 – JC2; CI1 – CI3 A Level Exam
	Polytechnic	3 years Diploma
	Institute of Technical Education (ITE); vocational	1 – 2 years after O or N Level Exam
(12+ to 17+)	Integrated Programme (IP) direct to A Level Exam	
19+	University or workforce	

in ages of students in Secondary 1 from these different streams, and a decision was made in early 1990 to standardise primary education to six years. Streaming was changed to Primary 5, but admission into GEP remains at Primary 4. The labels for the three revised streams became *EM1* (more capable), *EM2,* and *EM3* (weakest), where *E* stands for English and *M* stands for Mother Tongue, and the number refers to different levels of competency in the languages. In 2004, *EM1* and *EM2* were merged into a single course of study, and *EM3* was abolished with the 2008 Primary 5 cohort. In its place is a new system of subject-based banding at Primary 5 and Primary 6, where pupils will join classes at the *Standard* or *Foundation* level based on their strengths in specific subjects. Parents will have the final say which level their children will take the subjects. At the same time, the nine primary schools with GEP will allocate 50% of curriculum time for GEP pupils to have lessons with other high ability non-GEP pupils, but they continue to have their own separate lessons in English, Mathematics, and Science. This adjustment is in response to a concern about social division through academic elitism. These changes underscore the current priority to provide greater flexibility and differentiated learning experiences within the centrally monitored system (Ministry of Education, 2006c). The ways in which these events unfold over the past few years has useful lessons for engineering reforms in education: let some schools experiment with changes based on a policy shift, note the impacts, listen to the public and parents, announce another initiative at an opportune time, and provide additional resources to support the new initiative, preferably with parental participation. This works well in Singapore because of its political stability, ample funding allocated to education, and strong parental support for education.

At the end of Primary 6, all primary pupils sit for the Primary School Leaving Examination (PSLE). The examinable subjects are English (*Standard* or *Foundation*), Mother Tongue (*Standard* or *Basic*), Mathematics (*Standard* or *Foundation*), and Science (*Standard* only). The options in PSLE show a strong alignment between curriculum and assessment for pupils of different abilities. This examination is managed by the Singapore Examinations and Assessment Board (SEAB), a statutory board established in 2004 under MOE to provide quality assessment services to Singapore and the ASEAN region. The PSLE *Standard* Mathematics paper (2 hours 15 minutes long) consists of multiple-choice questions (20%), short answer questions (30%), and structured/long answer questions (50%). Many past PSLE questions are on sale to the public (Singapore Examination and Assessment Board, 2006). This open access to the examination questions avoids the angst faced by students, teachers, and parents of some other countries due to non-disclosure of test items in high-stakes examinations, but it also leads to some Singapore teachers spending substantial amount of time and effort to drill pupils on questions similar to these past test items, with less attention given to other desirable curriculum objectives that are not readily assessed by paper and pencil tests.

Pupils who pass PSLE are divided into four courses of study and the Integrated Programme (IP) in secondary schools based on their PSLE scores and the Direct School Admission (DSA) criteria available to *independent* and *autonomous* schools and schools with approved niches of excellence. These DSA criteria consist of school-based assessment of the pupils' special talents in academic and non-academic areas. Pupils who fail PSLE may repeat Primary 6; in 2005, about one thousand pupils failed PSLE. Those who fail PSLE three times may enter vocational training centres rather than continue with secondary schooling.

The four courses of study in secondary schools are: *Special, Express, Normal (Academic),* and *Normal (Technical).* The 2006 data show that 9.0% of all Secondary 1 to 4 students were in the *Special* course, 52.8% in *Express,* 25.5% in *Normal (Academic),* and 12.7% in *Normal (Technical).* Lateral transfers across certain courses of study are allowed up to Secondary 3, and these are based on overall school performance and/or strengths in English, Mathematics, and Science. At the end of four years of schooling, students in the *Special* or *Express* courses of study sit for the Singapore-Cambridge General Certificate of Education Ordinary Level (GCE O Level) Examination, and those in the two *Normal* courses of study take the GCE N Level Examination. *Normal* students who do well in the N Level Examination may undertake one more year of study and then offer the O Level Examination.

The results of the O Level Examination are used to select students into pre-university institutes (2 to 3 years) or polytechnics (3 years Diploma). Those who enter pre-university institutes will sit for the GCE Advanced (A) Level Examination. The A Level results are used as the main criterion for entry to local and overseas universities, and between 20 – 25% of each cohort enter universities (Soh, 2005).

About 1% of each cohort of secondary school students are in the GEP. However, with the introduction of IP and other enrichment programmes for high ability students in the top secondary schools, GEP for secondary schools will end in 2008. Students in the IP programme bypass the O Level Examination and after six years of secondary schooling will sit for the GCE A Level Examination or other qualifications, for example, the International Baccalaureate (IB). The first cohort of 357 students who took the International Baccalaureate at the end of 2007 did very well, with nine of them gaining the perfect score of 45; their average score of 39.4 was higher than the world average of 30.7 (Ng, 2008).

The above descriptions show that high-stakes national examinations have considerable influence on the academic paths of the students, and to excel in these examinations can become quite stressful for some students. Singapore teachers often cite these examinations as constraints why they do not use more "innovative" teaching strategies and they feel that the examination results of their students are the major yardstick used to judge their own teaching competence. "Teaching to the test" has been criticised by educators, yet aligning the test items to curriculum objectives can help to promote *some* of these objectives. For example, including "challenge problems" in PSLE is supposed to promote higher order thinking; this will be taken up again in Section 3.3.

3.2 *The School Curriculum*

The current desired outcomes of education are: "develops the individual and educates the citizen." This twin aim is more general than the three primary objectives stated in 1989: every child to excel to the limit of his/her ability; life-long learning; proper moral values for a useful and loyal citizenship (Ministry of Education, 1989). Indeed, the school curriculum has been revised regularly to achieve these important aims.

The school year consists of four 10-week terms beginning on 2 January each year and ending in mid November. There is a one-week vacation after the first term (March) and third term (September), a 4-week vacation in June, and a 6-week break at the end of the year. Not all the 40 school weeks are used for classroom teaching because curriculum times are "lost" through public holidays, school functions, formative tests, and summative examinations.

Each school week comprises five days of study from Monday to Friday. Many primary schools run double sessions with Primary 1 and Primary 2 in the afternoon and Primary 3 to Primary 6 in the morning. Secondary school sessions are in the morning, usually from 0730 to 1300, with afternoons allocated for remedial, enrichment, and other activities. Typical curriculum time is 24 hours per week, with variations among the schools.

The curriculum subjects in primary schools are English, Mother Tongue, Mathematics, Science (from Primary 3 onwards), Social Studies (from Primary 4 onwards), Civics and Moral Education, Arts and Crafts, Music, and Physical Education. However, pupils also participate in

Co-Curricular Activities (CCAs), Community Involvement Programme (CIP), activities related to National Education (NE) (Tan & Goh, 2003a), and Interdisciplinary Project Work (PW) (see Ho, Netto-Shek & Chang, 2004), all of which aim to inculcate social-emotional and communication skills. Curriculum times for the formal curriculum subjects vary across year levels and courses of study. A typical allocation of curriculum times at Primary 5 is given in Table 3.

Table 3
Typical Allocation of Curriculum Times of Subjects at Primary 5

Subject	No. of hours per week
English	6.5
Mother Tongue	6.0
Mathematics	5.5
Science	2.0
Others	4.5
Total (per week)	24.5

Secondary school students have a wider choice of subjects. *Special* and *Express* students offer 6 to 8 subjects at the GCE O Level Examination, comprising English, Mother Tongue, Mathematics, at least one Science subject (Combined Sciences, Physics, Chemistry, Biology), Combined Humanities, and other electives. At upper secondary level (Secondary 3 and Secondary 4), students begin to specialise broadly into the sciences, arts, or commerce field through the choice of electives. The number of elective subjects has multiplied in recent years to enable students to pursue their interests. Some schools are collaborating with the polytechnics to offer Advanced Elective Modules such as digital media and entrepreneurship at the O Level standards. The compulsory non-examinable subjects are Civics and Moral Education, Arts, Music, and Physical Education. Like their primary counterparts, they participate in CCA, CIP, NE-related activities (Tan & Goh, 2003b), and PW. A typical curriculum time allocation for Secondary 2 *Express* and *Normal (Technical)* is given in Table 4.

The *Normal (Academic)* students take a similar range of subjects as their *Express* counterparts but at the lower N Level. Selected students can offer up to two O Level subjects at the end of four years and the rest at the N Level. The *Normal (Technical)* course is more practice-oriented and covers similar subjects. Both groups of *Normal* students can take short, practical Elective Modules, such as digital animation and retail sales.

Table 4
Typical Allocation of Curriculum Times of Subjects at Secondary 2 Express and Normal (Technical)

Subject	No. of hours per week	
	Express	*Normal (Technical)*
English	4.0	4.6
Mother Tongue	3.5	1.8
Mathematics	3.5	4.6
Science	3.5	3.5
History & Geography	2.3	na
Social Studies	na	1.2
Interdisciplinary Project Work	1.8	1.8
Others	10.0	10.0
Total (per week)	28.6	27.5

The pre-university education is either two years at Junior Colleges (JC) or three years at Centralised Institutes. The percentage of Primary 1 cohort admitted to Junior College doubled from about 15% in 1980 to about 30% in 2006. The JC curriculum was revamped in 2002, and the revised curriculum took effect in 2006. This curriculum stresses thinking skills, greater breadth of learning, and different pathways. Most subjects are offered at three levels called H1, H2, and H3; each H1 subject covers 120 hours, H2 covers 240 hours, and H3 varies between these two limits. Every student will normally study seven subjects: four H1 subjects (General Paper, Project Work, Mother Tongue Language, and one elective) and three H2 subjects with at least one from a contrasting discipline, namely "Mathematics and Science" and "Humanities & The Arts." The more able students may study an additional H1 or H2 subject, or up to two H3 subjects, where the H3 subjects are at par with first year university standard. The main criterion of admission to local universities is based on an applicant's university admission score covering the best three H2 and one H1 subjects, with at least one subject from a contrasting discipline, Project Work and General Paper (or Knowledge and Inquiry). The applicants must also satisfy Mother Tongue Language requirement and special subject requirements. The H3 grade can be considered together with special talents under discretionary admission determined by the individual universities.

3.3 *Major Initiatives*

In the 1990s, globalisation and a knowledge-based economy mean that the workforce needs to be more flexible and creative rather than just be competent but acquiescent. Education is a powerful tool to prepare students for this purpose. Curriculum changes in the 1980s were discussed by Wong (1991), but in the 1990s, many more initiatives were announced by the prime ministers and ministers of education of the time, usually during the annual National Day Rally, Teachers' Day Rally (attended by several thousands of teachers), and MOE Work Plan seminars (when senior education officers and school principals and vice-principals discuss the latest projects). New policies are translated into specific programmes in the schools. A partial list of these initiatives is given in Table 5 (see Tan & Ng, 2005).

Since higher order thinking has been much emphasised and is relevant to mathematics thinking, it is worthwhile to give a brief description of the Thinking Programme to illustrate how an important educational theme was implemented from an addition to the curriculum to an infusion into standard curriculum areas. The Thinking Programme was first introduced in 1987 with de Bono's CoRT (Cognitive Research Trust) programme and later in 1997 using Marzano's dimensions of thinking. The 1997 Thinking Programme was a 2-year programme for lower secondary students. Core thinking skills were taught in two ways: (a) infused into the core subjects of English, Science, Mathematics, Geography, and History, taking up to 30% of curriculum time; (b) separate lessons in non-curricular contexts for about an hour fortnightly. Relevant to mathematics instruction were activities that develop deductive thinking, inductive thinking, and compare and contrast. Students were to apply these thinking skills to subject-related as well as everyday situations. For infusion of thinking into Mathematics, MOE produced a package consisting of detailed lesson plans for Secondary 1 and Secondary 2 (Ministry of Education, 1997). Providing concrete examples of pedagogical initiative is an important way to help teachers accept and implement the desired changes. Examples of how thinking skills can be inculcated across the curriculum are given in Chang and Cheah (2002). In year 2000, this programme was replaced by PW. PW has four learning outcomes, namely, knowledge application, communication, collaboration, and independent learning. PW becomes an infused part of the curriculum from primary to pre-university level. Some teachers were initially dismayed by these changes, but in-service training was available to help them modify their current practices to infuse these initiatives into their subject-based lessons. The infusion approach also has great appeal among UK teachers who work in a national curriculum similar to Singapore, which emphasises strong subject discipline (Moseley et al., 2005).

Table 5
Major Initiatives since the 1990s

Initiative	General Descriptions
NE (National Education), 1996	A shared sense of nationhood, six NE messages about homeland, racial and religious harmony, meritocracy and incorruptibility, make a living, defend Singapore, and have confidence in the future.
TSLN (Thinking Schools, Learning Nation), 1997	Life-long learning, collective tolerance for change, schools as learning organisations, students develop both lower and higher thinking skills and processes.
mp1 (IT Master Plan I), 1997	Equip schools with IT hardware, LCD projector in every classroom, whole school networking, IT use in 30% of curriculum time.
ADE (Ability Driven Education), 1998	Develop talents and abilities of every child to the maximum and emphasis on each child as an individual.
mp2 (IT Master Plan II), 2002	Integrate IT with curriculum design, student-centred learning environment, evaluation of the use of IT in education.
I&E (Innovation & Enterprise), 2003	Character development and focus on pupils' ability to create and seize new opportunities.
SAIL (Strategies for Active and Independent Learning), 2004	Active and reflective learning through knowledge of formative assessment using rubric, to nurture independent learning habits.
TLLM (Teach Less, Learn More), 2004	Engaged learning, better interactions between students and teachers.
SEED (Strategies for Effective Engagement and Development), 2005	Inculcate desired habits of mind, a more holistic approach towards learning at P1 and P2, a bridge between informal schooling at K1 and K2 (emphasising socialisation) and formal schooling at P1 and P2 (emphasising specific skill acquisition).
Subject-based banding, 2008	End of streaming in primary schools.

Despite the initiatives outlined above, the public examinations (PSLE and GCE) are still basically paper-and-pencil tests conducted under time limits. This form of examination has high reliability compared to school-based coursework. School-based coursework has not been included in national examinations for several reasons: it is difficult to manage at a large scale, such as heavy demand on time and effort for the students and teachers, unfair advantages when students' coursework may be completed by other people (e.g., tutors, siblings) outside class hours, sophisticated and expensive moderation across schools, and the possibility of cheating. As a result of these concerns, England has recently scrapped coursework for secondary mathematics (BBC, 2006). On the other hand, MOE recognises coursework and alternative types of assessment for formative purposes to improve teaching and learning. Examples of these modes of assessment for mathematics have been piloted and are discussed in Chapter 18.

Mathematics education is part of the national system, and it is obviously affected by some of these initiatives, global trends in mathematics education, and advances in instructional technology (Wong, Zaitun & Veloo, 2001). We will now turn our attention to the Singapore mathematics education.

4 SINGAPORE MATHEMATICS EDUCATION

4.1 *An Overview*

Mathematics is a compulsory subject in primary and secondary schools. By the end of ten years of general education, every Singapore student would have taken about 1600 hours of instruction in mathematics. Mathematics is taught in English at all levels.

The intended mathematics curriculum is centrally designed and the outcomes are also centrally assessed through public examinations. The curriculum is reviewed every six years with a mid-term review after three years by committees appointed by MOE (Soh, 2005). These committees gather feedback from school teachers and study trends in mathematics education in English-speaking, European, and Oriental countries. The latest official syllabuses for primary to pre-university levels are available at the MOE website. In the latest revision, MOE also produces, for internal circulation only, additional notes on teaching activities and assessment practices. These documents spell out the depth of coverage of the topics and schools are also given general guidelines about the number of hours per week for Mathematics. Nevertheless, teachers often work in teams to prepare the schemes of work for their classes to embed some of the initiatives mentioned earlier, such as use of ICT and infusion of National Education into mathematics lessons.

The main topics in primary schools cover standard work in whole numbers, fractions, decimals, percentages, ratio, money, measures, mensuration, statistical graphs, geometry, tessellation, speed, and algebraic substitution. Probability is not covered in primary schools. Scientific calculators are allowed starting from 2008, restricted to Primary 5 and Primary 6 only, and in the second paper of PSLE starting from 2009. The number of periods (one period = 30 minutes) per week on Mathematics increases from 7 at Primary 1 to 11 at Primary 4. At Primary 5 and 6, the *Standard* band has 9 or 10 periods of Mathematics, and the *Foundation* band has 13 periods per week. This is based on the principle that weaker pupils need more time to learn Mathematics.

The secondary mathematics topics are based on the Singapore-Cambridge GCE O Level examination syllabus for Elementary Mathematics. This syllabus includes topics on integers, real numbers, Cartesian geometry, algebraic equations and graphs, Pythagoras theorem, trigonometry, circle properties, transformation geometry, and statistics and probability. The GCE N Level examination syllabus for the *Normal* courses of study is a subset of the topics in the O Level syllabus. Scientific calculators have been used in secondary mathematics for many years, and are now allowed in all papers in the N and O Level examinations. The number of periods (one period = 35 to 40 minutes) per week is: 5 for *Special/Express,* 6 for *Normal (Academic),* and 8 to 9 for *Normal (Technical).* The same principle of giving more time for the weaker students to learn applies as for the primary schools.

In addition to the O Level Mathematics, the science-oriented students also take Additional Mathematics, which includes differential and integral calculus, more advanced topics in algebra, geometry, and probability and statistics.

Mathematics is not a compulsory subject at the pre-university level. However, around 95% of students at this level take at least one type of Mathematics. From 2006, the new curriculum for GCE A Level examinations includes Mathematics offered under the H1, H2, and H3 levels. The H2 level is normally taken by science-oriented students. The topics include complex numbers, functions and graphs, vectors, calculus, and probability and statistics (sampling, hypothesis testing, correlation). H1 covers half of the H2 topics (mainly functions and graphs, calculus, and statistics) but at similar level of rigour. This H1 subject is taken by students who wish to read business and social sciences in the university. The H3 level is for students who have strong aptitude for mathematics and includes a strong emphasis on mathematical modelling. H3 students must also offer H2 Mathematics. They may choose from the H3 modules approved by MOE and taught at pre-university institutes: Differential Equations and any two of Plane Geometry, Graph Theory, or Combinatorics. Non-programmable graphic calculators can be used at the A Level examinations for these modules. There are two alternatives to H3 mathematics: the students may take advanced modules offered by the universities or work on a research paper supervised by approved agencies. Since these three levels of Mathematics are designed by MOE to suit the Singapore contexts, the University of Cambridge Local Examination Syndicate (UCLES) will prepare the A Level examination papers to align with this new curriculum. Since 2002, MOE has gradually assumed greater control over the A Level examination and this control was extended in 2006 to the O Level examinations (Tan, Chow & Goh, 2008). With this change from previous practice, the Singapore papers are no longer linked to similar papers set by UCLES for other countries.

Ever since Singapore students achieved top performance in the TIMSS studies (see Chapters 19 & 20), there is an increasing interest among international mathematics educators to find out how this stellar result has been achieved. Singapore mathematics educators have written and presented papers about the Singapore mathematics education at conferences (Foong, 1999, 2004; Kaur, 2002, 2003a, 2003b, 2004; Lee N. H., 2004; Lee P. Y., 2005; Lee & Fan, 2002; Lim-Teo, 1998, 2002a, 2002b; Ng, 2007; Seng, 2000; Soh, 2005; Wong, Zaitun & Veloo, 2001). This section draws on their writings and attempts to trace some key issues historically to assist the readers to develop a meaningful mental framework to follow the discussions in some of the chapters of this book.

4.2 *Early Years*

During the past six decades, Singapore mathematics education has undergone changes in tandem with the evolution of the national education system, with influences from international reforms and changes in the Mathematics examination syllabuses offered by Cambridge University. Lee P.Y. (2005) divided these changes into five periods: *early days* (1945 – 1960); *first local syllabus* (1960 – 1970); *maths reform* (1970 – 1980); *back to basic* (1980 – 1995); *new initiatives* (1995 – 2005). Lee N.H. (2004) related three trends of national changes to mathematics education: *survival-driven* (1960s and 1970s); *efficiency-driven* (1980s and 1990s); *ability-driven* (present). Other ways of categorising these developments are also possible.

From 1940s to self-government in 1959, different types of mathematics were taught in the English medium schools administered by the colonial government and in the Chinese and Malay vernacular schools managed by the respective communities. After self-government, MOE developed a common curriculum that "knows no racial barriers" (McLellan, 1957). This first attempt to develop a national mathematics curriculum also unified different branches of mathematics (Arithmetic, Algebra, Euclidean Geometry, and Trigonometry) into a single subject (Ministry of Education, 1959). This document includes the topics to be taught, teaching notes, and recommended books for pupils, teachers, and school libraries. The number of hours per week to spend on arithmetic or mathematics increased with the grade levels: 4 (Year 1 to 5), 5 (Year 6 to 8), 6 (Year 9 to 10), and 8 or more (Year 11 to 13). Secondary mathematics from Year 6 to 10 was examined as the Elementary Mathematics (Syllabus B) under Cambridge O Level examination. The authors of this report recommended that mathematics lessons should take place early in the day. They also discussed several pedagogies, illustrated with relevant examples, that are quite progressive at the time: practical work, outdoor mathematics, discussion of common errors, group work, oral mental tests, individual reports on special topics, use of mathematics puzzles, emphasis on the "mathematics can be fun" theme, and links between mathematics and other subjects. They highlighted the general principle: "Where pupils can be made to understand, time is never wasted" (Ministry of Education, 1959, p. 3). Some of these techniques nowadays come under new labels, such as multiple representation and constructivism.

In 1971, the primary mathematics curriculum was revised to include a new section on "General Aim and Objectives of the Singapore Mathematics Curriculum" and a column on Outcomes in the form of "Children should be able to ..." (Ministry of Education, 1971a; 1971b; 1971c). This was in line with the outcome-based approach stimulated by Bloom's Taxonomy. Since 1983 the writing of specific instructional objectives (SIOs) has become an entranced part of lesson planning taught to trainee teachers. SIOs are also used to plan Tables of Specification of examinations to delineate the weightage of test items at different taxonomic levels (Fong, 1986). Mathematics textbook publishers have also included SIOs as supplementary materials for teachers to use and modify.

As the Modern Mathematics reform caught on internationally in the 1970s, the O Level examination introduced Elementary Mathematics (Syllabus C) to include modern topics such as commutative and associative laws, sets, transformation geometry, and vectors. In early 1980, some adjustments were made to balance traditional and modern topics, resulting in Elementary Mathematics (Syllabus D) for the O Level examination.

When streaming was introduced in 1980, differentiated syllabuses were required for different courses of study. The General Aims remained virtually the same, but a column called "Sample Exercises" was included to illustrate the different expectations of the various courses of study (Ministry of Education, 1980a; 1980b; 1981; 1982). A subset of the topics in the O Level Elementary Mathematics (Syllabus D) was selected for the N Level examination. These topics were assessed at a lower standard compared to O Level.

In the next section, we describe the theoretical framework underpinning the Singapore mathematics curriculum.

4.3 *The Singapore "Pentagon" Framework: Problem Solving as Its Focus*

In 1988, an MOE Committee reviewed several issues about the mathematics curriculum (Wong, 1991). It devised a framework for mathematics curriculum from Primary 1 to Secondary 2. This framework placed problem solving at its centre, surrounded by five inter-related factors deemed to contribute toward success in problem solving: *concepts, skills, processes, metacognition,* and *attitude.* This framework is frequently dubbed the "Pentagon framework," and it has been retained with minor changes when the curriculum was reviewed in 2000 and 2005. The original version (Ministry of Education, 1990a; 1990b) is shown in Figure 1.

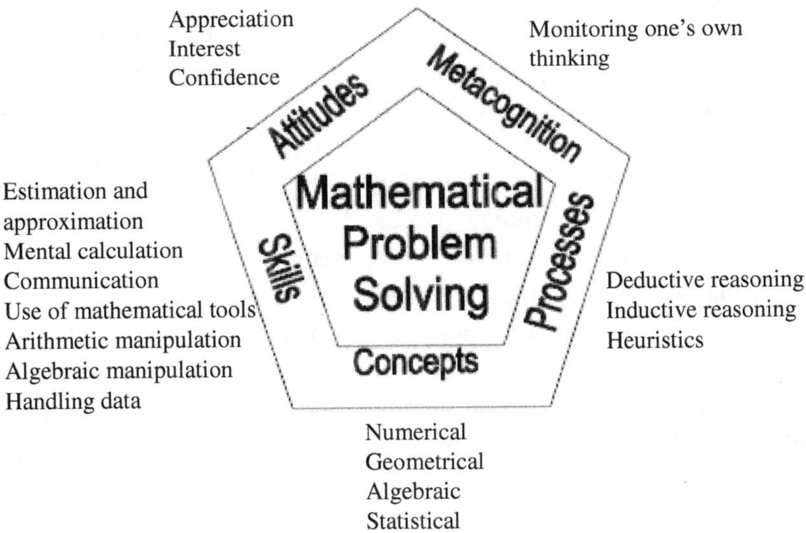

Figure 1. Singapore mathematics curriculum framework, 1990.

The 2001 version included "Perseverance" under *Attitudes* and combined "Deductive reasoning" and "Inductive reasoning" into "Thinking skills." The thinking skills include deduction, induction, spatial visualisation, classifying, and pattern seeking. Examples are given in the syllabus document. Many of these thinking skills are also covered in the generic Thinking Programme mentioned earlier.

NE, TSLN, and IT as listed in Table 5 were incorporated in the 2001 version. NE is to be "integrated into instruction by drawing examples from the prevailing national and current issues during mathematics lessons" (Ministry of Education, 2000, p. 17). Textbook authors and educators have also provided examples and pedagogical framework to promote this integration (e.g., Wong, 2003). IT can support the five factors of the Pentagon by providing opportunities for students to consolidate concepts and skills by linking concrete experiences to abstract ideas (*Concepts* and *Skills*), to enjoy mathematics lessons (*Attitudes*), to become independent learners (*Metacognition*), and to explore different ways of solving problems using higher level of competencies (*Processes*). In more recent years, some schools are experimenting with PC tablets, interactive whiteboard, and e-learning from home; see Chapter 12.

In the latest 2007 version (Ministry of Education, 2006d), a decision was made to extend the Pentagon framework to pre-university level. Thus, some of the items in the framework may not apply to lower levels. The additions are: "Analytical" and "Probabilistic" under *Concepts,* more suitable for upper secondary level*;* "Beliefs" under *Attitudes;* and "Self-regulation of learning" under *Metacognition.* The last item indicates a wider interpretation of metacognition to encompass both monitoring of thinking during problem solving and regulation of learning behaviours by the students themselves (Wong, 2002). The *Processes* component is reorganised as "Reasoning, communication and connections," "Thinking skills and heuristics," and "Application and modelling." At the pre-university level, mathematical modeling is given greater emphasis now than before. These changes are minor in order to provide some continuity from well established practices that the teachers are already familiar with, and yet, to include adjustments for new priorities, which require some teacher training.

The five factors in the "Pentagon" framework are very similar to the five intertwined strands of mathematical proficiency discussed by Kilpatrick, Swafford, and Findell (2001):

- *concepts* versus *conceptual understanding;*
- *skills* versus *procedural fluency;*

- *processes* versus *adaptive reasoning;*
- *attitudes* versus *productive disposition;*
- *metacognition* versus *strategic competence.*

This shows parallel thinking between mathematics educators in Singapore and the US; indeed, there is potentially productive cross fertilisation of ideas as mathematics educators all over the world address similar curriculum and instructional issues within their local contexts. The American Institutes for Research (AIR) report (2005) describes the Singapore national framework as "mathematically logical … lays out a balanced set of mathematical priorities centred on problem solving" (p. xi). On the other hand, Adams et al. (2000) compared the Singapore curriculum and textbooks against NCTM's *Principles and Standards* (2000) and two other US curricula. They found that the Singapore curriculum was rigorous but it did not engage students in higher order thinking skills (despite these being part of the "Pentagon") and did not recognise that students have different learning styles. However, their report was based on documents and textbooks prior to the 2001 syllabus. These criticisms touch on the implemented curriculum in actual lessons.

There is, however, one subtle yet important difference in how the key term "problem" is used. The international mathematics education community tends to differentiate "problems" from "exercises" (e.g., Skovsmose, 2001), with "problems" as open-ended or "non-routine questions, using a range of strategies to solve unfamiliar tasks" (Anderson & White, 2004, p. 127). However, the Singapore 1990 review committee defined it more inclusively to "cover a wide range of situations from routine mathematical problems to open-ended investigations that make use of relevant mathematics" (Ministry of Education, 1990a, p. 3). The same inclusive definition was retained in 2001 but included "unfamiliar contexts" and "thinking processes," while the latest version added "real-world problems" (Ministry of Education, 2006b, p. 12). The rationale for this is to strike a balance between applying standard procedures and algorithms to solve well-defined problems (after all, mathematicians have developed efficient techniques for this purpose, and these techniques need to be mastered) and using heuristics and thinking to tackle non-routines ones. With the introduction of scientific calculators in Primary 5 from 2008 and graphic calculators in pre-university mathematics since 2005, students will be able to solve real-world problems with messy data and to undertake mathematical modelling.

4.4 *Problem Solving and the Singapore Model Method*

Internationally the Modern Mathematics reform gave way to problem solving (National Council of Teachers of Mathematics, 1980) and investigative mathematics (Cockcroft, 1982) in the 1980s. These two themes generated much interest in Singapore about Polya's heuristics. These events coincided with the internal pressure to raise the difficulty levels of the PSLE Mathematics examination to further discriminate among the high proportion of pupils who were scoring the top grade in mathematics. This was done by introducing into PSLE a few "challenge problems" that required heuristic thinking. In 1987, the Curriculum Development Institute of Singapore (CDIS, later became the current Curriculum Planning and Development Division) under MOE published a series of challenge problem cards to provide practice on heuristics for Primary 4 to Primary 6 pupils. These heuristics include "make a systematic list," "look for pattern," "use before and after concept" and the "model method" or "model drawing" approach. The "model method", a "Singapore creation," was introduced to help pupils solve these challenge problems without algebra (Kho, 1987, 2008). Using this approach, pupils draw visual representations of the quantities and unknowns stated in a word problem so that they can see more clearly the structure of the problem and hence solve it. This method has been taught in a structured way beginning with simple problems at Primary 4 and ending with challenge problems involving fractions, ratios, and rates at Primary 6.

The model method has dominated the teaching of problem solving in Singapore primary schools for the past twenty years (Ferrucci, Yeap & Carter, 2003; Foong, 2007; see Chapters 7, 8, & 11). Its use may have contributed to the top performance of Singapore pupils in TIMSS. However, after several years of implementation in primary schools, several curriculum, assessment, and instructional issues have surfaced as outlined below.

A curriculum issue at the upper primary level is whether to teach the algebraic method instead of the model method because some of these challenge problems are more easily solved with algebra. This suggestion was not accepted because the priority is to reduce content coverage so that more curriculum time can be devoted to pupil-centred learning, higher order thinking tasks, and use of IT. In the last two curriculum revisions, the mathematics contents were reduced by about 30% to support IT use. In fact, algebraic methods were deleted from the primary syllabus in the 1970s in response to teachers' complain that there was insufficient time to teach algebra properly and that these methods were difficult for most Primary 6 pupils.

The assessment issue is whether the algebraic method can be used in PSLE instead of the model method to solve challenge problems. The official position has always been unequivocal: any mathematically valid method will earn full credits in PSLE. However, this becomes an issue only when some overzealous teachers forbid algebraic methods in school-based assessment in order to drill their pupils to use the model method. It is also known that some pupils have learnt algebraic methods from other sources, such as tuition teachers (Ho, 2006).

Finally, the instructional issues are harder to resolve. One issue is how to bridge the transition from the model method used in Primary 6 to algebraic methods taught in Secondary 1 (Fong, 1994). While learning the new algebraic methods, some Secondary 1 students revert to the model method, and this has caused some consternation to the secondary school teachers. The other issue is that with the model method as the predominant heuristic taught in primary schools, the other heuristics, such as work backward, simplify the problem, and guess and check, are not given adequate treatment in mathematics lessons. In a recent paper, Wong and Tiong (2006) reported that a sample of Primary 5 pupils (n = 221) and Secondary 1 students (n = 149) were not quite successful in using the heuristics of systematic listing, guess and check, equations, logical argument, and diagrams (model drawing is not suitable for the given problems) to solve challenge problems, and most of them did not use the same heuristic consistently to solve parallel problems. Secondary 1 students in the study preferred the algebraic method and were successful if this method was appropriate. Yet a third instructional issue is whether teaching a prescribed method is counter-productive to the promotion of creative thinking. Obviously, more local studies are needed to probe these instructional issues.

The model drawing method has inspired the development of online Thinking Blocks, an adaptation by a US group (http://www.ThinkingBlocks.com). An ongoing local study is examining how students can learn to use the model method through online manipulative (http://lsl.nie.edu.sg/m18.htm).

4.5 *Singapore Textbooks and Resource Materials*

In the 1980s, CDIS was the sole publisher of mathematics textbooks and teaching resources for primary schools. It published the series of textbooks under its Primary Mathematics Project (PMP) starting in 1983. This series was subject to three types of evaluation: intrinsic evaluation (Wong, 1986), classroom observations, and teachers' opinions. This series of textbooks was fully phased out in 2006, after MOE decided to open up the primary school textbook market to private publishers in late 1990. Currently there are six mathematics series for primary mathematics, including the *My Pal* series being marketed in the US.

The PMP series is one of the Singapore textbooks adapted into American editions (http://www.Singaporemath.com) and is recently approved for use in California elementary schools (Ho, 2007). Besides the United States, several countries such as Canada, New Zealand, Chile, Israel, and the United Arab Emirates are also interested in the Singapore mathematics textbooks. Comments about the Singapore mathematics textbooks are generally favourable. For instance, the American Institutes of Research (AIR) (2005) reported that Singapore textbooks "build deep understanding of mathematical concepts through multistep problems and concrete illustrations that demonstrate how abstract mathematical concepts are used to solve problems from different perspectives" (p. xii). However, Adams et al. (2000) found that "[t]he vast majority of student tasks in the Singapore curriculum is based on computation, which primarily reinforces only the recall of facts and procedures" (p. A-5).

This latter impression seems to agree with a local analysis that had classified 99% of the problems in the CDIS textbooks as "routine," 41.5% as "application," and 72.6% as "single-step" problems (Ng, 2002). However, what is "routine" or not depends on how a problem is introduced in the lessons, and challenge problems can become routine with sufficient practice to achieve mastery. As an illustration, take the well-known problem given to the young Gauss, namely, to find the sum 1 + 2 + 3 + … 100. If the teacher introduces this problem by asking the pupils to find their own ways to solve it, this would be classified as non-routine. After the initial exposure, the teacher may ask the pupils to do similar problems, which then become routine. Just looking at a problem in the textbook without knowing the kind of prior experience pupils already have with the problem or how it is used in the classroom is not sufficient to classify it as routine or non-routine. In 2004, PSLE set a similar problem with a twist: What is the digit in the ones place of the sum 1 + 2 + 3 + … 97? It is difficult to decide whether this problem is routine or non-routine in the Singapore context.

Textbooks for secondary schools have always been published by commercial companies. Zhu (2003) undertook a comparative study of the problem types found in lower secondary mathematics textbooks from Singapore, China, and the US. She found that most of the problems in the textbooks included in her study were routine and close-ended. The US textbooks had more non-traditional problems but these were least challenging as they involved few problem solving steps. Different definitions of problem types, consistency of classification, and knowledge of classroom practice may explain different findings reported by different researchers, and this suggests that more refined methodology is needed to analyse textbooks. Nevertheless, according to a local press report, "there are now over 200 American schools teaching maths the Singapore way, and about 10 American universities have begun showing trainee teachers how to use the Singapore books" (Low, 2005, p. 1). A pilot study in a US county found that students in schools that used Singapore Math "typically outperformed their peers in other district schools" (Garelick, 2006, p. 40).

Mathematics textbooks used in Singapore schools must be approved by MOE after they have been reviewed by panels of MOE officials and school teachers. Although the reviewers and authors may disagree in their interpretations of the intended curriculum, the published materials and suggested activities are largely aligned with the curriculum objectives and coverage. Teacher's resources, including samples of SIOs, often accompany the textbooks. Most textbooks are fairly thin consisting of brief explanations, suggested activities, and mathematics problems. Many activities at the primary level follow the concrete to pictorial to abstract (CPA) approach. CPA is based on Bruner's enactive, iconic, and symbolic modes (1964), but it is couched in everyday terms rather than abstruse jargons so that the teachers can understand them better. Harries and Sutherland (1999) noted the "consistency of use of appropriate representations" (p. 63) in the Singapore primary textbooks to link "the diagram/illustration to the symbolic notation" (p. 59). These three modes of representation (CPA) can be extended to six modes for secondary level, with each mode linked to an action verb: real things (*do*), words (*communicate*), numbers (*calculate*), diagrams (*visualise*), symbols (*manipulate*), and stories (*apply*) (Wong, 1999).

There is no comprehensive study about how Singapore teachers use textbooks in their lessons. Some teachers ask their students to carry out the activities given in the textbooks and assign homework from the textbook exercises, supplementing these with worksheets written by them or jointly with colleagues. Textbook explanations are usually quite brief and Singapore students seldom read the text to deepen their understanding, a disposition similar to what is reported elsewhere (Posamentier & Jaye, 2006; Toumasis, 2004). The students rely heavily on teacher's explanations and answers to their questions.

5 CONCLUDING REMARKS

For the past ten years, initiatives are introduced annually into the Singapore education system and at such a rapid pace that the main stakeholders (students, parents, and teachers) need detailed and clear guidance to make sense of them. Concerns have been raised whether this "rapid fire introduction of projects" is in the best interests of the students

(Siew, 2006). However, it is difficult to define what constitute "best interests" for the students because Singapore students nowadays are different from those in the past decade: they are IT savvy (*Clickerati kids,* see Harel, 1999); they seem to have shorter attention spans; many are brought up by foreign maids whose influences on the students' beliefs and behaviours have not been studied comprehensively; they have to find their own paths between Western influences and traditional practices, and so on. It might be difficult for some teachers to make allowance for these changes when teaching their students.

The attempts to align mathematics instruction with the ever-changing national priorities are not particularly drastic but can still be unsettling for some teachers. In the early years of implementation of the 1991 framework, teachers cited concerns about heavy workload, time constraints, lack of knowledge of the few new topics, and lack of teaching skills to help weak pupils with slow cognitive and language development (Foong, Yap & Koay, 1996). The subsequent revisions and in-service training have addressed some of these concerns. Emulation of the Singapore mathematics education is not feasible in other countries because of the very different compositions of racial, cultural, religious, political, economic, and geographical factors, as well as the quality of the teachers and how their mastery of a new mathematics curriculum can be adequately supported (Adams et al., 2000; AIR, 2005; Garelick, 2006).

Hopefully we have succeeded in preparing the readers to place some of the other chapters within a meaningful context. Much may be learnt from an in-depth study of another system and critical reflection of how this knowledge can be applied to personal situations.

References

Adams, L. M., Tung, K. K., Warfield, V. M., Knaub, K., Mudavanhu, B., & Yong, D. (2000). *Middle school mathematics comparisons for Singapore mathematics, Connected Mathematics Program, and Mathematics in Context (including comparisons with the NCTM Principles and Standards 2000).* Washington, DC: Department of Applied Mathematics, University of Washington. Retrieved December 10, 2004 from http://www.amath.washington.edu/~adams/sing.appendix.pdf

American Institutes for Research (AIR) (2005). *What the United States can learn from Singapore's world-class mathematics system (and what Singapore can learn from the United States): An exploratory study.* Washington, DC: Author. Retrieved on July 30, 2005 from http://www.air.org/news/default.aspx

Anderson, J., & White, P. (2004). Problem solving in learning and teaching mathematics. In B. Perry, G. Anthony & C. Diezmann (Eds.), *Research in mathematics education in Australasia 2000 – 2003* (pp. 127-150). Falxton, QLD: Post Pressed.

Barber, M., & Mourshed, M. (2007). *How the world's best-performing school systems come out on top*. London: McKinsey & Co.

BBC (2006, September 27). *GCSE coursework faces axe to curb Internet cheats.* Retrieved October 20, 2006 from http://www.thisislondon.co.uk/news/article-2336435-details/GCSE%20coursework%20faces%20axe%20to%20curb%20Internet%20cheats/article.do

Bruner, J. S. (1964). The course of cognitive growth. *American Psychologist, 19*, 1–15.

Chang, S. C. A., & Cheah, Y. M. (Eds.) (2002). *Teachers' handbook on teaching thinking skills across disciplines.* Singapore: Prentice Hall.

Cockcroft, W. H. (1982). *Mathematics counts*. London: HMSO.

Ferrucci, B. J., Yeap, B. H., & Carter, J. A. (2003). A modeling approach for enhancing problem solving in the middle grades. *Mathematics Teaching in the Middle Schools, 8*(9), 470-475.

Fong, H. K. (1994). Bridging the gap between secondary and primary mathematics. *Teaching and Learning, 14*(2), 73-84.

Fong, W. C. (1986). *The use of Specific Instructional Objectives for effective teaching and testing*. Singapore: Ministry of Education.

Foong, P. Y. (1999). Education in Singapore: Contributing factors to TIMSS results. Published in Swedish in *Namnaren, 3,* 40-45.

Foong, P. Y. (2004). Engaging mathematics curriculum: Some exemplary practices in Singapore primary schools. *Teaching and Learning, 25*(1), 115-126.

Foong, P. Y. (2007). Problem solving in mathematics. In P. Y. Lee (Ed.), *Teaching primary school mathematics: A resource book* (pp. 54-81). Singapore: McGraw-Hill Education (Asia).

Foong, P. Y., Yap, S. F., & Koay, P. L. (1996). Teachers' concerns about the revised mathematics curriculum. *Mathematics Educator, 1*(1), 99-110.

Garelick, B. (2006). Miracle math. *Education Next, 4,* 38-45. Retrieved on October 3, 2006 from http://www.hoover.org/publications/ednext/3853357.html

Harel, I. (1999). *Clickerati kids: Who are they?* Retrieved on September 9, 2006 from http://www.mamamedia.com/areas/grownups/new/home.html

Harries, T., & Sutherland, R. (1999). Primary school mathematics textbooks: An international comparison. In I. Thompson (Ed.), *Issues in teaching numeracy in primary schools* (pp. 51-66). Buckingham: Open University Press.

Ho, A. L. (2007, December 20). California endorses Singapore maths textbooks. *The Straits Times,* p. H17.

Ho, B. T., Netto-Shek, J., & Chang, S. C. A. (Eds.). (2004). *Managing Project Work in schools: Issues and innovative practices.* Singapore: Pearson Education.

Ho, R. P. Y. (2006, May 26). When rules foil creative students. *Today,* p. 38.

Kang, T. (2005). *Creating educational dreams: The intersection of ethnicity, families and schools.* Singapore: Marshall Cavendish International.

Kaur, B. (2002). Singapore's school mathematics curriculum for the 21st century. In J. Abramsky (Ed.), *Reasoning, explanation and proof in school mathematics and their place in the intended curriculum: Proceedings of the QCA International Seminar* (pp. 166-177). London: Qualifications and Curriculum Authority.

Kaur, B. (2003a, March). *Evolution of Singapore's secondary school mathematics curricula.* Paper presented at Talking It Through: A Cross-National Conversation about Secondary Mathematics Curricula, National Academy of Science.

Kaur, B. (2003b). Mathematics for all but more mathematics for some: A look at Singapore's school mathematics curriculum. In B. Clark, R. Cameron, H. Forgasz & W. Seah (Eds.), *Making mathematicians* (pp. 440-455). Victoria: Mathematical Association of Victoria.

Kaur, B. (2004, July). *Teaching of mathematics in Singapore schools.* Regular lecture presented at the 10th International Congress on Mathematical Education (ICME-10), Copenhagen.

Kho, T. H. (1987). Mathematical models for solving arithmetic problems. In *Proceedings of Fourth Southeast Asian Conference on Mathematical Education* (ICMI-SEAMS) (pp. 345-351). Singapore: Institute of Education.

Kho, T. H. (2008). The model-drawing method with algebra. In P. Y. Lee (Ed.), *Teaching secondary school mathematics: A resource book* (2nd ed.) (pp. 393-412). Singapore: McGraw-Hill Education (Asia).

Kilpatrick, J., Swafford, J., & Findell, B. (Eds.). (2001). *Adding it up: Helping children learn mathematics.* Washington, DC: National Academy Press.

Lee, N. H. (2004, July). *Nation building initiatives: Impact on school mathematics curriculum.* Regular lecture presented at the 10th International Congress on Mathematical Education (ICME-10), Copenhagen.

Lee, P. Y. (2005, November). *60 years (1945 – 2005) of mathematics syllabus and textbooks in Singapore.* Paper presented at the First International Mathematics Curriculum Conference, Chicago.

Lee, P. Y., & Fan, L. H. (2002, August). *The development of Singapore mathematics curriculum: Understanding the changes in syllabus, textbooks and approaches.* Paper presented at the Chongqing Conference, Congqing, China.

Liaw, W. C. (2006, November 5). Making sure no child is left behind. *The Straits Times,* p. 11.

Lim, L. C. (Ed.). (2007). *Many pathways, one mission: Fifty years of Singapore education.* Singapore: Ministry of Education.

Lim-Teo, S. K. (1998). Seeking a balance in mathematics education: The Singapore story. In *Proceedings of the First East Asian Regional Conference in Mathematics Education* (pp. 315-329). South Korea.

Lim-Teo, S. K. (2002a). Mathematics education in Singapore: Looking back and moving on. *The Mathematics Educator, 6*(2), 1-14.

Lim-Teo, S. K. (2002b). Mathematics education within the formal Singapore education system: Where do we go from here? In D. Edge & B. H. Yeap (Eds.), *Proceedings of the Second East Asia Regional Conference on Mathematics Education and Ninth Southeast Asian Conference on Mathematics Education* (pp. 29 – 37). Singapore: National Institute of Education.

Low, E. (2005, February 8). US schools take to Singapore maths. *The Straits Times,* p. 1.

McLellan, D. (1957). Foreword. In Ministry of Education (1959), *Syllabus for Mathematics in primary and secondary schools.* Singapore: Ministry of Education.

Ministry of Education. (1959). *Syllabus for Mathematics in primary and secondary schools.* Singapore: Author.

Ministry of Education. (1971a). *Revised syllabus for Primary 1 and 2, Mathematics (Draft).* Singapore: Author.

Ministry of Education. (1971b). *Revised syllabus for Primary 3 and 4, Mathematics (Draft).* Singapore: Author.

Ministry of Education. (1971c). *Revised syllabus for Primary 5 and 6, Mathematics (Draft).* Singapore: Author.

Ministry of Education. (1980a). *Mathematics syllabus for the New Education System, Part A: Primary 1 to 3 (Common Course).* Singapore: Author.

Ministry of Education. (1980b). *Mathematics syllabus for the New Education System, Part B: Primary 4 to 5 (Normal Course).* Singapore: Author.

Ministry of Education. (1981). *Mathematics syllabus for the New Education System, Part D: Primary 4 to 8 (Monolingual Course).* Singapore: Author.

Ministry of Education. (1982). *Mathematics syllabus for the New Education System, Part C: Primary 4 to 8 (Extended Course).* Singapore: Author.

Ministry of Education. (1989). *Education statistics digest 1989.* Singapore: Author.

Ministry of Education. (1990a). *Mathematics syllabus (Primary).* Singapore: Author.

Ministry of Education. (1990b). *Mathematics syllabus (Lower Secondary).* Singapore: Author.

Ministry of Education. (1997). *The Thinking Programme: Infusing thinking into Mathematics (Secondary 1 & 2).* Singapore: Author.

Ministry of Education. (2000). *Mathematics syllabus (Primary).* Singapore: Author.

Ministry of Education (2003). *Nurturing early learners: A framework for a kindergarten curriculum in Singapore.* Singapore: Author.

Ministry of Education. (2006a). *Education in Singapore.* Singapore: Author.

Ministry of Education. (2006b). *Mathematics syllabus: Primary.* Singapore: Author.

Ministry of Education. (2006c). *Nurturing every child: Flexibility & diversity in Singapore schools.* Singapore: Author.

Ministry of Education. (2006d). *Secondary Mathematics syllabuses.* Singapore: Author.

Moseley, D., Baumfield, V., Elliott, J., Gregson, M., Higgins, S., Miller, J., & Newton, D. P. (2005). *Frameworks for thinking: A handbook for teaching and learning.* Cambridge: Cambridge University Press.

National Council of Teachers of Mathematics. (1980). *An agenda for action: Recommendations for school mathematics of the 1980s.* Reston, VA: Author.

National Council of Teachers of Mathematics. (2000). *Principles and standards for school mathematics.* Reston, VA: Author.

Ng, J. (2008, January 8). ACS(I) among world's best in IB exams. *The Straits Times,* p. 1.

Ng, L. E. (2002). *Representation of problem solving in Singaporean primary mathematics textbooks with respect to types, Polya's model and heuristics.* Unpublished M Ed dissertation. Singapore: National Institute of Education.

Ng, S. F. (2007). The Singapore primary mathematics curriculum. In P. Y. Lee (Ed.), *Teaching primary school mathematics: A resource book* (pp. 15-34). Singapore: McGraw-Hill Education (Asia).

Posamentier, A. S., & Jaye, D. (2006). *What successful math teachers do, grades 6-12: 79 research-based strategies for the standards-based classroom.* Thousand Oaks, CA: Corwin Press.

Seng, S. H. (2000). *Teaching and learning primary mathematics in Singapore.* Paper presented at the Annual International Conference and Exhibition of the Association for Childhood Education International. Retrieved November 2, 2006 from http://eric.ed.gov/ERICDocs/data/ericdocs2/content_storage_01/0000000b/80/10/cd/34.pdf

Siew, K. H. (2006, October 25). For the students or for 'sir'? *Today*, p. 3.

Singapore Examinations and Assessment Board (2006). *PSLE Mathematics: Examination Questions 2001 – 2005*. Singapore: Hillview Publications.

Skovsmose, O. (2001). Landscapes of investigation. In L. Haggarty (Ed.), *Teaching mathematics in secondary schools: A reader* (pp. 115-128). London: RoutledgeFalmer.

Soh, C. K. (2005, November). *An overview of mathematics education in Singapore.* Paper presented at the First International Mathematics Curriculum Conference, Chicago.

Tan, J. (2005). The marketisation of education in Singapore: What does this mean for *Thinking Schools, Learning Nation*? In J. Tan & P. T. Ng (Eds.), *Shaping Singapore's future* (pp. 95-111). Singapore: Prentice Hall & Pearson Education South Asia.

Tan, J., & Ng, P. T. (Eds.). (2005). *Shaping Singapore's future.* Singapore: Prentice Hall & Pearson Education South Asia.

Tan, K. S. S., & Goh, C. B. (Eds.). (2003a). *Securing our future: Sourcebook for infusing National Education into the primary school curriculum.* Singapore: Prentice Hall.

Tan, K. S. S., & Goh, C. B. (Eds.). (2003b). *Securing our future: Sourcebook for National Education ideas and strategies for secondary schools and junior colleges.* Singapore: Prentice Hall.

Tan, Y. K., Chow, H. K., & Goh, C. (2008). *Examinations in Singapore: Change and continuity (1891-2007).* Singapore: World Scientific.

Toumasis, C. (2004). Cooperative study teams in mathematics classrooms. *International Journal of Mathematical Education in Science & Technology, 35*(5), 669-679.

UNESCO Institute for Statistics (2007). *Global education digest 2007: Comparing education statistics across the world.* Montreal: Author.

Wong, K. Y. (Ed.). (1986). *Summative evaluation of Primary Mathematics Project (PMP) curriculum package (P1-P3): Substudy: Intrinsic evaluation of instructional materials and proposed teaching methods.* Singapore: Institute of Education.

Wong, K. Y. (1991). Curriculum development in Singapore. In C. Marsh & P. Morris (Eds.), *Curriculum development in East Asia* (pp. 129-160). London: The Falmer Press.

Wong, K. Y. (1999). Multi-modal approach of teaching mathematics in a technological age. In E. B. Ogena & E. F. Golia (Eds.), *8th Southeast Asian Conference on Mathematics Education, technical papers: Mathematics for the 21st century* (pp. 353-365). Manila: Ateneo de Manila University.

Wong, K. Y. (2002). Helping your students to become metacognitive in mathematics: A decade later. *Mathematics Newsletter, 12*(5). Available from http://math.nie.edu.sg/kywong.

Wong, K. Y. (2003). Mathematics-based National Education: A framework for instruction. In S. Tan & C. B. Goh (Eds.), *Securing our future: Sourcebook for infusing National Education into the primary school curriculum* (pp. 117-130). Singapore: Pearson Education Asia.

Wong, K. Y., & Tiong, J. (2006, May). *Diversity in heuristic use among Singapore pupils.* Paper presented at the Educational Research Association of Singapore Conference 2006: Diversity for Excellence: Engaged Pedagogies, Singapore.

Wong, K. Y., Zaitun M. T., & Veloo, P. (2001). Situated sociocultural mathematics education: Vignettes from Southeast Asian practices. In B. Atweh, H. Forgasz & B. Nebres (Eds.), *Sociocultural research on mathematics education: An international research perspective* (pp. 113-134). Mahwah, NJ: Lawrence Erlbaum.

Zhu, Y. (2003). *Representations of problem solving in China, Singapore and US mathematics textbooks: A comparative study.* Unpublished PhD thesis. Singapore: National Institute of Education, Nanyang Technological University.

Chapter 2

Mathematics Teacher Education: Pre-service and In-service Programmes

LIM-TEO Suat Khoh

In Singapore, a city-nation, the education system is centrally run and, in general, all the teachers in the school system are employees of the Ministry of Education. There is only one institution, the National Institute of Education (NIE), which provides teacher education for all the teachers in the education system and thus, there is very close working relationship between the NIE and the Ministry of Education. The story of mathematics teacher education is primarily based on the teacher education programmes provided by the NIE. This chapter describes mathematics teacher education in Singapore in four sections. The first section sets the context of teacher education in Singapore, the second gives a description of the pre-service teacher education programmes together with a brief discussion of its evolution and issues, the third does the same for the professional development programmes and the fourth section summarises the research done on mathematics teaching and mathematics teacher education over the last decades.

Key words: teacher education, in-service, pre-service, pedagogical content knowledge, professional development of mathematics teachers

1 THE CONTEXT OF TEACHER EDUCATION

As a small but developed city-nation with no natural resources, Singapore puts a premium on education as it sees the development of human resource as absolutely necessary to its economic survival and success. Hence its official education system is a centrally managed and carefully planned system, designed to optimize resources to achieve desired outcomes.

Singapore's Ministry of Education recognizes that teachers play a primary role in developing their pupils and this belief is clearly articulated by the previous Minister of Education, Rear-Admiral Teo Chee Hean, when he said, "The teacher is central to everything that we do in education. Learning about new ideas and concepts is not achieved through textbooks alone, but when teachers engage their students by making ideas come alive. ... The continued strength of education in Singapore will therefore depend on our ability to produce good teachers, with the passion and commitment to inspire young minds and souls" (National Institute of Education, 2002). Teacher education in Singapore is thus a clear part of the official agenda, not to be left to independent institutions, and it is within such a context that the reader must understand the history and current state of Singapore's mathematics teacher education.

The development of teacher education in Singapore is embedded in the context of Singapore's development. While teacher education began during British colonial times, there was a surge in the training of pre-service teachers in the post World War 2 period of the 1950's with the establishment of the Teachers' Training College. When Singapore became independent from Britain and subsequently from Malaysia more than four decades ago, it was necessary to put its growing population through mass education and there was the corresponding need to produce teachers en masse. The Teachers' Training College was established as the Institute of Education in 1970 when teacher education was given due recognition for professional certification purposes. Subsequently, the Institute was reconstituted as the National Institute of Education (NIE) in 1991, a "universitation" phase where teacher education became located within a university, the Nanyang Technological University.

Throughout these decades, the NIE has been the sole provider of pre-service teacher education in Singapore as well as the main provider of professional development for teachers. It is thus not difficult for the

Ministry of Education (MOE) to have a strong input in the management of teacher education since there is only one institution to deal with.

At the governance level, while NIE is part of the university, it has a separate charter and the permanent secretary of the MOE is the chair of the NIE Council. This close working relation is necessary because the source of NIE's student teachers is the MOE which employs all teachers and then sends them to NIE for pre-service teacher preparation. The student teachers are paid salaries for up to two years of the pre-service programmes and MOE also pays their tuition fees. In exchange, the teachers are bonded to serve the MOE as school teachers and are deployed to schools by the MOE upon graduation. It is only in very recent years that NIE has been permitted to accept a small number of non-MOE-sponsored student teachers into its pre-service programmes. Because the student teachers of the NIE are already employees of the Ministry of Education when they enter training, the official descriptor of the programmes is "Initial Teacher Preparation" programmes rather than pre-service programmes. However, for the purpose of this chapter we shall use the international term "pre-service" which is more general and can apply to the non-MOE employees as well. While there may be a few instances where schools employ their own teachers, in general, the MOE is the single employer of NIE's graduates as well as its primary source of student teachers. The close relationship between the two organisations is illustrated in Figure 1.

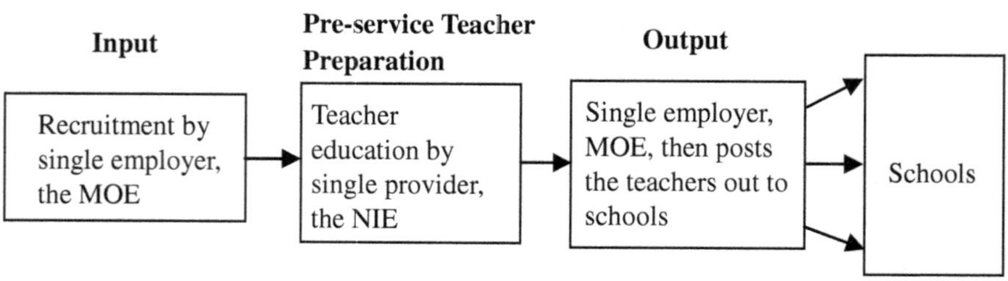

Figure 1. Model of pre-service preparation of teachers in Singapore.

Because of this relationship, the composition of NIE's pre-service student teachers in terms of subject combination is largely determined by MOE and is subject to fluctuations according to manpower planning at the ministry level. Moreover, since structural changes to NIE's pre-service programmes will have implications on deployment of the teachers upon graduation, the MOE is always consulted on such changes or on new programmes at the proposal stage.

At the in-service level, the MOE's staff training branch is the main sponsor of teachers' professional development although there is a move towards decentralization of funds for staff development to schools and school clusters. The NIE is not the only provider of in-service teacher education although it is the main provider in terms of the range of courses provided and the total numbers enrolled in such courses. The relationship between the two organisations is thus extremely strong as between a main client and a main supplier.

While the relationship between the MOE and the NIE is an unequal one since the MOE continues to have central governmental control of educational policies and of funding for all tertiary institutions including the NIE, this dynamic relationship has evolved to one of collaborative partnership in recent years where the NIE is seen as holding the professional and academic expertise which can contribute knowledge and research evidence for decision-making.

2 PRE-SERVICE MATHEMATICS TEACHER EDUCATION

2.1 *Pre-service Teacher Education at the National Institute of Education*

Since the NIE is the sole provider of pre-service teacher education in Singapore, a description of pre-service mathematics teacher education will necessarily be a description of NIE's pre-service programmes and the mathematics components therein. Since 1991, there are four programmes which prepare teachers for teaching in the formal education system from primary to secondary and junior college levels, i.e., from ages 6 to 18. A summary of the programmes is given in Table 1.

Since the NIE is the sole teacher education institution for pre-service preparation, the intakes of student teachers for the pre-service programmes is very large for a single institution. Figure 2 shows the intakes of the four programmes from 2002 to 2006. The PGDE(S) intake figures are the combination of two intakes, the main intake being in July with the intakes of all the other programmes, and a supplementary intake in January.

Table 1
Pre-service Programmes

Program	Entry Requirements	Duration in years	Level for teaching
PGDE(S)	University graduate	1	Secondary
PGDE(P)	University graduate	1	Primary
Dip Ed	A Level or polytechnic graduates	2	Primary
B.A./B.Sc. (Education)	A Level or polytechnic graduates	4	Primary or Secondary

2.2 *Pre-service Preparation for Secondary Level Mathematics Teaching*

The Postgraduate Diploma in Education (Secondary) (or PGDE(S) in short) is the oldest programme for preparing secondary school teachers although it was called the Diploma in Education prior to 1991. In fact, except for 2002 and 2003, it was the only programme graduating teachers for teaching at secondary level and thus has the highest intake numbers among the four programmes. Prior to enrolling for the PGDE(S) program, these potential teachers receive their university education in various disciplines other than education. Almost all the PGDE(S) student teachers come from two universities in Singapore while some come from overseas universities.

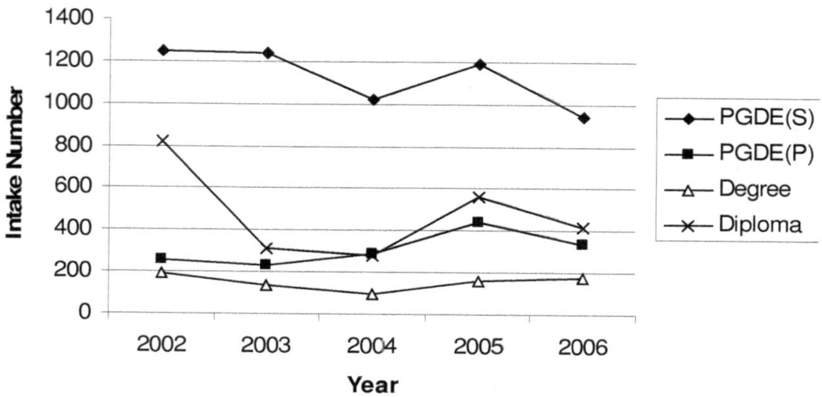

Figure 2. Student teacher intake numbers of pre-service programmes.

During the programme, which focuses on pedagogy rather than content, each student teacher is prepared to teach two main subjects, called Curriculum Studies (CS) subjects. The eligibility for admission to a particular CS subject is based on pre-agreed criteria between MOE and NIE. The first teaching subject, denoted CS1, is the stronger teaching subject for the student teacher while the second teaching subject, denoted CS2, is usually based on less stringent criteria. MOE recruits the student teachers based on projected manpower needs of schools and assigns the CS subjects based on their undergraduate backgrounds at the point of admission. For example, a Science graduate who majored in Mathematics and minored in Chemistry at his/her university will likely be required to take two methodology courses, to prepare him/her to teach Mathematics and Chemistry respectively. However, if he/she also meets the criteria for teaching English and there is a shortage of English teachers, he/she may be assigned English instead of Chemistry.

In the particular case of Mathematics, there is an option for CS2 Mathematics called CS2 Lower Secondary Mathematics. This option is for student teachers who meet lower criteria than for CS2 Mathematics and they will only be prepared to teach Mathematics at lower secondary level because of their lack of mathematical background. This option was created to meet MOE needs for recruitment of teachers in shortage areas where the teachers may not meet criteria for any second subject at CS2 level.

Since 1993, the number of student teachers in this programme who have taken Mathematics as one of the two CS subjects is about slightly more than half of the intake cohorts. In recent years, the NIE annually graduates about 500 to 550 teachers from the PGDE(S) programme who have been prepared to teach secondary Mathematics. Whereas those taking CS1 outnumber those taking CS2 in the early 1990s, this trend has reversed in the past decade. Also in the last decade, there has been an increase of non-Mathematics majors doing CS1 Mathematics, and these are usually engineering graduates. Also, because most graduates would have taken Mathematics at A Level or have taken some service mathematics modules at university, many of the PGDE(S) are assigned Mathematics as their second teaching subject, albeit at CS2 Lower Secondary Mathematics level.

In the current PGDE(S) programme, each of the CS courses takes up about a quarter of the total credit units in the one-year programme, with another quarter going to general education courses such as educational psychology, classroom management, instructional design and the social context of education as well as a basic communication module. The remaining and largest component of the program is Teaching Practice, occupying 10 weeks of the program.

The CS Mathematics (methods) course of the PGDE programme will be described in the next section while the Teaching Practice component which cuts across all pre-service programmes will be described in detail in a later section.

The only other programmes which also prepares student teachers for teaching mathematics at secondary level is the B.A./B.Sc. (Ed) programme. The programmes were introduced in 1991 as programmes for preparing primary school teachers and were then known as the B.A./B.Sc. with Diploma in Education. However, due to a shortage of secondary teachers, several cohorts in the 1990s were given additional curriculum studies modules to enable them to teach two subjects at secondary level in addition to their primary school teaching preparation. Subsequently, the heavy curriculum was modified in 1998 with separate primary and secondary tracks so that adequate preparation for the different levels could be given to the student teachers. The intakes of 1998 and 1999 saw around a total of 40 graduates in 2002 and 2003 from the B.Sc with Dip Ed programme who were prepared for secondary mathematics teaching. However, just after three years, the programmes reverted to their original objective of preparing strong primary school teachers, were structurally changed and were renamed the B.A./B.Sc. (Ed) in 2001. Within the context of overall curriculum review of NIE programmes in 2003 - 2004, the programmes were once again changed for the 2005 intake, with the current structure again carrying two tracks, one for the preparation of primary teachers, and the other for secondary teachers. These programmes, B.A/B.Sc with Dip Ed and B.A/B.Sc (Ed) will be collectively referred to as the degree programme in the subsequent discussion.

One of the constants of the degree programmme through these years has been the inclusion of at least one academic subject at university level. In the earlier versions of the programme and in the secondary track of later versions, a teacher being prepared to teach at secondary level had to master academic content in two selected teaching subjects.

2.2.1 The Secondary CS Mathematics Course (The Secondary Mathematics Methods Course)

The Mathematics Methods course has evolved over the last twenty years from being rather theoretical and psychology-based in the late seventies to a combination of theory and practice in the late nineties. There is a need for an effective balance in the distribution of time given to the understanding of theories and principles of learning and time for practical ideas and teaching strategies so necessary to the beginning teachers' immediate survival in the schools. Thus, while student teachers were given basic understanding of the learning theories of Piaget, Bruner, Dienes, Gagne, Ausubel and Skemp, a large part of the course would discuss the applications of these in teaching strategies for various topics in the secondary mathematics curriculum. In the last decade of the twentieth century, with the emphasis on problem-solving, there was a movement towards more independent learning by the trainees and a deeper understanding of the mathematics they should teach. Thus in 1993, the pedagogy course was revised quite extensively with the purpose of including a very large component of independent learning where the student teachers had to read up, discuss collaboratively and present various aspects of the different strands of Arithmetic, Algebra and Graphs, Geometry, Trigonometry and Statistics (Lim & Yap, 1996). One of these aspects was the study of the deeper mathematical points of certain secondary curriculum topics such as the classification of subsets of real numbers, the different concepts of letters in algebra and so on. Other aspects were particular learning theories and ensuing learning difficulties e.g., Van Hiele Theory in Geometry and motivational strategies for teaching a sample of topics in each of the strands. However, in response to feedback from the student teachers, the component of such independent learning, requiring much extra work outside class time, was substantially reduced in the following years.

The current methods course is done over two modules totaling 120 hours. Despite the large numbers of student teachers in recent years, only 6 to 8 hours of the course are delivered in a large lecture mode and these would be for efficient dissemination of information. The rest of the course is taught in groups of around 25 to 30 student teachers, the smaller group size being more conducive to discussion, reflection and collaborative learning. Each group is taught by one assigned instructor who takes the group throughout the course.

The course begins with a general overview of the aims and structure of the mathematics curriculum in Singapore education system. General psychological theories for learning mathematics as mentioned in the above paragraph as well as lesson planning in general are also covered towards the beginning of the course. The main bulk of the course is spent on the teaching of the various topics in the school curriculum. Within broad strands of Arithmetic, Algebra and Graphs, Statistics, Geometry and Mensuration, Trigonometry and Calculus, student teachers will examine understanding of some topics, specific learning difficulties, strategies and teaching approaches, use of Information Technology, enrichment tasks, methods for teaching difficult topics, etc. In addition, topics such as problem solving heuristics and inculcation of thinking skills may be interwoven into the mathematical areas as well as taught separately and explicitly due to the importance of problem-solving in the curriculum. The problem-solving approaches and heuristics to be used are based on Polya's model and approaches developed by mathematics educators such as Charles, Lester Jr, Krulik, and Rudnick. Although error analysis of pupils' work may be done within the topics, the general procedures of setting mathematics examination and test papers and developing marking schemes are covered in general rather than topic by topic. The course also includes practice sessions where trainees present mock lessons or parts of lessons to practice their planning and communication skills in explaining or developing a particular mathematics concept. The planning of these lesson segments could be discussed with instructors or course-mates prior to the presentation and reflection, feedback and more discussion would be given after the sessions. The reader is referred to the main textbook used for this course (Lee, 2006) for more details on the coverage of the course.

2.2.2 Mathematics Content Knowledge for Teaching at Secondary Level

It was mentioned earlier that the number of pre-service teachers being prepared for teaching secondary mathematics had increased tremendously in the 1990s and moreover, the profile of these teachers has been changing. Two decades ago, pre-service Mathematics teachers would be mainly graduates from Science degrees majoring in Mathematics and the CS1 Mathematics teachers would outnumber the CS2 Mathematics teachers. When the CS2 Lower Secondary Mathematics was first offered in 1996, there were fewer than 15 student teachers in that year and only 17 in the following year. Since 2001, however, the majority of teachers doing CS1 Mathematics are not mathematics majors but engineers. The number of CS2 mathematics teachers have grown tremendously and in current years, the CS2 Lower Secondary Mathematics teachers form between 20 – 25% of all those doing mathematics. With this changing profile of pre-service mathematics teachers, there was some concern over the mastery of mathematics content for teaching at secondary levels among the teachers.

However, it was impossible to introduce extra content courses to the already heavy curriculum of the PGDE(S) programme. As a measure to encourage student teachers to improve their mastery of mathematics content in terms of understanding concepts, a School Mathematics Mastery Test which examined the student teachers' understanding of O Level Mathematics was introduced in 2003. Those who did not pass the test at mastery level at the first try were required to re-take the test before they completed the programme. During the time in between the tests, the student teachers were given online materials for revision and also given teaching help from mathematics teaching staff of the NIE. Although this test has produced some initial sense of insecurity among the pre-service student teachers, there has been positive feedback that they felt more confident about their content mastery after they were "shocked" into acknowledging their lack of mastery and had undergone the extra learning. More details of this test can be found in Toh, Chua & Yap (2007).

2.3 *Pre-service Preparation for Primary Level Teaching*

There are three programmes which prepare teachers for teaching at primary level. The discussion below will not include specialist tracks of these programmes for preparation of mother tongue language teachers as these teachers do not teach mathematics.

The oldest of the programmes is the Diploma in Education (or Dip Ed) programme which was named the Certificate in Education programme prior to 1991. This two-year programme had high intakes prior to 2003 and enrolment again increased after a one year dip in 2003. The student teachers of this programme are, in general, academically weaker than those of the other two programmes. The programme was revised prior to 1991 to include academic subjects but these were later removed so that the programme would focus more strongly on pedagogy and subject content knowledge for teaching rather than normal tertiary level content.

The one-year Post-graduate Diploma in Education (Primary) programme, (or PGDE(P) for short), was introduced in Singapore in 1991. The programme takes in graduates with a vast range of different degrees from other universities. They are not selected for their teaching subjects based on university courses since Singapore's primary teachers are expected to be generalists as mentioned above. It was considered that graduates of all disciplines should have sufficient content to teach at primary levels.

The degree programme which was introduced in 1991 had undergone the greatest number of structural changes during its 16-year history. This was due to policy changes regarding its role in preparing either primary teachers or both primary and secondary level teachers. The student teachers of this group meet normal university entry requirements. The intake sizes have varied between 200 and 100 but the number which graduate 4 years later exceeds the intake numbers quite substantially. This is because Dip. Ed. student teachers who graduate in the top 20–30% of their cohort are allowed to enter the degree programme and normally take two to two-and-a half more years to complete the degree.

In all the main tracks of these three programmes, teachers are prepared to be generalist teachers to meet the deployment requirements of primary schools. Known as English medium teachers, they are deployed to teach the main examinable subjects of English, Mathematics and Science, as well as the non-examinable subjects of Social Studies, Art, Music, Physical Education which are all taught in English. There are schools which do deploy some of their primary teachers as subject teachers to teach just one or two of the examinable subjects. However, these are likely experienced teachers whose strengths in particular areas have been identified after their initial teaching years.

In 2004, NIE carried out a major curriculum review of its pre-service programmes and modified the programme structures with agreement from the Ministry of Education to allow more flexibility in choice of subjects as well as a reduction in the number of teaching subjects. These modified programmes were implemented with the July 2005 intakes of all the programmes. Table 2 shows the current three pre-service general programmes for preparing primary teachers and the subjects which the programmes prepare them to teach. In addition to greater choice of teaching subjects, the number of teaching subjects in the 4- year degree programmes was reduced to three from the previous 4 teaching subjects. Although the PGDE(P) and Dip. Ed. programmes still prepared the student teachers to teach 3 subjects, some selected student teachers in these programmes are given the option to be prepared to teach two subjects instead but these must be chosen from English, Mathematics and Science. The curriculum space freed up from this reduction is used for additional modules which would deepen their knowledge in subject pedagogy and subject matter knowledge of the two subjects.

In the more flexible programme structures since 2005 as given above, Mathematics is no longer compulsory although most of the student teachers still opt to offer Mathematics since the subject forms a large component of the primary school curriculum and they will very likely be expected to teach the subjects on graduation. Historically, when student teachers in the programmes are prepared to teach particular subject, it means that they undergo pedagogy courses in those subjects in addition to their general education courses. These courses are known as

Curriculum Studies (CS) courses, as in the case of secondary level teaching mentioned earlier, and the various CS subject areas form the CS component of each programme. In Table 2, the last column specifies the CS courses to be taken within the programme; for example, a degree programme student teacher would take three CS courses, for instance CS English, CS Mathematics and CS Science, to form the CS component of his/her programme. It is these courses that would begin the development of pedagogical content knowledge of the subject in the pre-service teacher. These courses are not to be confused with content courses and the requirement for offering a particular CS subject area in the primary programmes is almost negligible beyond general educational qualifications.

Table 2
Teaching Subjects of Pre-service Programmes (2005 onwards)

Programme	Entry Requirement	Duration	Teaching Subjects
Postgraduate Diploma in Education (Primary)	University graduate	1 year	English, Mathematics, Science with possible replacement of one of these with Social Studies, Art or Music
BA/BSc (Ed)	A Level or Polytechnic Diploma holders	4 years	Any three from English, Mathematics, Science, Social Studies, Art or Music
Diploma in Education	A Level or Polytechnic Diploma holders	2 years	English, Mathematics, Science with possible replacement of one of these with Social Studies, Art or Music

In particular, for mathematics, due to the need for all primary teachers to teach mathematics, there was no requirement for those who undertake these mathematics pedagogy courses to have any background in mathematics beyond O level mathematics as part of the national O level examinations taken at age 16. In recent years, this condition has been raised to a higher grade in the same examination but the student teachers may enter and complete their pre-service preparation without having done any post-secondary mathematics.

2.3.1 The Primary Mathematics Methods Course

The mathematics methods courses have been established as long as Singapore's teacher education but have undergone constant review and revision. Due to the different lengths and structures of the three programmes and the introduction of the programmes at different times, prior to 2005, the CS Mathematics courses in the three programmes used to be quite different in emphasis, content and assessment although the core content would be largely common across programmes. In the curriculum review of 2004, the CS courses in each subject area were streamlined across programmes, the reason being that the student teachers are being prepared for the same teaching functions in the same schools upon graduation regardless of programmes. In fact, when Professor Jeremy Kilpatrick of the University of Georgia was invited to review the CS mathematics curriculum, his advice was that the contents of the CS Mathematics courses in different primary programmes should in fact be similar for that same reason. Thus in the current curriculum implemented in 2005, each CS subject, regardless of programme, would consist of the same number of courses with the same course content. The only difference would be the placement of these courses within the semestral structure of the programmes since the PGDE(P) programme had to accommodate the courses within 2 semesters whereas the same courses could be spread across semesters in different years for the 4-year degree programme.

While the contents of the pedagogy courses for the three programmes are similar, some consideration is given for the different backgrounds of the student teachers from different programmes. However, the differences in mathematical background are more likely to be found between individuals within a programme rather than on a programme basis as the PGDE(P) student teachers tend to come with non-Science degrees and may not have done any post secondary mathematics. Although many in the Singapore education system do tend to offer mathematics in the A Level National examination at age 18, there is a growing number of graduates who take up polytechnic courses after O levels at age 16 and thereafter proceed to universities. This group of graduates may have no post-secondary mathematics at all. Some of the student teachers are also mature students who decided on a career switch and it would have been a long time since they last encountered school mathematics. The adjustments for student background are therefore done at class level by individual faculty rather than on a programme basis.

Over the years, the CS Mathematics course has moved from a more generic psychological orientation to a more practically-oriented course. While student teachers are exposed to theories, they are encouraged to re-think their own learning of mathematics and to explore teaching approaches based on an understanding of the concepts which they will be teaching. The CS Mathematics course covers an overview of the Singapore mathematics curriculum, general learning theories, pedagogical aspects of various primary mathematics strands (whole numbers, fractions, decimals, space, etc.) including the use of technology in the teaching of these topics, the diagnosis of errors and practical aspects of setting assessments in mathematics. The curriculum content is constantly reviewed in the light of international and local mathematics education research and findings could be brought in where relevant. The course content will not be too different from mathematics methods courses in the United States and in fact some American textbooks are used as references in these courses although the NIE has recently developed its own textbook for the CS Mathematics course (Lee, 2007). There are only two additional localized topics and these are the use of the abacus (Japanese soroban) for addition and subtraction and the use of the "model method" in solving arithmetic problems, a method used in Singapore schools which primary mathematics teachers must be able to teach to their pupils. During these methods courses, there is much discussion, learning activities, experimentation and presentation of teaching ideas. Assessment of these courses include presentations, assignments and group projects. The total contact time for the modules which constitute the mathematics pedagogy courses is 96 hours, slightly lower than the 100 or 140 hours in the United States or France as reported in Comiti and Ball (1996) although the American or French cases may also include time for developing the student teachers' own mathematical understanding which is built into a separate component of Singapore's programmes.

Having undergone the pedagogy courses, the student teachers will put what they have learned into practice during their practicum. There are two periods of practicum for the two longer programmes and only one for the PGDE(P) programme. One of the shortcomings of the programmes is the lack of specific subject mentoring by faculty during the student teachers' practicum in schools. Due to the large numbers of student teachers in every intake, much of the responsibility for mentoring student teachers lies with the mentor teachers in the schools rather than with faculty. Moreover, because the subject specialists are needed for supervision in the secondary programmes, student teachers in primary schools are mostly supervised by faculty from general education departments rather than subject specialists. Thus, faculty members who specialise in primary mathematics education are only able to see their student teachers in action in real classrooms through private and individual arrangements, sometimes through the avenue of research. Nevertheless, student teachers do freely consult their methodology teachers during their practicum if they need special assistance in addition to what their supervisor or school mentor teacher can offer.

2.3.2 Mathematics Content Knowledge for Primary Level Teaching

There has always been an issue of teachers' subject knowledge in the subjects which they are to teach. While laymen may assume that anyone with post-secondary education is able to handle primary level mathematics, teacher educators realize that to develop strong pedagogical content knowledge, a good grounding and deep understanding of the mathematical content which is being taught is a necessary though not sufficient condition. Moreover, while the methods courses may cover a wide repertoire of teaching approaches and skills, unless the student teacher has a strong mastery of the concepts behind mathematical procedures, he or she will not be able to appreciate the need or usefulness of such teaching approaches and may in fact apply pedagogically sound approaches either superficially or in wrong contexts.

In Mewborn (2003), a survey of American research on teacher knowledge paints a "dismal picture of teachers' conceptual knowledge of the mathematics they are expected to teach" compared to their "strong command of the procedural knowledge of mathematics." While research has established links between a teacher's conceptual understanding and her choice of classroom activities, it is less clear about the links between a teacher's characteristics and the achievement of her pupils. Nevertheless, the research of Ma (1999) indicates that the strong content knowledge of China's mathematics teachers is an important factor contributing to their teaching. It is noted that primary level mathematics teachers in China taught mathematics only and had far less contact time each day, spending much time in observing each other's classes, discussing and planning their lessons. Mewborn (2003) points out that Chinese teachers are thus "likely to encounter both the opportunity and the necessity to deepen their understanding of the mathematical content they are teaching" in contrast to generalist US teachers who teach a whole range of subjects and are occupied with teaching throughout the day. Within the Singapore situation where elementary teachers are generalists, while the situation may not be as "dismal" as that shown by American research, the requirement that almost all primary teachers must teach mathematics regardless of proficiency or interest in Mathematics does not give confidence that our primary mathematics teachers will be able to bring out deeper mathematical thinking in their children which goes beyond procedural knowledge.

With the concern about primary teachers' subject understanding, a subject knowledge (SK) component was introduced in the degree programme in 1998. For each curriculum studies (CS) subject, a corresponding set of SK courses would be required of the student teachers. The SK courses were aimed at building up the student teachers' subject knowledge for teaching at primary level and the content was closely linked to the Singapore primary school curriculum for that subject but going beyond providing basic knowledge of the topics to inculcate deeper understanding of structures under-girding these topics. Subsequently, in 2001, the SK component was also introduced to the Dip. Ed. Programme. The SK component could not be included in the 1-year PGDE(P) due to the short duration of the programme and the heavy curriculum to be completed. However, after a curriculum review of all initial teacher preparation programmes, the modified PGDE(P) programme of 2005 had an option where the student teachers are allowed to "specialise" in two teaching subjects instead of the normal three teaching subjects. The curriculum space freed up from this allowed the student teachers to supplement their subject knowledge with SK courses in the two CS subject areas. In fact, with the same curriculum review, the Dip Ed modification implemented in 2005 also has an option where the student teachers are prepared to teach two subjects instead of three but had one additional CS and SK module in those two subjects compared to the option where the student teacher had three CS subjects.

The SK mathematics courses comprise topics which will help build up the student teachers' mathematical foundations for the primary mathematics topics. The content thus included the topics: Problem-solving heuristics, historical numeration systems, algorithms on operations in a place-value system, divisibility, ratio and proportion, mathematical processes of deduction and induction, the study of 2-dimensional space (geometrical entities, properties of geometrical figures, concepts of similarity and congruency, processes of constructions and proofs), measurement of space in 2 and 3 dimensions, motion geometry and tessellations, statistical investigations.

While the student teachers appreciate the pedagogical principles learnt in their CS classes, they were unable to see such teaching methods put into practice in the learning of mathematics content within these classes since the content of these classes is pedagogy and not mathematics. It is thus in the SK Mathematics course that the teacher educators are able to demonstrate the pedagogical methods of helping the student teachers make sense of mathematics through discussion, examining mathematics concepts, deductive and inductive thinking, and so on. Taken in its best sense, the SK Mathematics course has thus two objectives, firstly, to raise the level of understanding in the mathematical foundations of the student teachers so that they have a deeper understanding and appreciation of the seemingly simple mathematics they need to teach at elementary level and secondly, to see the pedagogical principles they learn in their methods course put in practice in their own learning of mathematics. In most of the programmes, wherever possible, the SK module runs before the CS module covering the same topic so that the pedagogy class can build methods upon deeper understanding of mathematical structures. The teaching methods employed in the SK course can then provide current experience of learning mathematics which acts as a springboard as well as example for further discussion of mathematical learning and teaching in the CS course.

The description of the links between the CS and SK Mathematics courses given in the preceding paragraph is the ideal state and in reality, there are problems in achieving these links very effectively. Some of the reasons are logistical: due to large cohorts and manpower constraints, the SK and CS courses are taught by different teacher educators who may not be able to link the two courses as well as they are intended to be. Moreover, as the two courses are separated by a semester break, student teachers do not appreciate the usefulness of the SK modules which are taught first. Some possible measures to be taken to improve the situation could be the amalgamation of the two courses which could allow immediate links of topics as well as pairing of the teacher educators who could work to improve the synergy between what is learnt in the two courses.

3 PROFESSIONAL DEVELOPMENT FOR MATHEMATICS TEACHER

3.1 *Professional Development Infrastructure in the Singapore Education System*

In the Singapore Education System, the Ministry of Education's commitment to the professional development of teachers can be seen in its provision of policies, measures and more than adequate resources to encourage all teachers to constantly update and upgrade themselves in their knowledge and skills. Every teacher in the school system is eligible to 100 hours of professional development a year. When teachers attend professional development courses, their fees are met by the Ministry of Education, either directly or through school/cluster funds. There is a whole range of professional development courses, ranging from one session workshops, whole day conferences/seminars to long certification programmes. Mathematics teachers in schools may choose to attend generic professional development courses or those pertaining to mathematics teaching and learning. The only issue is that unless the teacher is overseas attending higher degree programmes on a full-time basis or is on full-time attachment to other institutions, all other professional development work is done without any reduction in school duties. Since formal school curriculum time runs for slightly more than half a day in Singapore, teachers are expected to attend courses during the other part of the day and still do their normal day's work, a practice which does not ensure optimal learning. However, in a society where examination performance is highly emphasised, the teachers themselves would not be comfortable for their classes to be taken over by substitutes so that they can attend professional development courses.

Short half-day, one or two day workshops and seminars as well as many professional development courses which are stand-alone short courses ranging from 12 hours to 39 hours are popular with teachers who seek short learning stints with practical results. While the National Institute of Education offers the majority of the courses which exceed 12 hours, there are also many other institutions or organisation providing the shorter workshops and seminars for professional development. There are many commercial providers for short generic courses but the main providers of mathematics professional development courses are as follows:

- The National Institute of Education offers short workshops as well as stand-alone courses of up to 39 hours run by her own staff or specially engaged foreign academics
- Professional bodies such as the Singapore Mathematical Society and the Association of Mathematics Educators organise conferences and seminars with a judicious mix of foreign experts, local experts and practicing teachers. Such conferences and seminars provide a platform for teachers to learn from foreign and local academics as well as for them to present their own experiences for feedback and critical review.

- The Ministry of Education's Curriculum Planning and Development Division which run courses conducted by their curriculum specialists or other experts. These courses usually coincide with curriculum initiatives and are meant to provide teachers with the skills and knowledge to implement the initiatives.
- School or Cluster organised customized sessions with their own teachers sharing or tapping on the expertise of master teachers, local university or foreign experts.

3.2 *Structured Programmes and the Professional Development Continuum Model*

While teachers have greater flexibility in developing themselves according to their needs by taking short courses, structured programmes could result in deeper understanding in specific subject areas or specialisations. Other than doctoral programmes which are largely research based, the National Institute of Education offers Masters degree programmes by coursework and dissertation for graduate teachers as well as Advanced Diplomas for primary level teachers.

Each Advanced Diploma consists of a coherent selection of related courses and for mathematics teachers, the Advanced Diploma in Primary Mathematics Teaching provides generalist primary school teachers with a programme which deepens their knowledge and pedagogy in teaching mathematics so that they can become specialist mathematics teachers. The courses offered are mainly in the areas of fostering mathematical thinking in the various primary level topics such as Numbers, Fractions and Decimals, Data topics, developing geometrical thinking or algebraic thinking. An additional course which is becoming increasingly important in Singapore is a course on Action Research as teachers are encouraged to carry out Action Research.

To meet non-graduate teachers' aspirations of obtaining a degree, the National Institute of Education has made the Advanced Diploma an alternative route into the degree programme for serving teachers. Teachers who do not qualify to cross into the degree programme on completion of their pre-service Diploma in Education are able to take relevant advanced diploma courses and, should they show good academic performance in the advanced diploma, they are eligible for admission to the degree programme.

The Professional Development Continuum Model (PDCM), implemented by the National Institute of Education together with the Ministry of Education in 2003, allows serving teachers who are admitted to the bachelor degree and master degree programmes to accredit relevant courses taken under professional development towards the programme requirements of the respective bachelor/master degrees. In fact, the Institute offers many of the bachelor or master degree courses as professional development courses and the teachers can take them as stand alone courses. Taking these courses prior to admission into the master/bachelor degree programmes and the subsequent accreditation thus reduce the full-time candidature necessary for completion of the programme, thereby reducing the opportunity cost for serving teachers who need to take no-pay leave to enter the full-time programme.

However, in the case of mathematics, for the non-graduate primary teachers, the PDCM accreditation advantage is not substantial for shortening the full-time candidature of the degree programme. Because of the close alignment of the pre-service diploma and degree programmes, the main accreditation comes from the diploma programme courses and this can reduce the degree requirements to almost half the programme. Figure 3 below explains visually the types of mathematics courses for the pre-service diploma and degree programmes and for the in-service programmes.

In the case of other subjects such as Science, the in-service courses offered are more aligned with the content courses of the degree programme and thus can be accredited towards the degree. In Mathematics, however, the university degree-type of content courses are unpopular with in-service non-graduate primary teachers and are therefore not offered. For the primary teachers who obtained their pre-service diploma before the subject knowledge component was introduced in 2001, they could take the subject knowledge in-service modules and accredit them towards the degree programme. However, for the in-service programmes, the large majority of the courses are advanced pedagogy courses which are not part of the pre-service degree programme and thus cannot be accredited towards degree courses.

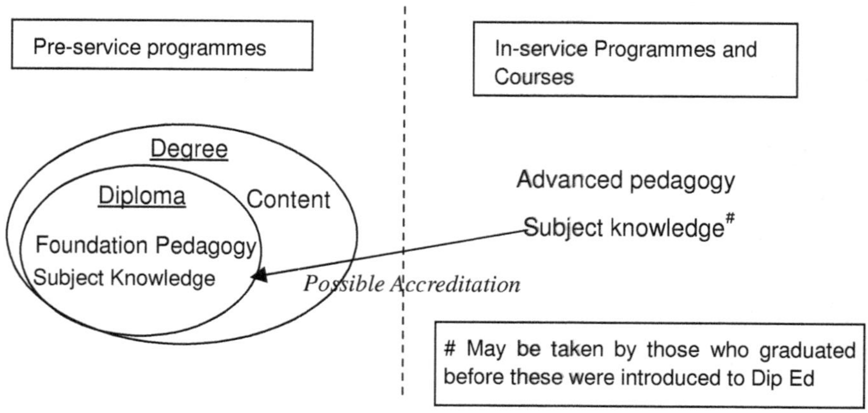

Figure 3. Types of mathematics courses in programmes for non-graduate primary teachers.

For the master degree programmes, graduate teachers (mainly from secondary schools) are benefiting from the PDCM in accrediting courses towards the M.Ed (Mathematics Education) programme. By taking the courses under professional development, the Ministry of Education sponsors the teachers by paying for the fees whereas in the past, those who enrolled as part-time candidates in the master's programme had to pay for their own fees. Teachers enrolled in such coursework masters' programmes typically complete their course in three years on a part-time basis.

This M.Ed (Mathematics Education) coursework programme is one of the strands of the M.Ed programme offered by the National Institute of Education. The programme seeks to build up the teachers' understanding of issues in mathematics education but is rather different from many M.Ed programmes in its basic philosophy that the mathematics teachers must also have a deep understanding of mathematics content. The courses thus involve a mix of courses which are educationally inclined such as *Educational Inquiry*, *Curriculum Studies in Mathematics* and those which are pedagogy-content oriented such as *Geometry and the Teaching of Geometry. Number Theory and the Teaching of Arithmetic.*

A more content-oriented coursework masters has also been offered since 2006. This is the M.Sc (Mathematics for Educators) programme. The programme emphasises the acquisition of wide and in-depth content knowledge in mathematics as well as its linkage to mathematics teaching and is especially designed for mathematics teachers who wish to strengthen their mathematical understanding. This programme admits non-mathematics majors who have mathematical aptitude but lack content knowledge or depth of understanding due to their university backgrounds. It has a compulsory core module *Mathematical Inquiry* and in general focuses on university type mathematics courses such as *Advanced Calculus* and *Abstract Algebra* but with applications for educators. It also shares three common courses with the M.Ed (Mathematics Education), all of the content-pedagogy type. There are also higher level mathematics courses such as *Real Analysis, Functional Analysis* and *Statistical Methods* to cater to diverse interests and needs of teachers.

3.3 *Some Issues in Professional Development*

It was mentioned in Section 3.1 that the Ministry of Education has been extremely supportive of teacher professional development and has put in place policies and resources to develop teacher expertise. While Singapore seeks to upgrade the professional practices of her teachers through professional development courses which could ultimately lead to higher degrees, there are two concerns with regards to the actual implementation practicalities.

The first concern is the teachers' ability to manage longer and higher level courses such as those leading to higher degrees. Since a large majority of teachers will not be able to attend courses on a full-time basis, it is difficult for them to balance unreduced school duties with course commitments and family life. This can result in attrition when they cannot meet the rigours of the courses, especially those at master's level.

The second concern, especially in the area of content knowledge, is that teachers prefer to take generic courses or pedagogical courses which interest them rather than content courses which address their understanding in areas of weakness. There seems to be a perception that after years of teaching, they are adequately knowledgeable in the content which they teach. Just like school pupils who avoid their weak subjects, teachers likewise do not wish to enroll for courses whose content is deemed difficult to master. The perception of teacher educators in general is that teachers prefer courses in practical pedagogy which they can immediately apply rather than those which stretch their mathematical muscles even if these courses can benefit them if they master such processes and transfer the learning to their students. The new initiative to appoint staff development officers in schools who can counsel the teachers on the needed areas for improvement would do much to develop teachers in areas of needs rather than strengthen already strong areas.

4 RESEARCH ON MATHEMATICS TEACHING AND MATHEMATICS TEACHER EDUCATION IN SINGAPORE

In the last two decades, there have been many research studies in Singapore on Mathematics Education, primarily in the areas of mathematics learning and achievement in various topics, pupil attitude towards mathematics, and effectiveness of various pedagogies including the use of ICT in mathematics learning. Research in the area of mathematics teaching and teachers is rather more limited. This area encompasses teachers' attitude towards mathematics and mathematics teaching and learning, teachers' knowledge and applications of such knowledge to their own teaching, general pedagogical practices preferred by teachers, and relationships between teacher factors and student learning. This section will summarise most of the research done in mathematics teacher education and teaching and the reader should refer to the publications for greater detail. There is also a broader collection of research on teacher education in general of which mathematics teacher education is a subset. The reader is referred to Deng and Gopinathan (2001) for a summary of general teacher education research in Singapore from 1989 to 1999.

Broadly, the research done on Mathematics Teacher Education in Singapore can be organised into the following four broad categories:

(a) Mathematics Pedagogical Content Knowledge of teachers
(b) Pedagogical Practices of Mathematics Teachers
(c) Initial preparation of mathematics teachers
(d) Professional Development for Mathematics Teachers

4.1 *Mathematics Pedagogical Content Knowledge of Teachers*

The term Pedagogical Content Knowledge (PCK) was coined by Shulman (1987) to be the special kind of knowledge combining content and pedagogy which is peculiar to teachers and necessary for teaching. The issue of Mathematics Pedagogical Content Knowledge (MPCK) has been of concern in teacher education research of the United States of America. Review of research studies in this area can be found in Mewborn (2003), Ball, Lubienski, and Mewborn (2001), Grouws and Schultz (1996) and Ma (1999). However, the links between teacher characteristics and student outcomes is far from clear and although it is generally agreed that teachers with strong MPCK will be able to teach "better" lessons, the actual measurement of the abstract concept of MPCK is far more difficult than the measurement of mathematical knowledge which is usually done through achievement tests. Studies which look at teachers' understanding of mathematical concepts are also included in this section since the teachers' content knowledge is an integral part of their MPCK although strong mathematics content knowledge is a necessary but not sufficient condition for strong MPCK.

Four studies specifically investigated teachers' understanding of mathematics rather than MPCK. Anderson and Wong (1989) compared the basic mathematical skills of pre-service teachers being prepared for primary school teaching from Singapore's Institute of Education and from a similar institution in Australia. They found that the Singapore pre-service teachers scored significantly better although majority of both groups (73% of the Singapore group and 64% of the Australian group) attained mastery level of at least 75% of the test. Kaur (1990) adapted a mathematics test covering several topics in the primary school syllabus from one used by an Australian teacher education institution and used it to test pre-service teachers in the Certificate in Education[1] programme who were being prepared for teaching at primary level. She found that almost 19% of the student teachers did not meet the 75% mastery level criterion. At the secondary and junior college level, Lim (1991) analysed the understanding of Calculus concepts of student teachers in the Diploma in Education[2] programme. The survey of 31 student teachers found that while they were competent in manipulative skills, they were lacking in the understanding of concepts (such as concept of limits) and were unable to explain these in a clear and correct way. In the fourth study, Liu of the Centre for Research in Pedagogy and Practice (CRPP) at the National Institute of Education, Singapore, studied the junior college teachers' personal understanding and pedagogical understanding of statistics. The study interviewed 18 teachers in two parts: first to understand their instructional practices and their understanding of pedagogy and the second was for the teachers to solve 25 statistics problems to gauge their understanding of concepts such as population and sample, probability, sampling distribution, hypothesis testing, confidence interval and so on. However, the findings of the study have not been published.

As mentioned earlier, while there is a general understanding of what constitutes MPCK, it is far more complex to define it exactly and to measure it. The study of Ong (1987) was aimed at developing and validating an Observer Rating Scale of Teaching Competence in Mathematics (ORSTCM). The actual format of the ORSTCM was based on four areas of teaching competency, namely, (i) command of mathematical content, (ii) preparation as including an understanding and application of methodological principles and thoroughness and detail in lesson preparation, (iii) lesson presentation in terms of motivation, flair and originality, effectiveness, use of appropriate teaching materials, and (iv) pupil learning which included promotion of mathematical thinking, awareness of individual learning rates, level of rapport and interaction and encouragement of problem solving. The study found the ORSTCM to be a reliable and valid instrument based on data gathered through the

[1] This programme was renamed Diploma in Education programme in 1991.
[2] This programme for university graduates was renamed PGDE (Sec) in 1991.

rating of video lessons as well as through the lesson observation of 34 student teachers during practicum.

In more recent years, a funded research project was initiated in 2003 to study the development of MPCK in novice teachers. This project, known as the MPCK project, first developed a test instrument for measuring MPCK for primary mathematics teachers based on four constructs. The instrument was first piloted with two groups of student teachers in 2003 and subsequently used to measure the development of MPCK in one Diploma in Education cohort and one PGDE(P) cohort. The findings were that the student teachers in both programmes showed significant improvements in their MPCK after their programme although these improvements varied across constructs and across topic areas. The reader is referred to Lim-Teo, Ng, and Chua (2006), Lim-Teo et al. (2007) and Cheang et al. (2007) for the detailed results.

The MPCK project also sought to understand how beginning teachers developed MPCK after they graduate from pre-service programmes. Four beginning teachers from two programmes were observed and video-taped teaching a unit of mathematics lessons a few months after they graduated from their programmes and again a year later. Through coding the transcriptions of the video tapes, the study is seeking to identify MPCK-in-action observable outcomes which are illustrated by episodes in the mathematics lessons and whether these outcomes have developed in the one year period. This work of the project is still ongoing.

Another study, Wong and Ng (2007), used a case study method to investigate the PCK of three established primary mathematics teachers in teaching the concept of area. Based on lesson observations and interviews, the findings showed that teachers had a perceptive understanding of pedagogies, pointing to good pedagogical content knowledge. They were observed to use practices such as proceeding from simple to complex knowledge, using of real life examples to illustrate abstract concept of area and made links to other topics. Some of these practices coincide with those identified in the MPCK study as desirable practices.

On a more specific competency area, Yeo (2004) investigated the problem-solving frameworks of twenty prospective secondary mathematics teachers. He found that fifteen of the teachers had frameworks which consisted of Polya's four phases of problem-solving and that they were able to be very detailed and comprehensive in reflecting upon their own practices in problem-solving.

While the above studies investigated the actual MPCK of teachers as determined by tests, observations and interviews, the perceptions of stakeholders on what pedagogical practices are effective for learning mathematics present another angle from which MPCK-in-action outcomes could be studied. Lim and Wong (1989) sought to identify the characteristics of an effective mathematics teacher through a 40-item questionnaire with a 4-point Likert scale. The findings showed that the characteristic most highly rated by both pre-service and practicing teachers was the "ability to explain concepts or methods clearly."

These observable outcomes are also studied in the third strand of the MPCK project, which sought to investigate what practices school middle management valued in their mathematics teachers. Mathematics Heads of Departments in primary schools were surveyed to find their perceptions on various MPCK-in-action outcomes, whether they regarded them as important in contributing to student learning. Selected department heads were subsequently interviewed to expand on their perceptions. In general, the findings (Lim-Teo, Ng & Chua, 2006) showed that Singapore department heads valued practices which contributed towards conceptual understanding and pupil motivation rather than those which developed procedural skills. While the department heads strongly felt that the given MPCK-in-action practices were desirable for effective teaching, there was some concern with the ability of their teachers to carry out these practices effectively.

4.2 *Pedagogical Practices in Singapore*

While the MPCK studies in the previous section could include some pedagogical practices of teachers as putting MPCK into action, studies in this section investigate the actual classroom practices of teachers in general. As mentioned earlier, studies on effectiveness of particular pedagogical innovations would not be part of the scope of this chapter.

A large-scale research study managed by the Centre for Research in Pedagogy and Practice (CRPP) in NIE investigated the pedagogical practices of teachers in Singapore using the Singapore coding system developed by the CRPP. The general findings based on 220 mathematics lessons show that the pedagogical practices of Mathematics teachers tend to be teacher-centred, didactic, patterned and consistent, compounded by an emphasis on logical and algorithmic problem solving tasks with convergent outcomes. These findings are reported in Nie, Lau, and Hogan (2005) and in Venthan and Abd Rahim (2007). A further analysis of the official curriculum documents and the data from the lesson observations by Abd Rahim and Liu (2007) show a tightly coupled system of curriculum, pedagogy and assessment and proposes that this strong alignment could be the stimulus for Singapore's success in international tests but may present challenges to efforts to foster high peaks of excellence.

Another study which includes a strong strand in pedagogical practices in Singapore is the Learner Perspective Study, an international project with home base at the University of Melbourne. The Singapore contribution to this project is entitled Student Perspective on Effective Mathematic Pedagogy whose objective is to study the relation between teacher practices and learner practices across a sequence of lessons. The study explores the practices in three secondary two mathematics classrooms, the pedagogical flow with respect to the nature of teachers' instructional approaches, the role of textbooks and homework, the questioning used by teachers etc. Publications related to this study are Kaur, Low, and Seah (2006), Seah, Kaur, and Low (2006), Kaur, Low, and Benedict (2007), and Benedict and Kaur (2007).

The last study in this section is a case study examining the balancing act of teaching mathematics by Leong and Chick (2006). It examines the complexities involved in the actual work of classroom instruction by looking at the interactions among the goals of teaching during a series of geometry lessons in an intact secondary one class. The method of research is through analysing video-taped lessons and post lesson reflection data and the findings showed that teaching carried problems of trying to achieve multiple goals which may conflict with each other in the reality of carrying out the lesson.

4.3 *Initial Teacher Preparation Programmes*

While the studies in Section 4.1 on MPCK of teachers may include pre-service teachers, this section will focus on research concerning pre-service teacher education programmes from perspective of the programme.

The Teacher Education Study in Mathematics (TEDS-M) project, begun in 2006, is part of an international study on the preparation of mathematics teachers for elementary and (lower) secondary schools in about 20 countries, examining the links between teacher education policies, practices and outcomes. Led by Wong Khoon Yoong, the local research team consists of teacher educators at the NIE. The objectives of this international study are covered under 3 components of the study:

1. To understand teacher education policies for elementary and secondary mathematics teachers and the cultural and social contexts of these policies. The theoretical framework developed by the international team will enable local teacher educators to reflect on Singapore policies and contextual factors from a different perspective.
2. To study the various routes, programs, standards and expectations of the learning of future mathematics teachers. This includes an investigation of the perceptions of mathematicians, mathematics educators, teacher educators, and practicum supervisors.
3. To study the mathematics content knowledge and pedagogical content knowledge of future mathematics teachers.

Since NIE is the only pre-service teacher education institution in Singapore, the study covers all pre-service programmes at the NIE. The first component was carried out through a series of meetings and electronic communication in 2006. At the time of writing (2007), the study in Singapore is at the data gathering stage with respect to the second and third components. Component 3 of the study will be more aptly classified under Section 4.1 and the findings will strengthen the understanding of pre-service teachers' MPCK and further the work of the MPCK project.

One other study to be included in this section is a small-scale study by Chua (2005) on twenty-two pre-service teachers. She found that by using observation, daily logs and critical reflection during their one-week of school experience in the earlier part of their pre-service programme, the teachers were able to set goals and chart their own professional development during their practicum.

4.4 *Professional Development of Mathematics Teachers*

The underlying objective of professional development is to continue to develop teachers so that they become more effective in their practice. Compared to pre-service teachers who are novices, in-service teachers have more classroom experience and thus learn differently. The fact that they attend professional development courses on top of their teaching duties also places constraints on their learning. The studies outlined here therefore examine various approaches and models to professional development to find out which approaches are more effective in achieving the goals of developing the teachers concerned.

The first study by Dindyal and Tan (2005) discussed the issues in the implementation of a three-month long school-based professional development project. Premised on the empowerment paradigm, this approach was designed based on several characteristics such as perceived need by the teachers, initiation by the teachers, support of school management, being situated at the school premises, no predetermined content and no formal assessments. The project team was considered a part of the group rather than having an outside expert role. The findings were on the whole positive although there were some problems such as on-site distractions, initial perception of project team as outside experts and some lack of PCK.

Along the same beliefs about effective professional development being school-based, and based on the findings on pedagogical practices, a new project "Enhancing the Pedagogy of Mathematics Teachers" was begun in 2006. This CRPP project, led by Berinderjeet Kaur, will work with communities of mathematics teachers in five primary schools and five secondary schools to help teachers examine their existing pedagogies, acquire the know-how of suggested pedagogies, use their learning in classroom practice and reflect on their learning journeys. This project is still in its beginning stage.

Ng (2005) also examined her approach in an in-service module entitled Enrichment Activities for Primary Mathematics. She used a "Replicate-Reflect" teaching learning model which encouraged teachers to replicate what was learned in workshops immediately afterwards in their own teaching, reflect upon their practice and share their experiences with their coursemates. They were also encouraged to conduct in-house workshops for their own schools. As learners, it was found that the teachers understood mathematical structures better and had improved in their mathematical thinking. As reflective practitioners, the approach enabled them to improve their articulation and communication of their mathematics teaching and to regard different teaching approaches with greater confidence and respect for complexities.

The study by Ho (2007) sought to provide a professional development pathway for teachers to widen their repertoire of heuristics and problem-solving approaches after findings from his research showed that some teachers leaned towards a limited teaching for problem-solving stance.

5 CONCLUSION

This chapter has sought to paint a broad picture of pre-service and in-service teacher education of mathematics teachers in Singapore. It has also summarised the research in mathematics teacher education up to the time of writing. It would appear that almost all the research cited has been carried out by NIE faculty. This is because, although there is much research work done by graduate students and teachers, such work tends to be on pupil learning and effectiveness of specific teaching strategies and not on teacher education which is of less interest to practicing teachers.

One area yet to be mentioned is the work by the Mathematics and Mathematics Education faculty at NIE in the organisation of mathematics education conferences locally and assisting in such conferences overseas. The intention of such work is not only to promote research sharing among academics but also the dissemination of research findings, pedagogical and research skills to teachers, thus contributing towards the professional development of teachers.

Due to the smallness of Singapore and the existence of one teacher education institution, this chapter is also a story of NIE's journey in preparing and developing mathematics teachers for the Singapore education system in the past twenty years. The scope of the story will widen in the next few years to the region and this has already begun as NIE builds upon its work of the past to reach out to impact educators and education beyond Singapore's shores.

References

Abd Rahim, R., & Liu, Y. (2007, April). *Mathematics pedagogical practice in Singapore.* Paper presented at the Annual Meeting of the American Educational Research Association, Chicago.

Anderson, A., & Wong, K.Y. (1989, July). *A cross cultural study of basic numeracy skills of primary education student teachers at Singapore Institute of Education and St George Institute of Education.* Paper presented at the Mathematics Education Research Group of Australasia/Mathematics Education Lecturers Association Conference, Bathurst, Australia.

Ball, D.L., Lubienski, S.T., & Mewborn, D.S. (2001). Research on teaching mathematics: The unsolved problem of teachers' mathematical knowledge. In V. Richardson (Ed.), *Handbook of research on teaching* (4th ed.), (pp. 433 – 456). Washington: American Education Research Association.

Benedict, T.M., & Kaur, B. (2007, May). *Using teacher questions to distinguish pedagogical goals: A study of three Singapore teachers.* Paper presented at Redesigning pedagogy: Culture, Knowledge and Understanding Conference. National Institute of Education, Nanyang Technological University, Singapore.

Cheang, W.K., Yeo, K.K.J., Chan, C.M.E., Lim-Teo, S.K., Chua, K.G., & Ng, L.E. (2007). Development of mathematics pedagogical content knowledge in student teachers. *The Mathematics Educator, 10* (2), 27 – 54.

Chua, K.G. (2005). Effective observation and critical reflection. *Journal of Science and Mathematics Education in South East Asia, 28*(1), 1 – 15.

Comiti, C., & Ball, D.L. (1996). Preparing teachers to teach mathematics: A comparative perspective. In A. Bishop (Ed.), *International handbook of mathematics education* (pp. 1123-1153). Netherlands: Kluwer Academic Publishers.

Deng, Z., & Gopinathan, S. (2001). Research on teaching and teacher education in Singapore (1989 – 1999): Making a case for alternative research paradigms. *Asia-Pacific Journal of Education, 21*(2), 76 – 95.

Dindyal, J., & Tan, J.L. (2005, May). *Professional development of mathematics teachers: A school-based approach.* Paper presented at the Redesigning Pedagogy: Research, Policy, Practice Conference. National Institute of Education, Nanyang Technological University, Singapore.

Grouws, D.A., & Shultz, K.A. (1996). Mathematics teacher education. In J.P. Sikula (Ed.), *Handbook of research in teacher education: A project of the Association of Teacher Educators* (pp. 442 – 458). London: Macmillan.

Ho, K.F. (2007, May). *A professional development pathway for teachers to teach mathematics for, about and through problem-solving.* Paper presented at Redesigning Pedagogy: Culture, Knowledge and Understanding Conference, National Institute of Education, Nanyang Technological University, Singapore.

Kaur, B. (1990). A study of the basic numeracy skills: Primary education student teachers at the Institute of Education. *Singapore Journal of Education, 11*(1), 74 – 81.

Kaur, B., Low, H.K., & Seah, L.H. (2006). Mathematics teaching in two Singapore classrooms: The role of the textbook and homework. In D. Clarke, C. Keitel & Y. Shimizu (Eds.), *Mathematics classrooms in 12 countries: The insider's perspective* (pp. 99-115). Rotterdam: Sense Publishers.

Kaur, B., Low, H.K.C., & Benedict, T.M. (2007, May). *Some aspects of the pedagogical flow in three mathematics classrooms in Singapore.* Paper presented at Redesigning Pedagogy: Culture, Knowledge and Understanding Conference. National Institute of Education, Nanyang Technological University, Singapore.

Lee, P.Y. (Ed.). (2006). *Teaching secondary school mathematics: A resource book.* Singapore: McGraw-Hill.

Lee, P.Y. (Ed.). (2007). *Teaching primary school mathematics: A resource book.* Singapore: McGraw-Hill.

Leong, Y.H., & Chick, H. (2006). *An insight into the balancing act of teaching.* Technical Report ME 2006-01, Mathematics and Mathematics Education Academic Group, National Institute of Education, Singapore.

Lim, S.K. (1991, September). *Do pre-service students (teachers) understand calculus concepts?* Paper presented at the Fifth Annual Conference of the Educational Research Association of Singapore.

Lim, S.K., & Wong, K.Y. (1989). Perceptions of an effective mathematics teacher. *Singapore Journal of Education, Special Issue,* 101 – 105.

Lim, S.K., & Yap, S.F. (1996). Restructuring a mathematics methodology course. In R. Zevenbergen (Eds.), *Mathematics education in changing times: Reactive or proactive* (pp. 103-110). Proceedings of the Mathematics Education Lecturers' Association Conference, Melbourne, Australia.

Lim-Teo, S.K., Chua, K,G., Cheang, W.K., & Yeo, K.K.J. (2007). The development of Diploma in Education student teachers' mathematics pedagogical content knowledge. *International Journal of Science and Mathematics Education, 5*, 237 – 261.

Lim-Teo, S.K., Ng, L.E., & Chua, K.G. (2006, November). *MPCK variables valued by schools' mathematics department heads.* Paper presented at Asia-Pacific Educational Research Association Conference, Hong Kong.

Ma, L. (1999). *Knowing and teaching elementary mathematics: Teachers' understanding of fundamental mathematics in China and the United States.* Mahwah, NJ: Erlbaum.

Mewborn, D.S. (2003). Teaching, teachers' knowledge and their professional development. In J. Kilpatrick, W.G. Martin & D. Schifter (Eds.), *A research companion to the Principles and Standards for School Mathematics* (pp. 45 – 52). Reston, VA: National Council of Teachers of Mathematics.

National Institute of Education. (2002). *Moulding lives, shaping tomorrow: the NIE story.* A commemorative volume on the official opening of the NIE. Singapore: Author.

Ng, S.F. (2005). Teachers as learners, teachers as reflective practitioners: A model for professional development. In P. Singh & C.S. Lim (Eds.), *Improving teaching and learning of mathematics: From research to practice* (pp. 1 – 26). Kuala Lumpur: Universiti Teknologi MARA.

Nie, Y., Lau, S., & Hogan, D. (2005, May). *Patterns of pedagogical practices of mathematics and English teachers in Singapore classrooms.* Symposium presented at the Redesigning Pedagogy: Research, Policy, Practice Conference. National Institute of Education, Nanyang Technological University, Singapore.

Ong, S.T. (1987). Validation of an observer rating scale of mathematics teacher classroom performance. *Singapore Journal of Education, 8*(2), 56 – 66.

Seah, L.H., Kaur, B., & Low, H.K. (2006). Case studies of Singapore secondary mathematics classrooms. In D. Clarke, C. Keitel & Y. Shimizu (Eds.), *Mathematics classrooms in 12 countries: The insider's perspective* (pp. 151-165). Rotterdam: Sense Publishers.

Shulman, L.S. (1987). Knowledge and teaching: Foundations of the new reform. *Harvard Educational Review, 57*, 1 – 22.

Toh, T.L., Chua, B.L., & Yap, S.F. (2007). School mathematics mastery test and pre-service mathematics teachers' mathematics content knowledge. *The Mathematics Educator, 10*(2), 85 – 102.

Venthan, A. M., & Abd Rahim, R. (2006, April). *A large scale study of Singapore's science and mathematics pedagogy.* Paper presented at the Annual Meeting of the American Educational Research Association, San Francisco.

Wong, S.O., & Ng, S.F. (2007, May). *Pedagogical content knowledge and the teaching of area.* Paper presented at Redesigning Pedagogy: Culture, Knowledge and Understanding Conference. National Institute of Education, Nanyang Technological University, Singapore.

Yeo, K.K.J. (2004). Problem-solving frameworks of prospective secondary mathematics teachers. *Journal of Science and Mathematics Education in South East Asia, 27*(1), 54 – 64.

Chapter 3

Learning Communities: Roles of Teachers Network and Zone Activities

CHUA Puay Huat

The purpose of this chapter is to unravel the roles of Teachers Network (TN), the school clusters and zones in promoting mathematics learning among teachers in Singapore. Specifically, the author draws on the concept of Learning Communities to discuss teachers' continuing learning in mathematics in these communities. The case study on TN in particular looks at how, at the national level, teachers are trained to create an environment for mathematics sharing and learning through the practice of the Learning Circles (LC) methodology in their schools. At the zone level, the author describes the work of a Mathematics Centre of Excellence (COE) in its creation of scaffoldings for collaborative learning and sharing of mathematics for schools in the East Zone. Through their activities and programmes, both the TN and the COE spawn learning communities within schools. Issues challenging the roles of TN and of COE in teachers' continuing professional development and the implications of these issues for policy makers are also discussed.

Key words: teacher education, learning communities, Learning Circles, Teachers Network, action research

1 TEACHER EDUCATION IN GENERAL

In Singapore, formal mathematics teacher training is provided by the National Institute of Education (NIE). Besides the various NIE mathematics teacher education programmes described in Chapter 2, teachers also go through some informal collegial learning in schools. Teachers can also be involved in sharing of classroom practices within their mathematics departments. Others may be engaged in sharing outside schools at the Teachers Network (TN) which is a centre set up by the Ministry of Education for the further professional development of its teachers. Teachers can also be engaged in professional exchanges at school clusters and school zone events. Specifically within a school zone, there is a Centre of Excellence (COE) for mathematics which is set up to foster teacher learning and sharing. Such on-going sharing plays an important role in promoting the professional growth of teachers. They provide a social milieu for inquiry into teachers' practices in a non-formal environment. Besides, these episodes of learning also provide opportunities for mathematics teachers to cross-pollinate their localized knowledge of their practices across schools which otherwise would remain insulated within their departments. Fan and Cheong (2002) describe such teachers' "own teaching experience and reflection", informal sharing with colleagues as the more important sources by which teachers can develop their pedagogical knowledge. This is what Cochran-Smith and Lytle (1999) term the "knowledge-of-practice" (learning through working within "contexts of inquiry communities") in their conception of teacher learning. Elsewhere, Llinares and Krainer (2006) use the term "mathematics teachers as learners" to emphasise the importance of inquiry in the professional growth of practicing teachers. Shulman (2004) takes the view that teachers must be in "communities where they can actively and passionately investigate their own teaching" (p.498) so that they can build a knowledge base that goes beyond what they would have in isolation in their departments.

The purpose of this chapter is to explicate the roles of the TN and the mathematics COE in the East Zone in contributing to this non-formal aspect of teacher development through their spawning of learning communities in schools.

2 LEARNING COMMUNITIES

Increasingly there has been recognition that the professional learning community plays an important role in teacher development (Beck, 1999;

Day & Sachs, 2004; Stein, Silver & Smith, 1998; Toole & Louis, 2002). The term "learning community" is used variously to mean different things. On the one hand, there are teacher communities of practice where members come together to solve problems of practice, share knowledge and learn from each other. Such a community is characterized by having a shared purpose, having evidence of broad networking, sharing of craft knowledge and having an enquiring collaborative culture. Martin-Kniep (2004) describes such communities as being versatile in terms of their sizes and in terms of their structures. Their sizes can vary and their setups can vary from loose informal groupings to those with fixed formal structures. On the other hand, there are also university learning communities. Among universities in America, a learning community comprises a small group of students from different faculties enrolling in two or more courses together with the intention to build academic as well as social connections between what "otherwise would be discrete academic and social experiences" (Tinto, 2002). It also requires faculty and staff members to collaborate so that the learning experience provides for an academic coherence that is interdisciplinary in nature. In addition to other qualities, such a community (where membership is voluntary) typically focuses on shared learning, shared leadership and the valuing of diversity of perspectives among its members. Yet there are also virtual learning communities which Lewis and Allan (2005) characterize as having discussions and learning taking place through active on-line collaboration. In this chapter, references are made on the commonalities across these various conceptions of learning communities. A learning community is conceived to be one that is characterized by voluntary membership, a leadership that is facilitative, a sense of a shared purpose to solve a problem or to resolve an issue, a spirit of collaboration and the valuing of diversity of views. The roles of TN and the COE in bringing about the growth of learning communities will be discussed in terms of their organisational structures, the nature of their programmes and activities and their impact on teachers professional learning.

3 TEACHERS NETWORK AND LEARNING CIRCLES

3.1 *Organisational Structure*

To support the Ministry's vision of producing thinking schools, it set up the Teachers Network (1998) to provide for the professional development of its teachers. Launched on 30 April 1998, TN is part of the Ministry's efforts to produce thinking teachers who must have the interest to improve their professional practice through continuous learning and sharing. Whilst TN has in place various structures and programmes (http://sam11.moe.gov.sg/tn/abouttn.html) to promote the professional and general well-being of teachers, the focus in this chapter will be on a study of its Learning Circles (LC) methodology where teachers are trained to facilitate their own group learning using tools of reflection and dialogue. Such a focus will help to unravel the role of TN in the continuing education of mathematics teachers. LC, which is recognized by the Ministry as one of the three innovation tools that schools could use for their work improvement, promotes collaboration, reflection, sharing and learning among the teaching fraternity. This is in line with TN's vision "to build a fraternity of reflective teachers dedicated to excellent practice through a network of support, professional exchange and learning." TN therefore aims to be a catalyst to spawn teacher collaboration and learning. TN operates on an invitational mode by publicizing its LC methodology and activities through its website. It also uses the Ministry's Training and Development Division platform in reaching out to schools.

TN is supported by an administrative team and a team of 12 PDOs (Professional Development Officers) who serve schools in the four zones. Each PDO is an LC activist who champions and propagates LC methodology to schools which have indicated their interest to embark on the LC journey. When invited, a PDO team from TN will give teachers an introductory workshop on the skills of LC reflection and dialogue. A PDO then follows through with the LC with an action plan. A typical LC comprises four to eight teachers. It has a lifespan of six to nine months. Members of an LC may come from the same school or they may come from different schools. An LC is formed to serve a specific purpose of looking at teaching and learning issues like mathematics teaching and students learning. An LC is facilitated by a member who is trained and is partnered closely throughout the LC process by a PDO. Each PDO is a "critical friend" to the LC in terms of guiding the group in its methodology by being a professional critic, being a sounding board and by lending emotional support to the team. The PDO facilitates the processes during which teachers explicate their practices and create new knowledge. The PDO then helps to record, publish and disseminate such learning so that knowledge of practice can be replicated elsewhere in schools. If necessary, a PDO may also get expert help in a specific area.

3.2 *Nature of Programmes and Activities*

TN's core beliefs about personal mastery, shared vision, mental models, team learning and systems thinking are drawn from Senge et al. (2000)'s five inter-related disciplines about a Learning Organisation. In a Learning Organisation, members continually expand their learning capacities through learning together, freeing their collective aspiration and through nurturing "new and expansive patterns of thinking" (Infed, 2006). TN's five core beliefs form the working principles under girding the working of an LC. The concept of the LC closely mirrors Action Research (AR) which incorporates reflection, journal writing, dialogue, collaboration and the 5 step RPAOR cycle (R: initial reflection; P: plan, A: act; O: observe; R: critical reflection). As an example, teachers can use the LC as scaffolding to surface their mental model of students' picture of a mathematical concept, to challenge their practices about teaching that particular concept and to draw an action plan to improve those practices that can translate to better students learning. Ball (1996) takes the view that such approach of contextualizing their teaching practices and reflection through the on-going interactions with colleagues is an effective professional development model for teacher learning.

Unlike learning from some mathematics in-service courses, an LC does not serve as a disseminator of specific content knowledge. It draws heavily on the members' sense of inquisitiveness into the group's practices. It fosters an attitude of self-critique and of drawing upon contextual knowledge to resolve a workplace issue. In this sense, the theory behind an LC is also linked to constructivism (knowledge creation) and to andragogy (adult learning). An LC also has a shared and facilitative leadership where the leader leads without dominating and behaves in a way that enhances the collective ability of the group to adapt, solve problems and to improve performance.

3.3 *Impact of Learning Circles on Teachers' Learning*

In this section, the author is most grateful to TN officer, E. Chong (personal communications, August 24, 2006, September 6, 2006, September 11, 2006) for providing information that had helped to illustrate the impact of LCs on teachers' learning.

Much of the learning outcomes of LCs are captured through personal journals, group reflective journals, TN publications, TN teacher-led workshops and TN conferences. As an example, a mathematics LC in School A (a government-aided primary school), reflected in their group reflective journal that "being part of the LC has enabled us to become more conscious of the needs of our pupils" and that "too often, we have made sweeping assumptions about why our pupils are not able to learn." In another group's reflective journal, a team from School B (a government primary school) shared about how Primary 4 (aged 10) students can be helped to improve on their problem solving skills in mathematics by supplementing the demonstration of problem solving strategies with affective actions like encouragement and extension of questions (posing new questions based on the original question). They reflected that they had become more empathetic towards their students and more focused, patient and reflective in their teaching. This reflection on teaching and student learning as an interactive and consultative process was also shared by the LC from School C (a government secondary school). In their project on how to use "five-senses" to help weaker students remember the basic formulae in solving problems about mensuration, they used a combination of strategies involving pair work, chorus/reciting, students presenting on the whiteboard and students proposing their own ways of remembering the formulae. The team believed that students performed better if they engaged themselves in "at least three of out of the five senses", namely, listening, visualizing and verbalizing. The team were surprised to find when asked to come up with strategies for remembering, students were able to come up with their own ideas of remembering formulae in the topic on mensuration. Although their suggestions may not be new to the teachers or pedagogically high in value, the team noted that students' pride in taking ownership of these strategies appeared to have worked in improving their performance in solving problems involving such formulae. Two of the students' suggestions were $V = (1/3)\,Ah$ (to be read as "One-third *Ah*") for remembering the formula for volume of a pyramid and $A = \pi rl$ ("the π is ready to be loved") for helping them to recall the surface area of a cone. The team also reflected that the high level of cooperation, openness and a tolerance of diverse views among the team members were key factors to the success of their LC.

TN has also published two handbooks for mathematics teachers (one for primary teachers and the other for secondary teachers). In these publications, teachers share their successful practices and useful strategies and innovative ideas for more effective classroom teaching. As an example, there is a sharing of common errors made by students in upper secondary Elementary Mathematics and strategies to help students in this area from the mathematics LC team of School D (a government secondary school). Another team from School E (a government secondary school) looked into the motivating and monitoring strategies in Additional Mathematics. In another publication, "Challenging Maths Problems Made Easy", Wan (2006) shared resources about the Primary 6 Mathematics model methods of problem solving.

TN teacher-led workshops and TN conferences are the other platforms which serve to promote communities of learning in schools. In TN teacher-led workshops which incorporate LC methodology, classroom teachers share their teaching ideas to fellow practitioners. The contents of such workshops are therefore contextualized to be directly relevant to the local classroom. As examples, there were workshops on "Designing Effective Student Centred Tablet PC Lessons" by two teachers from School F (a government secondary school) and on "Eureka Maths" (problem solving for non-routine questions) by a teacher from School G (a government primary school). Participants reflected that they had benefited in terms of picking up good ideas. Although such workshops do bring about the professional development of the presenters when they shared their implicit knowledge of their practice, not much is known about the follow-through learning in terms of how participants actually translate these into practices and how such knowledge influences their professional views of a topic.

Teachers also share their learning on the TN website. Tripp (2004) in his study of TN describes this website as an important repository and an archive of teacher knowledge. TN website archives contain links like the TN Workshop, TN Conference, LC method and the Reading Room (for resources on general reading but not mathematics specific topics). It also has its online catalogue of resources at Read@TN link. Besides promoting professional growth of teachers through its LC methodology, TN also has in place a recognition structure to honour teachers who have contributed to developing the community of teachers as learners. A teacher will be recognized as a "TN Associate" if he or she has either contributed more than 20 hours to TN or have written or edited at least three publications that are the direct outcomes of TN activities. This recognition structure has contributed to the teachers' sense of professionalism across the fraternity. The number of LCs in schools is shown in Table 1. But anecdotal evidence seems to suggest that LCs are more popular with primary schools. The number of new LCs in 2006 shows that 84 are from primary schools, 35 from secondary schools and only three are from junior colleges.

Table 1
Number of LCs (2003 – 2005)

Year	Number of LCs
2003	196
2004	156
2005	220
2006	122

Although the impact of LCs may not be fully quantified by the number of sessions offered at TN conferences, the number may be taken as a proxy indicator. Across the three TN Conferences (Table 2), the number of mathematics sharing sessions remains small. These concurrent sessions by teachers involved topics like how to help teachers infuse National Education in mathematics classroom teaching, how to create interesting classroom games in the teaching of bearing, algebra and rate of change and how to build strategies to motivate students. Although the LC methodology fosters teachers' learning and reflection into their classroom practices, there have been concerns about its viability in the further professional development of teachers. With increasing demands on teachers to deliver results and to complete administrative tasks, membership to an LC may be construed as an add-on layer of work for teachers if the intent of the LC methodology is not properly communicated.

Table 2
Teachers Conferences (2001 – 2006)

	2001	2004	2006
Number of Sessions	105	93	48
Number of Math Sessions	7	7	4
Percentage of Math Sessions	6.7	7.5	8.3

The sustainability of an LC and the keeping of the intent of an LC therefore may present a challenge to school leaders. Teachers need to believe that the LC methodology is a viable and meaningful platform for them to develop professionally.

4 CENTRE OF EXCELLENCE (COE) FOR MATHEMATICS AT EAST ZONE

4.1 *Organisational Structure*

Since 1998, schools in Singapore are grouped into clusters according to their geographical locations (*Cluster Schools and Its Challenges*, 1998). Several clusters are then grouped into a zone. There are four zones: North, South, East and West Zones. Each cluster is facilitated by a Cluster Superintendent. Cluster Superintendents help to establish networking, sharing and collaboration among schools within the zone in order to raise the capacity of the school leadership teams and the level of performance in each school. Within a cluster, there are special interest groups that serve particular needs. For example, there are mathematics groups comprising Heads of Departments from different schools who organise enrichment activities like Mathematics Trail and Mathematics

Camp for students within their cluster (*Cluster Teamwork* x *Maths Camp*, 1999; *Maths Trail*, 2001). There are also various mathematics workshops and discussion forums organised by mathematics cluster support groups to take care of teachers' learning needs in specific topics of common interest and to foster professional networking among the teachers. These learning communities serve the needs of schools within the clusters. At the zone level, there are Centres of Excellence. The purpose of each Centre is to promote teacher professional development in a specific subject area. For example, there are Centres of Excellence for Arts and for Physical Education. Each Centre serves as a platform for active discussion and sharing of best practices among teachers within the zone. The focus of this chapter is to look at the East Zone Centre of Excellence (COE) for mathematics which was set up on 6 September 2003 and hosted at Tao Nan School (Chua, 2003; *Towards Maths Excellence*, 2004). It is one of the more established mathematics COE among the four zones. It serves 45 primary schools, 40 secondary schools and four Junior Colleges.

Like the TN, COE is a formal grouping initiated by the Ministry's Schools Division. The primary purpose of the COE is to promote good mathematics classroom teaching and learning among schools in the East Zone so as to bring about higher standards of mathematics in these schools. From its Work Plan documents for 2006, the core business of COE is to "lead in the pursuit of excellence in the teaching and learning of mathematics." Its vision is to be the Centre of Excellence in mathematics pedagogy and practices in the East Zone. It does this by enhancing the capacity of its mathematics teachers and by leading the drive for innovation and creativity in the teaching and learning of mathematics. It actively promotes Action Research (AR) as one of the thinking tools for teachers. To engender a more sharing and learning culture, it organises the annual EZ Mathematics Sharing Day for mathematics teachers in the East Zone. Besides inviting NIE's MME (Mathematics and Mathematics Education) staff and partners from other agencies to share their professional knowledge, the day is also a celebration of teachers' work and an occasion to showcase good teaching practices in the classrooms.

The COE has a recording studio (Maths Studio) which is designed as a classroom setting with video-taping and editing capability. It is used for recording good teaching practices and for teachers' evaluation of their own teaching. The COE also has a mathematics resource room known as MATELL (Mathematics Teaching and Learning Lab) which houses teaching resources from schools. The Centre has its own e-portal which is hosted at Tao Nan School (http://www.eastzone-coemath.moe.edu.sg /default.asp). E-portal accounts are given to all mathematics teachers in the zone. The e-portal serves as a one-stop resource that produces and promotes good teaching and learning ideas and resources for teachers. Besides functioning as a channel for communication between schools and the COE, it also serves as a repository of good lesson plans, a clearinghouse of successful classroom practices, learning ideas, resources and mathematics assessment papers from schools. Using its "Matell" link, teachers can have access to pedagogical strategies that are posted by fellow practitioners. In its "Event" link, schools can publicize their mathematics programmes and activities so that others may be able to link up to schools for collaboration. Likewise, from its "Forum" link, teachers can engage in online discussions about specific topics in the mathematics curriculum.

The COE co-opts cluster representatives who are mainly Heads of Departments or Senior Teachers from the seven clusters within the East Zone. This administrative setup places the cluster groups in a complementary position within the COE. Each cluster representative acts as a spokesperson for schools within the cluster. Information dissemination about activities and initiatives from the COE and the ground feedback from schools and clusters about the COE's programmes are actively promoted. Having cluster representatives also facilitates 2-way communications between the COE and schools. It complements the COE's e-portal which is used as a platform to link schools to the COE. The Principal of Tao Nan chairs the COE Executive Committee. The Executive Committee comprises cluster representatives and three school Vice Principals who help in the running of the Centre. A panel of School Superintendents provides advice and guidance to the COE in its work. Much of the decision making within the Executive Committee are participatory in nature. The Principal leads in its strategic planning for the Centre every year. Before mapping up plans for the year, due attention is paid to the current issues on the educational horizon. In-depth discussions and considerations are made to align the Centre's work with the current Ministry's key initiatives for schools and the learning needs of mathematics teachers in schools. Partnerships with NIE's MME department and other external agencies are also made so as to add value to the work of the COE.

4.2 *Nature of Programmes and Activities*

The COE functions as a decision making body that organises training and programmes to teachers to bring about learning. Its members being representatives from clusters are in a good position to understand the professional learning needs of the schools. The activities and initiatives taken by the COE serve as catalysts to promote learning communities within schools and clusters. Its AR initiative, its annual mathematics Sharing Day, its outreach through the e-portal and its workshops are targeted at fostering teachers' continuous learning and reflection.

Whilst TN champions the LC methodology, COE adopts AR as its strategy to promote teachers as reflective practitioners. Unlike the PDOs in TN that promote their LC methodology to schools, AR training is conducted by external agencies. Like TN, teachers' involvement in AR training is on a voluntary basis. Mathematics teachers are invited to attend such training if they are keen to use AR as one of the tools for their professional reflection. Other than these external agencies, the COE does not have the capacity to provide its own follow-up continuous support to schools in their AR journey after the training. This is due in part to the nature of the COE setup where members are not full time staff.

The COE offers a spread of teacher-led sharing and training to cater for the different learning needs of its teachers. Teachers can learn about the varied classroom pedagogical practices from their fellow colleagues from different schools. This cross-pollination of ideas is actively promoted during the course of the sharing. There is also training on specific skills which are of interest to teachers. A course on item calibration and item banking, for example, has been conducted by an external agency so that teachers can learn how to set quality Primary mathematics questions. Some courses are designed for the dissemination of pedagogical content knowledge. Teachers can also learn specific knowledge through presentations by MME lecturers during the Sharing Day. But there is no active attempt to fully match schools' learning needs and the training that is organised during the Sharing Day. This difficulty is a natural consequence of COE's varied composition of schools within the East Zone and as such it is difficult to meet all their learning needs within a single platform.

4.3 *Impact of COE Activities on Teachers' Learning*

In 2005, the COE engaged an external agency to conduct its AR training to 89 teachers from 23 primary schools and 21 secondary schools. As a follow up from the course, teachers implemented small scale projects relating to their areas of work. The completed projects which addressed specific issues in mathematics teaching and learning are housed in the COE e-portal. In a study of this training, Wong and Chua (2006) found that the participants had gained insights into the process of AR and most of the participants found that the AR training was useful and that they could apply it in their own schools to bring about work improvement. As an example, a team from a primary school in teaching pie-charts did an AR project on how to better engaged students in this topic. They used a mini survey activity with the students and asking them to compile data on their drinks preferences. The team found that students' interest and performance in this topic had improved as a result of this different approach to teaching the topic. The team members reflected that they had "gained insights into the thinking of their pupils" and that they had a closer bond with pupils (by engaging the passive students in lessons). This was a result of having a more relaxed classroom atmosphere because of the informal activities. Ong, Ho, and Wong (2006) in their AR on investigating the impact of mathematics journal writing, found that besides being able to better understand their students' feelings towards the subject they were also able to identify students' misconceptions. The team were able to "reflect and modify" their teaching to address the misconceptions expressed by their students. Such an ownership of work issues and the capacity of teachers to reflect on their work are important learning outcomes which the COE is trying to propagate. There are also teams who reflected the value of working together as a team to solve a common problem. They also learnt to appreciate alternative views from members during the process of collaboration. These are essential attributes of learning communities which the COE hopes to create. To further increase the number of teachers initiated into AR, another workshop was organised in 2006 from March to September with 35 participants coming from six primary schools and 10 secondary schools.

Since 2003, COE has been organising its annual Sharing Day. In each session, mathematics teachers from the East Zone are invited to share their best classroom practices. External partners such as NIE's MME department, AME (Association of Mathematics Educators) and the TN are also invited to give presentations on that day.

Table 3
Mathematics Sharing Day (2003 – 2006)

	Number of Teachers Attended	Number of Concurrent Sessions
2003	300	12
2004	439*	33*
2005	235	16
2006	210	22

** 2 sharings in 2004 because of the official opening of COE*

Table 3 shows the number of participants and the extent of the sharing from 2003 to 2006. Although participants' feedback have always been positive, the extent of the follow up learning by teachers after these sessions have not been fully studied. A typical course evaluation form after each session only captures what the teachers had learned about the sharing at that point in time and the feedback about the presentation. But if learning is conceived as going beyond just picking up new techniques, then course evaluation may have to address questions like what exactly it is that they had learnt, what types of follow up that can be taken and to what extent had their conceptions of the topic changed after the learning. The subsequent reflection after the course is much left to the mathematics departments within their respective schools.

Although the e-portal is a platform for facilitating exchange of good teaching ideas and for serving as a springboard for innovation in classroom teaching, the number of teachers actively involved in online discussions in the "Forum Board" link has not been significant given that COE serves 89 schools. This is reflected in the low number of online topics registering hits (only eight in 2006) at various levels. Perhaps this could indicate the teachers' reluctance to participate in such exchanges because of the lack of time due to other work commitments or that it is not a popular culture for teachers to engage in online mathematical discussions. Another possible reason (technical in nature) is that the keying in of mathematical symbols and equations during an online discussion can be cumbersome and this may discourage teachers from contributing. Because of this lack of online collaboration, it is difficult to construe that the e-portal has spawned a virtual learning community for the zone. But this is not to be taken that teachers do not desire to share their teaching ideas. A total of 38 teaching resources have been uploaded in the Matell link within the e-portal and there has been at least 158 downloading of these resources by teachers. But it is difficult to describe the type of learning that takes place and the impact on classroom practice after a teacher has downloaded a teaching resource. Like the AR workshop and the Sharing Day, there has been no comprehensive study on how the e-portal has impacted on teachers learning.

The COE also organises workshops by teachers for sharing to a wider audience within the zone. In the 3-day workshop on Problem Solving and Fun Revision Strategies for Primary school teachers held on 24, 31 March and 7 April 2006, participants reflected that they found the training useful and that they could apply the ideas gleamed for their teaching. Similarly in another 2-day workshop on Assessment Strategies for Primary school teachers held on 12 and 19 April 2006, teachers also reflected that they could improve on their own scoring rubrics when setting their school assessment papers. But again, to what extent does such knowledge change teachers practice has not always been easy to fathom.

5 CONCLUDING REMARKS

Although there is no deliberate policy to create learning communities in mathematics education in Singapore, what can be inferred is that learning communities are growing as a result of a general policy towards teacher engagement in reflection and sharing. There is a move away from a centrally directed model of educational management towards one that encourages initiatives from the ground. Shanmugaratnam (2005), the previous Minister of Education, talked about the policy of a "bottoms-up approach with a top-down support" where teachers are encouraged to take ownership of workplace issues with support from the Ministry. Through this policy, a more thinking and learning culture can be encouraged and this is consistent with the Ministry's vision of "Thinking Schools Learning Nation." TN and COE are examples of groups created by the Ministry that help fan such bottoms-up engagement in workplace issues.

TN and COE by themselves are strictly not learning communities. They are more like activists of change for schools. They help to spawn learning communities within schools. TN has its LC methodology, its teacher-led workshop structure and its TN conferences. Its activities touch across a range of subjects. It is not mathematics specific. Full time PDOs in TN serve as agents of championing the LC methodology. TN targets schools at the national level. It propagates its ideology of a Learning Organisation approach towards teacher learning and development. The COE has its AR methodology, its annual Sharing Day programme, the e-portal framework, its support facilities and its teachers' workshops to meet the needs of schools in the East Zone. The COE targets specifically at fostering learning and innovation for mathematics teachers. Members of the COE are not full time staff. They are co-opted from schools and are representatives from different school clusters. Although significantly different in their setups, both TN and COE share some commonalities.

Participation in both TN and COE activities and programmes is on an invitational mode. They do not "roll out" (make mandatory) their activities and programmes to schools. Instead schools are invited to join them in their journey to bring about teacher learning. Both endorse sharing across schools and actively encourage collaboration. TN's Teachers Conference and the COE's annual Sharing Day are examples. Both TN and COE recognise the need to archive good practices and they have taken steps in this direction. TN has its publications of teachers' work and its e-portal repository. The COE has its e-portal archives of teaching ideas and the MATELL room which houses teaching artifacts contributed from schools. Above all, both share a passion towards promoting the non-formal aspect of teacher training which is important because teacher learning is situated in communities. Besides self-reflection, there is a need for teachers to have the desire to be part of a community of learners. It is important for teachers to extend the conversation about learning and collaboration beyond the close circle of their colleagues within the department. In partnering this aspect of teacher training, both the TN and the COE play important roles as advocates of learning communities and as catalysts of change by giving teachers their different scaffoldings to build learning communities in schools. But how much of teacher learning (and in particular, their learning in communities within their schools) is actually translated into improving or changing their classroom pedagogies so as to impact on students learning is not yet clear. However, it is clear that although training hours can be mandated, much of teacher learning is a complex and an individual experience and cannot be regulated. This is an area that would probably warrant further work.

References

Ball, D.L. (1996). Teacher learning and the mathematics reform: What do we think we know and what do we need to learn? *Phi Delta Kappan*, *77*, 500-508.

Beck, L.G. (1999). Metaphors of educational community: An analysis of the images that reflect and influence scholarship and practice. *Educational Administration Quarterly*, *35*(1), 13-45.

Chua, G. (2003). *A synergy of mathematical minds*. Retrieved on July 13, 2006, from http://www.moe.gov.sg/corporate/contactonline/pre2005/vol51/in_around/minds.html

Cluster Schools and Its Challenges (1998). Retrieved on July 13, 2006, from http://www.moe.gov.sg/corporate/contactonline/pre-2005/glance1/web/glance3.htm

Cluster Teamwork × Maths Camp = Fun^3! (1999). Retrieved on July 13, 2006, from http://www.moe.gov.sg/corporate/contactonline/pre-2005/vol1499/page8.htm

Cochran-Smith, M., & Lytle, S. (1999). Relationships of knowledge and practice: Teacher learning in communities, *Review of Research in Education, 24*, 249-305.

Day, C., & Sachs, J. (2004). Professionalism, performativity and empowerment: Discourses in the politics, policies and purposes of continuing professional development. In C. Day & J. Sachs (Eds.), *International handbook on the continuing professional development of teachers* (pp. 3-32). Maidenhead: Open University Press.

Fan, L.H., & Cheong, N.P.C. (2002). Investigating the sources of Singaporean mathematics teachers' pedagogical knowledge. In D. Edge & B. H. Yeap (Eds.), *Mathematics education for a knowledge-based era (Vol. 2)* (pp. 224-231). Singapore: Association of Mathematics Educators.

Infed Encyclopaedia Archives (2006). *The learning organisation*. Retrieved on October 10, 2006, from http://www.infed.org/biblio/learning-organization.htm

Lewis, D., & Allan, B. (2005). *Virtual learning communities: A guide for practitioners* (pp.5-15). New York: Society for Research into Higher Education & Open University Press.

Llinares, S., & Krainer, K. (2006). Mathematics (student) teachers and teacher educators as learners. In A. Gutiérrez & P. Boero, (Eds.), *Handbook of research on the psychology of mathematics education: Past, present and future* (pp. 429-459). Rotterdam, The Netherlands: SensePublishers.

Martin-Kniep, G.O. (2004). *Developing learning communities through teacher expertise* (pp.1-14). Thousand Oaks, California: Corwin Press.

Maths Trail Takes Off At Changi Airport (2001). Retrieved on July 13, 2006, from http://www.moe.gov.sg/corporate/contactonline/pre-2005/vol28/inaround-frameset.htm

Ong, I., Ho, W.Y., & Wong, H.P.C. (2006). Impact of math journal writing on students' performance and attitude: An action research project at St. Hilda's Primary School. *Maths Buzz, 7*(2), 3-5.

Senge, P.M., Cambron-McCabe, N., Lucas, T., Smith, B., Dutton, J., & Kleiner, A. (2000). *Schools that learn: A fifth discipline fieldbook for educators, parents, and everyone who cares about education* (pp. 59-93). NY: Currency Doubleday.

Shanmugaratnam, T. (2005). Speech delivered at the MOE Workplan Seminar, MOE 22 September 2005. Retrieved on September 29, 2006, from http://www.moe.gov.sg/ speeches /2005 /sp20050922_print.htm

Shulman, L.S. (2004). Communities of learners and communities of teachers. In L.S. Shulman & S.M. Wilson (Eds.), *The wisdom of practice: Essays on teaching, learning, and learning to teach* (pp. 485-500). San Francisco: Jossey-Bass.

Stein, M.K., Silver, E.A., & Smith, M.S. (1998). Mathematics reform and teacher development: A community of practice perspective. In J.G. Greeno & S. Goldman (Eds.), *Thinking practices in mathematics and science learning* (pp. 17-52). Mahwah, NJ: Lawrence Erlbaum.

Teachers Network (1998). Retrieved September 5, 2006, from http://www.moe.gov.sg/press/1998 /980430a_print.htm

Tinto, V. (2002). *Learning better together: The impact of learning communities on student success in higher education.* Retrieved on October 25, 2006, from http://www.mcli.dist.maricopa.edu/events/lcc02/presents/tinto.html

Toole, J.C., & Louis, K.S. (2002). The role of professional learning communities in international education. In K. Leithwood & P. Hallinger (Eds.), *Second international handbook of educational leadership and administration* (pp. 245-279). Dordrecht, the Netherlands: Kluwer Academic Publisher.

Towards Maths Excellence (2004). Retrieved July 13, 2006, from http://www.moe.gov.sg/corporate/contactonline/pre-2005/vol61/in_around/in.html

Tripp, D. (2004). Teachers' networks: A new approach to the professional development of teachers in Singapore. In C. Days & J. Sachs (Eds.), *International handbook on the continuing professional development of teachers* (pp. 191-214). Maidenhead: Open University Press.

Wan, A. C. H. (2006). *Challenging maths problems made easy.* Singapore: Marshall Cavendish Education.

Wong, K.Y., & Chua, P.H. (2006, May). *Promoting action research through the zonal centre of excellence (COE) for mathematics in Singapore.* Paper presented at the Conference on "Preparing teachers for a changing context," London.

Chapter 4

Lesson Study in Mathematics: Three Cases from Singapore

FANG Yanping Christine Kim-Eng LEE
SHARIFAH THALHA Bte Syed Haron

This chapter reports mainly the mathematics research lessons component of a two-year intervention project (2006 and 2007) funded by the Centre for Research in Pedagogy and Practice (CRPP). The project team worked closely with a government school to implement Lesson Study as a teacher-directed form of instructional improvement. The chapter introduces the conceptual framework of cultural-historic activity theory and Wenger's community of practice and how they guide our intervention. It then examines the continuous improvement processes and teacher learning through three cases of mathematics research lessons conducted in three Lesson Study cycles. The topics cover *long division, area and perimeter,* and *equivalent fractions* in Primary 3 and 4. As the cases highlight, Lesson Study has become not only a powerful tool to bring together knowledge from diverse communities but also a rich site for the induction and mentoring of novice teachers. Researchers' learning from the implementation is equally powerful.

Key words: Lesson Study, instructional improvement, community of practice, activity theory, mathematics research lessons, mentoring of novice teachers

1 INTRODUCTION

Lesson Study is a cycle of activities in which teachers design, implement, and improve one or more research lessons and make positive changes in instructional practice and student learning (Stigler & Hiebert, 1999). From the beginning of this cycle, Lesson Study groups set long and short-term goals, plan, and conduct lessons over a period that can last beyond a year. Planning is done collaboratively. One of the teachers carries the planned lesson out with team members observing the lesson and taking careful notes on how pupils respond to the lesson and how well they learn. Observers and the team of teachers then meet to review the evidence gathered during the lesson, discuss it, reflect upon ways to improve the lesson, revise it, and then teach it in another class. Observations and findings are documented and shared with other teachers.

For over five decades in Japan, teachers have used Lesson Study, a form of teacher-led instructional change cycles, to improve their practices. Lesson Study has not only been recognised as playing a part in the development of effective teaching of Mathematics and Science in Japan (Lewis, 2002a, 2000b, 2000c; Lewis & Tsuchida, 1998; Stigler & Hiebert, 1999), but it has also been acknowledged as contributing to outstanding student achievement. This ground-up teacher-led form of professional development contrasts with the typical top-down approach to reform and instructional improvement in many educational systems around the world.

In recent years, Lesson Study has been introduced to many states in the U.S. (Chokshi & Fernandez, 2004; Fernandez, 2002; Lewis, 2002 a, b, & c), implemented widely in Hong Kong in the form of Learning Study (Elliot, 2004; Lo, Pong & Pakey, 2005), in China in the form of Action Education (Gu, 2005), and in many APEC member countries (Fang & Lee, 2007; Isoda, Shimizu, Miyakawa, Aoyama & Chino, 2006). When introduced to these systems, they undergo a contextualising and adapting process to fit the local contexts and to address inherent problems of education beyond borders. This adaptation process provides an important site of inquiry for education researchers to test the vitality of Lesson Study and understand what conditions and factors lead to successful implementation in different educational systems (Lewis, Perry,

Hurd & O'Connell, 2006). The spread and impact of Lesson Study internationally have triggered our interest in finding out whether it works for Singapore and what it takes for it to work.

Teachers in Singapore attend various workshops as a major professional development avenue with limited transfer to classroom practices. An extensive two year research project by the Centre for Research in Pedagogy and Practice (CRPP), National Institute of Education (NIE), found that pedagogical practices in Singapore were dominated by traditional forms of teacher-centred and "teacher as authority" approaches with little attention to the development of more complex cognitive understanding (Luke, 2005). Lesson Study as a process and tool for teacher learning is a new phenomenon that could constitute an innovation to deal effectively with the long-standing problem of poor transfer of learning to real practice.

This chapter presents selected mathematical component of our pilot experience of Lesson Study in one primary school. We draw on Vygotskian and Engeström's cultural-historic activity theory and Wenger's notion of community of practice to frame our study and design experiment to guide our intervention. This conceptualisation underpins our design methodology, data collection, and analysis. More importantly, it enables us to examine the contexts and local education practices through direct participation, learning, and theorizing from the intervention to build sustainability (Barab & Squire, 2004). We present here three cases of mathematics research lessons across the two years to illustrate the process and features of continuous improvement that characterises Lesson Study and teacher learning. We conclude the chapter with implications for future adoption of Lesson Study by more schools as a tool to build teacher-directed instructional improvement in Singapore.

2 CONCEPTUAL FRAMEWORK OF THE STUDY

2.1 *An Activity System Mediated by Tools and Communities of Practice*

In studying human labour activity as developmental work in social institutions, Engeström (1990) brought Vygotsky's idea of the tool- and sign-mediated nature of human activity together with Leont'ev's (1977) view of the community-mediated nature of collective human activity and put together the following composite model of a human activity system as represented in Figure 1.

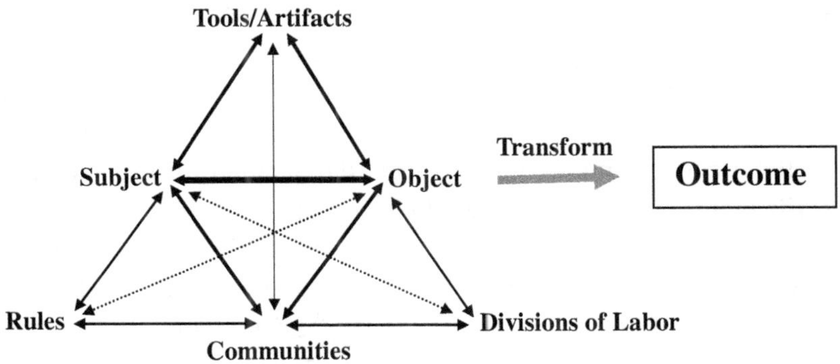

Figure 1. The Human Activity System (Engeström, 1990).

In the activity system, the human subject's (can be an individual or a group) actions and activities are mediated by both tools and artifacts (represented in the upper triangle in the above figure) and communities of practice (represented in the lower triangle) at the same time. Therefore, these actions and activities are also mediated by rules and norms governing the communities and the participation of the members is mediated by division of labour. In the process of participation, subjects are transformed (Wertsch, 1981) and such transformation represents learning from social practice or praxis.

The school as a social institution functions like an activity system. With teachers, pupils, and school administrators as the major subjects, the central actions and activities of teaching and education are directed at the object of pupils' development and learning. This is mediated by the curriculum, instructional discourse, and other tools and artifacts at different levels of the activities. Ideally, such a system should be organised as a site for teachers learning from one another and from their own daily practice.

2.2 *Lesson Study as a Tool for Building an Activity System for Professional Learning in Schools*

As Lave and Wenger (1991) posited, "A community of practice is an intrinsic condition for the existence of knowledge…." (p. 98). Facing new technologies and globalisation, schools today have been subject to constant reforms demanding knowledge and skills that they have to consistently learn and unlearn. Often, such knowledge resides in different communities such as universities, research institutions, policy makers, as well as school communities. To meet the demand for the knowledge and skills required in reform-minded practice, a community of practice needs to be formed with members coming from diverse communities but sharing common tasks (Cobb, McClain, Lamberg & Dean, 2003). Within this community of practice, knowledge, and learning are co-constructed through distributed expertise (Brown, 1992; Collins, Joseph & Bielaczyc, 2004).

Inspired by the examples of the Japanese and Chinese traditions and our own intervention experiences, we view Lesson Study as a tool that brings together the knowledge from different communities to mediate teacher learning in a local school community. Researchers brought knowledge of Lesson Study, curriculum theories and knowledge and skills in research and development; teacher educators and content area resource persons or specialists from academic groups at the National Institute of Education contributed disciplinary knowledge, pedagogical knowledge and knowledge about learning of the related content areas; and a school teaching community possessed knowledge of pupils, teaching materials, and instructional knowledge and experience. Each of these partners plays an important and indispensable role in the knowledge building processes. This arrangement has allowed us to examine the distributed expertise and boundary-crossing across the different communities in the design and implementation of our intervention. The boundary-crossing remains a key challenge in harnessing resources from both internal and external institutions to sustain such intervention work.

By introducing Lesson Study as a new set of mediating tool, we envision the school as an activity system for teacher professional learning. Ideally, in this interwoven system, the organisation of teachers' work and resources is all directed towards the central object of research lessons that aim at systematically inquiring into and bringing about the outcome of pupils' learning and development (Sato, 2006). Mediated by Lesson Study and a community of key stakeholders of education, the object is to be owned eventually by teachers and pupils. Supported by time, space, and resources, teachers are engaged in continuously improving and refining the research lessons as part of their routine practices, which will enable them to develop their pedagogical content knowledge, sensitivity towards pupil learning, and pedagogical decision-making based on evidence. Working in such an activity system, teachers are empowered as professionals who develop professional knowledge and skills in communities of practices.

Nevertheless, school institutions are activity systems operating around a set of their own established organisational routines and agendas. When a new activity system mediated through Lesson Study is introduced, it takes tremendous negotiation and strong will on the part of school leadership and teachers in restructuring resources, building new norms, and redefining roles and responsibilities. This re-negotiation process has provided numerous challenges as well as learning opportunities for researchers who have learned to do Lesson Study together with the school and its teachers for the past two years.

2.3 *Continuous Cycles of Improvement across Cycles and beyond Research Lessons*

As we implement Lesson Study, we find ourselves making adjustments and adaptations at the end of each cycle to address new issues arising from existing structures and norms. Over the four cycles in the past two years, we have gone through iterative cycles of continuous improvements in trying to make teacher inquiry of pupils' development and learning central to teachers' professional development.

Lesson Study itself is made up of a set of activities: studying curriculum and formulating goals, planning, implementing research lessons, reflecting on the implementation, and improving the research lessons. Through weeks of careful planning, each team produces a carefully crafted set of learning experiences for pupils. These sets are then used for the teaching and observation. When the research lesson is taught at one team member's class and observed by the rest of the team, it becomes a rich site for data collecting, collective reflection for continued improvement, aiming at further enhancing pupil learning. The power of Lesson Study lies within this iterative cycle of continuous improvement. When teachers go through an activity in well-articulated ways, they produce solid outcome that goes to mediate the next activity. Therefore, it is valuable to examine the trajectory of the tools and artifacts generated at each activity level and their mediating roles in the activity system. How they are produced and transformed carries the messages of how they mediate teachers' actions and reasoning. Such mediations embody teachers' participation, engagement, reification of meaning, and identity building (Wenger, 1998) through Lesson Study.

2.4 *Methodology*

Working closely with a local school to develop a community of practice through Lesson Study, we followed the tradition of design-based research. We have been deeply involved not only in designing and implementing an intervention but also studying the implementation process itself. As mentioned earlier, the multiple roles played by the researchers have brought uncertainty and challenges as well as valuable learning opportunities (Barab & Squire, 2004; Wilson & Berne, 1999).

Viewed from the activity system level, Lesson Study brings a new object for this system. Viewed from an intervention project level, the new object for the project integrates the object of design, the object of implementation, and the object of research. These objects are interrelated but need to be differentiated at the methodological level.

2.4.1 Design Experiment

Design experiment constitutes iterative cycles of improvement (Bannon-Ritland, 2003). For Lesson Study, "this approach of progressive refinement in design involves putting a first version of a design into the world to see how it works. Then the design is constantly revised based on experience, until all the bugs are worked out" (Collins, Joseph & Bielaczyc, 2004, p. 18-19). As Collins and his colleagues further noted, such continuous improvement culture not only exists in the Japanese car industry in its design technology but also in its schools' instructional designs through Japanese Lesson Study (Stigler & Hiebert, 1999). Lewis, Perry, and Murata (2006) commented, "Japan's elementary education system provides a provocative example of the power of a systematic focus on refinement of ideas ...such as the ideas of John Dewey, George Polya ..." (p. 9). For countries where teacher collaboration has not been part of the school culture, when Lesson Study is introduced for teacher development, it becomes a purely design-based effort for culture building. As Wenger (1998) reminded us, designing is for participation not just use. The relationship between the practices of design and practices of use is crucial in design experiment.

2.4.2 Training and Implementation

In addition to our researcher role, we played a training role and introduced to teacher participants the fundamentals of Lesson Study, and the knowledge and skills involved in planning, observing lessons, team work, and reflection. To enhance disciplinary knowledge and pedagogical content knowledge of the teacher participants, we acted as brokers to negotiate for instructional resources and get faculty support from the academic departments at NIE. The video cases of research lessons developed, for instance, by Mill's College Lesson Study Group headed by Catherine Lewis, provided the technology for our own training and learning.

In many ways, the training has gone beyond generic Lesson Study knowledge and skills to address both emergent and urgent needs that surfaced during each cycle. For instance, teacher's lack of curriculum knowledge for unit-based (versus lesson-based) planning and knowledge and skills for analysing teaching materials were both addressed through additional training workshops. As part of the design, some of these newly added elements and tools proved useful but needed continued refinements.

2.4.3 Research on the Design and Implementation

Our research takes both the design and its implementation as objects of systemic examination to study the processes and their interrelations in order to better inform further design and future implementation. Key design questions that drive our current and future research then become: what are entailed in designs for successful implementation and teacher learning, and what counts as teacher learning; how to design to maximise teacher participation and pupil learning; how to manage the relationship between implementation for teacher learning and research on teacher learning; and how to balance the roles of designer, implementer, trainer, researcher, and so on.

3 THREE MATHEMATICS RESEARCH LESSONS

In this section, we illustrate the continuous improvement model of Lesson Study, the knowledge co-construction in a community of practice and the processes of teacher learning through three cases of mathematics research lessons on three different topics conducted across the two years of project implementation. The three data sources for these cases are given in Table 1. Triangulation with teacher interviews and planning conversations were also done wherever needed.

Table 1
Data Sources of Mathematics Research Lessons

Planning sessions	Audio recordings
	Observation notes
	Verbatim transcriptions of audio record
Research Lessons	Video recordings
	Observation notes
	Verbatim transcriptions of video record
Post Lesson Discussions	Video recordings
	Observation notes
	Verbatim transcriptions of video record

A chronological description of the cases was deemed most appropriate in the elucidation of the teachers' learning trajectories. For preliminary analysis, notes taken by the researchers were reviewed to recollect the details of the planning sessions, the research lessons, and the post-lesson discussions. References to the transcripts were then made to check and confirm the accuracy of the notes in order to produce a clearer understanding of the processes in the Lesson Study cases. Teachers' verbalised thoughts and ideas as captured in the planning and interview data were also utilised to back up researcher interpretations. In the case report on equivalent fractions below, teachers' involvement as co-authors in conference presentations further corroborates and validates the findings.

There were six research lessons in mathematics as shown in Table 2, and three lessons are highlighted for subsequent discussion in this chapter.

Table 2
Research Lessons in Mathematics

Grade	Topic	Cycle
Primary 3	*Long Division with Remainder*	Cycle 1
Primary 1	*Introduction to Multiplication*	Cycle 2
Primary 3	*Equivalent Fractions*	Cycle 2
Primary 4	*Area & Perimeter*	Cycle 3
Primary 5	*Addition of Mixed Fractions*	Cycle 3
Primary 3	*Equivalent Fractions*	Cycle 4

3.1 *Case 1- Developing Teachers' Pedagogical Content Knowledge through Research Lessons and Videos of Teaching: Primary 3 Lesson on Long Division in Cycle 1*

This case took place from February to April 2006. Primary school teachers in Singapore have heavy teaching loads and seldom observe colleagues' classrooms. Lesson Study was thus able to bring to the teachers a set of entirely new experiences. Through a first cycle of collaborative team planning, observing research lessons, and conducting post-lesson discussions, teachers felt strongly about the professional learning opportunities created by such teacher learning communities in their school (End-of-cycle-three survey).

This Lesson Study team consisted of six teachers from Primary 1 to 3 who were also involved in another school-based curriculum innovation project. The idea was to use Lesson Study to support the school-based curriculum making processes. Right from the beginning, the demands created by two projects on the teachers were tremendous. They needed time to understand the school-based innovation which is a radical departure from the norm of classroom organisation. At the same time, they were committed to the Lesson Study process which involved weekly meetings of collaborative planning. From teacher journals, it was obvious that such competing demands created stress for teachers. In spite of this, they still regarded the collaborative planning helpful in "putting the many minds together" and "get every one's perspectives." They chose "Global Citizens" and "Independent Learners," two of the three major school goals as their long term goals to guide their planning and choice of teaching tools. The research lesson focused on the use of manipulatives to help pupils visualise the abstract procedures of long division, use of authentic problem situations, and group problem solving.

Traditionally, planning has been lesson-based and these teachers were found to lack a sense of unit and cross-grade interconnectedness of the same topic. The team chose mathematics because "Mathematics is progressive" and they wanted to "link the Mathematics Syllabus from P1 to P3." Long division became the chosen topic because it was difficult for pupils' learning and the Primary 3 teacher volunteered to teach the research lesson to a small group of 10 pupils in her class.

To diagnose pupils' prior knowledge and misconceptions of division, the team designed a 6-item pretest with 4 items of division sums with no remainder and 2 items of division sums with remainder. Of the 10 pupils in the group, 4 pupils were not able to solve division sums with no remainder and 9 out of 10 pupils were not able to solve division sums with remainder. During the teaching, the pupils also had problems with multiplication. On hindsight, the pre-test should have been used to diagnose pupils' knowledge of multiplication so that revision could be provided.

To prepare the observers for the research lesson, the team prepared pupils' profiles and assigned pairs of pupils to different observers to focus on. An observation template was designed to focus on pupils' thinking and learning in terms of what the pupils said and did as evidence of their thinking and learning. At the same time, observers were also expected to pay attention to teacher actions and evidence of teacher facilitation and support. A post-lesson conferencing was conducted in the afternoon of the same day with the support of a resource person from

NIE. Of the entire Lesson Study process, the teachers found observing live classroom and sharing observations during post-lesson discussions the most helpful. The team learned to give "non-threatening and constructive feedback to colleagues." They became more aware of the importance of basing their decisions for lesson improvement on evidence derived from observation.

In this lesson, multi-links and story books were used to help pupils link mathematics to daily life. By engaging with hands-on activities, pupils were able to make sense of the long division procedure and concept of remainder. Even though these were clear in the teachers' minds during planning, bringing the ideas on paper to life in the classroom was entirely something else. Similar to what is often reported in the Lesson Study work in other countries, implementing the first research lesson tends to suffer from being over ambitious or having too many goals. For this lesson, seven different activities were planned: introductory quiz and review, teaching the new concept of remainder with an authentic problem situation (7 story books to be shared among 2 pupils), practicing with manipulatives, teaching the Long Division Method, testing understanding (doing long division sums in pairs), creating story problems, and independent seat work.

The execution of the plan provided opportunities for the team to reflect on the failures and build their desire to improve the lesson. In doing so, the videotape became an essential tool for reflection, particularly for the research lesson teacher herself. "Frankly speaking," she wrote in her journal, "when I conducted the research lesson that Tuesday, I was not aware of what actually took place. That was why I needed to watch the video." Her struggles during the lesson was brought back to her, "…I too found it difficult to go through questions with them as each group had different questions to solve and I could not help them to see where they have gone wrong. At that moment, I felt helpless." She also noticed the problem in choosing group tasks and the demonstration problem: "I had to teach the pupils how to divide the tens value with remainder when they could not even divide correctly a single digit number with remainder…" For the use of manipulatives, she not only saw what was problematic but also realised that the effective use of manipulatives is to seize the right moment to help pupils form a clear mental image of pictorial representation of the concept. In her words, "…I should have immediately proceeded to show them how the vertical method of long division is carried out using what they had done for me earlier instead of waiting till the end thereby losing focus and meaning for the pupils."

Watching the videotape of the revised lesson which was also taught by her, the RL teacher demonstrated a deeper understanding of the subject matter involved in division and how to make it understood by pupils, the relevant pedagogical content knowledge. She realised that, "…teacher should not say start with 1 first but help pupils to see that 1 ten could not be divided into 3 groups as a whole, thus they had to change it to 10 ones and add it to the four ones to get 14 ones." She also became reflective of her own role to involve pupils actively rather than to merely pass on to them the knowledge, a traditional one she had played for many years. Understanding not only constitutes being able to solve problems in worksheets, but also involves being able to explain the how and why of their solutions to pupils. She wrote, "For the next few sums, I could have asked different pairs of pupils to explain how they derived their answers to check their level of understanding of the concept taught." Through teaching, getting feedback from colleagues, and spending hours to improve the lesson, she also became conscious of using evidence from observation to support decision-making. Having the courage to volunteer and teach the research lessons at the beginning of the project, the RL teacher went through a powerful learning experience, even though the process proved to be quite tough.

During the post research lesson discussion, the resource person from NIE analysed the inappropriate choice of the demonstration problems and tasks and pointed out the complication of the tens and ones caused by going beyond the times table and thus making it too difficult for the pupils. When this resource person missed the second lesson, she watched the videotaped teaching of both research lessons. She commented on the use of manipulatives and explained why the use of manipulatives was better carried out in the revised lesson.

3.2 *Case 2 - Lesson Study as a Site for Mentoring New Teachers: Primary 4 Lesson on Area of Rectangles in Cycle 3*

In January 2007, a Lesson Study team of five teachers in the Primary 4 Group decided on the topic "Area and Perimeter" for cycle 3. These five teachers were new to the grade level and valued the opportunity to collaboratively research on teaching the topic so that the pupils could derive the area formula on their own.

Over a period of 11 weeks, the teachers studied the curriculum materials and analysed pupils' understanding based on the pre-test administered to pupils and pupil interviews. They also examined curricular resources from Shanghai and Japan to develop a unit plan for "Area and Perimeter." The collaborative setting enabled the team members to dwell deeper into their planning and each teacher contributed in his or her own unique way. The collaboration also enabled the younger teachers and the more experienced ones to learn from each other. What is noteworthy, in this case, is the presence of the significant "others" – the facilitator and the mathematics resource person – that allowed the team to co-construct and learn the pedagogical content knowledge at a higher level.

To get a sense of pupils' prior knowledge, the teachers administered a pre-test and interviewed a sample of six pupils. Although the pupils were taught the topic, Area and Perimeter, in Primary 3, not everyone was able to distinguish between the two concepts. The pre-test also indicated that the pupils who understood the meaning of area were familiar with and able to use grid-squares in finding the area of composite figures. The teachers also found, to their dismay, that few could recall and apply the formula for the area of a rectangle, which had already been taught in Primary 3. For instance, in the interview, pupils were asked to find the area of the figure in Figure 2. A pupil applied the area formula and obtained 4 square centimetres.

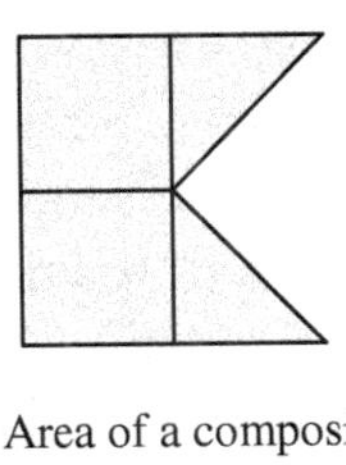

Figure 2. Area of a composite figure.

From the above information, the teachers saw the importance of revisiting the Primary 3 learning goals and content. With the help of the resource person, they decided on a tuning-in activity in which pupils were asked to draw an outline of their palms and compare the sizes so that they could distinguish between the perimeter of one's palm and its area. The team also planned to scaffold pupils' learning and guide them to derive the mathematical formula for area of rectangles on their own through the use of manipulatives. Although the team's vision was commendable, soon after carrying out their first research lesson, they realised that their main goal was unclear and that they were overly ambitious in trying to cover many concepts in just one lesson. The main goal (to enable pupils to derive the formula) was relegated to the very end of the lesson and was thus not addressed with much success.

The lesson was observed by the teachers in the planning team, the facilitator, a Primary 5 mathematics teacher, and two resource persons. While the live lesson was a useful learning platform in itself, it was the post-lesson discussion that enabled the research lesson teacher and the teacher-observers to try to improve the research lesson collaboratively. The discussion started with a reflection by the research lesson teacher who was able to identify the weaknesses of the lesson. The teacher-observers also contributed useful insights on how to further improve the lesson and gave feedback on teacher's questioning technique and students' responses during the lesson. The resource person elevated the discussion to a higher level in terms of both subject matter understanding and specific pedagogical strategies. She elaborated on why the major goal of the research lesson was unclear when the team aimed to achieve too many purposes in one lesson. She urged the team members, who were young and beginning teachers, to pay attention to the precision of using instructional language in order to highlight mathematical meaning and facilitate pupils' understanding. She pointed out that to help pupils differentiate between the concepts of area and perimeter, there should be a prior lesson focusing on knowing the terms. The research lesson could then concentrate on developing a deeper understanding of Area. This observation sparked a discussion of the spiral nature of the mathematics curriculum in Singapore. Teachers saw the importance for stronger communication, collaboration, sharing, and partnership among them across different grade levels.

During the same post-lesson discussion, some team members believed that the research lesson was successful in providing the scaffolding for pupils to derive the formula. According to their observation, some pupils were able to articulate the relationship between breadth, length and area of the rectangles in the given worksheet. The resource person, however, explained that those pupils who may be able to work out number patterns in the worksheet do not necessarily understand the relationship between length, breadth, and area. To move the team to focus more on concrete strategies, she proposed that the team use repeated addition of rows of squares (arrays) to enable pupils to observe the relationship between the changing dimensions of the length and breadth of the rectangles and how they relate to the area of the rectangles. She noted that to actively involve pupils in learning, what is important is that teachers need to ignite pupils' sense of wonderment and eagerness to problem solving by posing challenging questions to them; at their age level, they are expected to deduce the area formula.

In addition to the contribution by the resource person, the facilitator tried to shift teachers' attention onto the teacher-pupil relationship. It was observed that the teacher's voice dominated the lesson and drowned out pupils' thinking and voice. She noted the importance of giving pupils the space and time to analyse, consult with their peers, and articulate their thoughts. She asserted that the inclusion of these types of activities into the lesson will deepen pupils' learning. She showed the team how to carefully plan the utilisation of the whiteboard to record the trajectory of the lesson and facilitate pupils' viewing and learning. For teachers who were used to blithely writing and erasing on the whiteboard, they started to rethink how a teacher's writing on the board can help pupils learn during a lesson.

Towards the end of the post lesson discussion, the research lesson teacher suggested breaking up the lesson plan into three separate lessons as follows, and the team started to view the lesson from the perspective of a unit which uses the repeated arrays to induce pupils to derive the formula:

> **Lesson 1:** Review the definition of perimeter and area. Use the activity of Measuring the Palm to assist pupils in discerning the differences between the two concepts.

Lesson 2: Derive the formula for the area of rectangles ***(Research Lesson)***

Lesson 3: Understand relationship between Area and Perimeter ***(Follow-up lesson)***

During the two weeks that followed, the teachers met to modify and refine their research lesson. A novice teacher, with only nine months of teaching experience, took the challenge to teach the lesson. To better prepare her for the teaching, the team observed her in two mock lessons in which the team members played the role of pupils and responded to her teaching with anticipated pupil responses. The mock lessons helped the team in further refining the instructional strategies, particularly in how to improve whiteboard writing.

Two resources from the original research lesson were preserved for use in the revised research lesson: the 32 square centimetres rectangle and 36 square centimetres square. The transparent grid-square sheet was utilised for the tuning-in activity. Without the dimensions of the figures, pupils were challenged to identify the figure with the larger area. The grid-square sheet was also used for counting area in one square centimetre unit. Based on the advice from the resource person, the team used arrays of square units as a scaffold to support pupils to arrive at the formula. Magnetic squares were used for teaching demonstration and pupils used arrays of unit squares.

During the second post-lesson discussion, the team shared their observations and opinions openly. They encouraged the young research lesson teacher and commended on her good use of the whiteboard, clear instructions to the pupils, and good classroom management demonstrated through her good rapport with the pupils. At this culminating moment, she said with excitement, "I don't have much experience to say what is the best way to teach this topic. So through lesson study, in a way, I learn a lot faster instead of, like, maybe teaching area and perimeter for five years then I realised, oh, I should use the concrete to teach them. But now, I almost immediately know." Everyone clapped at hearing her words.

During the discussion, it was pointed out that the lesson may be too simple for some of the high achieving pupils. The facilitator pointed out that the repetition of the two arrays followed by the three arrays and then the four, five, and six arrays, was too monotonous and teacher-directed. She suggested challenging the pupils to try out on their own instead of the teacher demonstrating on the board while directing pupils to copy the teacher's actions using their set of manipulatives. The team noted that the worksheet was challenging and the lack of time did not permit the teacher to give the assignment. The facilitator also suggested that the team should improve the questioning technique. During the revised research lesson, pupils were asked to answer what the lengths and breadths of the rectangles were when arrays of squares were consecutively added. These questions merely got them to give correct answers but could not get them to visualise the pattern of the relationship between area, breadth, and length. She suggested tweaking the question from "what is the length?" or "what is the breadth?" to "what changes do you see?"

The discussion also led the team to realise that while teaching tools are meant to support pupils' learning, they may be counter-productive if used without careful thought. While the preceding paragraphs illustrate the pedagogical content knowledge that the teachers have learnt in the course of designing and improving the research lessons on Area and Perimeter, their learning was not limited to that mathematical topic per se. How to use the whiteboard, for instance, transcends the teaching of mathematics and goes to inform their teaching of other subjects as well. As far as the planning is concerned, the teachers in the team were increasingly more attuned to the importance of planning for a unit.

3.3 *Case 3 - Continuous Improvement in Using Manipulatives: Primary 3 Equivalent Fraction Lessons in Cycle 2 and Cycle 4*

The research lessons on Equivalent Fractions (EF) took place over two cycles in the second semester of 2006 and 2007 respectively. Fifteen teachers were involved in the first cycle in 2006 and three teachers followed up on previous cycle's effort in the second cycle in 2007 to continue exploring the teaching and learning of EF.

In July 2006, seven teachers (three Primary 3, two Primary 4 and two Primary 5) formed a cross-grade-level Lesson Study team focusing on Primary 3 mathematics. They first examined carefully Singapore's spiral curriculum to better understand how the fractions unit was designed in the national syllabus across the primary school. They found that, in the Japan and Shanghai textbooks, EF is briefly taught in Primary 4 while it is taught in Primary 3 in Singapore as an independent unit made up of 7 lessons. In the Singapore curriculum, the mastery of equivalent fractions is believed to be foundational for learning more advanced fraction topics at Primary 4 and 5.

To get a picture of pupils' knowledge of fractions in general and EF in particular, diagnostic tests were administered to all the pupils from Primary 3 to Primary 5. The team analysed various pupils' difficulties and misconceptions and became convinced that a lack of conceptual understanding in Primary 3 impeded the teaching and learning of advanced fraction topics in Primary 4 and 5. In light of this finding, they sought to gain a better understanding of what constitutes the conceptual understanding of EF and how to help pupils attain such an understanding.

The team reviewed the literature on the teaching and learning of fractions. They saw more clearly that size relations in fractions contradicted what was learnt with whole numbers, causing confusion among the pupils. Understanding fraction sizes and their relationships is an important and often difficult concept for pupils. To help pupils overcome the difficulty, the teachers needed to reconstruct the number system in a pupil's mind with the help of concrete manipulatives. The team agreed that to achieve conceptual understanding of EF, they must help pupils to move beyond calculations based on algorithms and enable them to apply the knowledge in problem solving.

The teachers were aware of the Concrete-Pictorial-Abstract (CPA) model but they felt that time constraints discouraged them from using it in their regular lessons. Conducting a series of four research lessons on EF across two cycles were opportunities for them to develop a deeper appreciation of the value of the CPA model. This case exemplified how the team sustained their effort in improving the design and use of the manipulatives in four research lessons across two cycles.

Research Lesson One

In the first research lesson, the team used rectangular paper strips as concrete manipulatives to provide students with hands-on experience. Each pupil received one paper strip at a time. The teacher asked pupils to make the number of folds and shaded the fraction of the whole for each strip. They included one-half, two-quarters, and four-eights of each given strip one after another. The teacher spent a considerable amount of time asking pupils to share how they folded and shaded their strips and the number of folds they produced. It was not until the 45th minute into the lesson that the pupils were probed to examine and articulate any observable patterns that they noticed. So the first research lesson became a folding, shading, and telling lesson, with limited space to guide pupils to observe the patterns and articulate the mathematical relationships embedded in the concept of EF.

Research Lesson Two

In revising the research lesson, the teachers decided to eliminate the folding activity and provided a set of paper strips in equal length, pre-divided into different numbers of equal parts: a two-part strip, a four-part strip, and an eight-part strip. Clearly demarcated lines highlighted that the same whole can be divided into different equal parts. Pupils were tasked to shade their paper strips one at a time: shade one part of two equal parts on the first strip, shade two parts out of the four equal parts on the second strip, and shade four equal parts out of eight equal parts on the third strip. They then aligned the strips together and placed them evenly on their desks below another given strip which indicates one-whole. They were given time to observe and articulate what was the same and different about the shaded parts and the wholes until they could find out that they were all halves of the same whole but were written in different fraction numbers. Pupils were given time to practise articulating this finding before the teacher told them that these were called equivalent fractions. Using the transitive law, the teacher led the pupils to see why they could be written in the form of a chain connected by equal signs. This tightened activity with improved questioning techniques produced a marked improvement as pupils spent less time on manipulating and had more time to think about the product of their "manipulation." Yet, the pupils were able to offer the comparative observations only after much probing by the teacher. There is a long way to go before the quiet pupils in this class were able to articulate and communicate the mathematical ideas through the use of manipulatives.

Research Lesson Three

When the teachers continued to explore the teaching of EF in 2007, they wanted to move away from teacher-centred instructions to give the pupils space to struggle in order to make sense of the abstract mathematical concepts through the use of carefully designed manipulatives. Thus, they designed a problem-based activity for research lesson three.

In planning this research lesson, the teachers first set out to prepare an investigative worksheet to allow the pupils to discover the concept of EF at their own pace. This daunting task was not fully achieved even after three weeks of discussions. Following an electronic consultation with their resource person, the team decided to include a pattern identification and sorting activity, with sets of eight pre-shaded circles as manipulatives. Pupils were tasked to examine, identify, and sort the one-halves from the non-halves. Working individually and then through pair discussions, the pupils were given time to engage in self-discovery as the teachers had intended. However, the outcome was less than satisfactory. The most obvious drawback was that the instructions posed to the pupils were too open. Pupils were asked to sort their circles into two groups and later to share with each other and the teacher how they did their sorting and what they observed from their sorting. The non-structured instruction was meant to invoke pupils' curiosity and was open to two different sorting patterns: halves versus non-halves and one-part shaded versus the other shaded circles. The lack of structure in the activity took a huge amount of the lesson's time before the teacher narrowed the focus of the pattern identification to concentrate on noticing and articulating the mathematical ideas surfaced through manipulating. Several defects in the design of manipulatives also caused some confusion among the pupils. All these stumbling blocks became learning points that informed the preparations for research lesson four.

Research Lesson Four

To facilitate the pattern identification through manipulatives and with a focus on articulating the key mathematical concept, the team made the task easier by choosing six circles shaded one-half, one circle shaded two-thirds, and one circle shaded three-fifths. For white board demonstration and sharing, four sets of bigger circles were prepared and magnetised. To avoid confusion, the colour of the circles and the shaded parts used by teachers and the class were the same. Attention was paid particularly to the use of the whiteboard to record all the pupils' responses. The team also made a special effort to carefully choose the wording of the instructional questions. They used "how" and "why" questions to lead pupils to higher level mathematical thinking. At times by feigning ignorance, the teacher successfully challenged the pupils to articulate their observations and understanding. Taken together, the outcomes of research lesson four had drawn the pupils closer to the desired learning outcomes.

Across the four research lessons, the improvement made in the manipulatives had enabled the teachers to have a stronger sense of using manipulatives to serve different instructional needs. It took the team a tremendous amount of time and commitment to move away from a risk-free teaching habit. When they started to shift away from "teaching as telling," they saw the importance of challenging pupils and giving them the space to think and articulate their mathematical observations. Teaching is indeed a complex enterprise. The collaborative Lesson Study setting could stimulate a concerted effort in continuously improving the teaching of a topic.

4 CONCLUSION

What weaves these three cases is the gradual unpacking process of the CPA model through ten research lessons created in teaching the three mathematics topics. By putting this abstract theoretical model into practice through the actions of planning, teaching, reflecting, and revising, the teachers had also formed a tangible image in their minds of what the model embodies pedagogically and in terms of pupil learning vis-à-vis the subject matter. The teachers seem to have improved their knowledge and skills in these areas. More importantly, with support from the mathematics educators, they had developed a well-blended form of pedagogical content knowledge that is directly applicable to improve pupils' deep understanding of these topics. This set of experiences attests to the power of Lesson Study in bringing different expertise together from diverse communities to develop the knowledge and skills that teachers need to teach for deeper and enduring understanding of key mathematical concepts.

Embedded in these three cases is another theme that pertains to how Lesson Study has become a rich site for the induction and mentoring of new teachers. With an increasing number of teachers retiring from the service, the teaching force in Singapore has become quite young, where the average age of the teachers is around 35 (Ministry of Education, 2006). When the mentoring practices that pair up experienced teachers with novices are no longer feasible under current circumstances, Lesson Study has become a promising pathway for effective induction of new teachers into the teaching profession. Among the Lesson Study teams in this study, a majority of the members were novice teachers. Out of the three cases, four research lessons and the revised lessons were taught by novice teachers. Having taught the research lessons, they reported that they were able to learn what could have taken them for years to acquire without taking such challenges. Such processes have helped them accomplish the demanding tasks of teaching while being observed and critiqued by colleagues. They gained confidence and satisfaction when they saw that they could help pupils learn. In our future research, we will continue to explore how to support more new teachers to grow on the job by learning to inquire, reflect, and problem solve in communities of practice through Lesson Study.

Across the cycles, however, as illustrated by the three cases above, a sense of progression through continuous improvement is evident; yet such progression could, once in a while, present itself as a kind of regression with little or no evident improvement at all. Even though this essentially characterises the complexity in initiating any change process in the real world, particularly in instructional improvement, the gradual and slow pace can often discourage both practitioners from action and policy-makers from funding such work. Against the backdrop of fast-paced reform culture in Singapore, we will continue our work with the valuable lessons we have learned so far; yet we are fully aware of the huge challenges facing us on the way ahead.

As we end our two-year pilot of Lesson Study, we see ourselves as learners of Lesson Study through learning by doing in the first year and consolidating our experience in the second year. As designers, implementers, and researchers, we have benefited from conceptualising the three levels of work as an Activity System mediated through Lesson Study as a tool in building teacher professional learning. This framework is comprehensive in capturing the actions, interactions, and dynamic discursive patterns depicting the nature of continuous improvement. It organises the three levels of our work in ways that allow us to stay alert of the challenges we have encountered at each level and of the adaptations we have sought to address the tensions and contradictions that constantly arise in the course of implementation.

We appreciate the contribution of Ng Swee Fong, Lim-Teo Suat Khoh, Yeap Ban Har, Jaguthsing Dindyal, and Foong Pui Yee as resource persons of the project.

References

Bannon-Ritland, B. (2003). The role of design in research: The integrative learning design framework. *Educational Researcher, 32*(1), 21-24.

Barab, S., & Squire, K. (2004). Design-based research: Putting a stake in the ground. *Journal of the Learning Sciences, 13*(1), 1-14.

Brown, A.L. (1992). Design Experiments: Theoretical and methodological challenges in creating complex interventions in classroom settings. *The Journal of Learning Sciences, 2*(2), 141-178.

Chokshi, S., & Fernandez, C. (2004). Challenges to importing Japanese lesson study: Concerns, misconceptions, and nuances. *Phi Delta Kappan, 85*(7), 520-525.

Cobb, P., McClain, K., Lamberg, T., & Dean, C. (2003). Situating teachers' instructional practices in the institutional setting of the school and district. *Educational Researcher, 32*(6), 13-24.

Collins, A., Joseph, D., & Bielaczyc, K. (2004). Design research: Theoretical and the methodological issues. *The Journal of the Learning Sciences, 13*(1), 15-42.

Elliot, J. (2004, November). *Conceptualizing and implementing learning studies in Hong Kong primary schools: Some views.* Paper presented at the annual meeting of the Australian Educational Research Association, Melbourne.

Engeström, Y. (1990). *Learning, working and imagining: Twelve studies in activity theory.* Helsinki: Orienta-Konsultit.

Fang Y., & Lee, C. (2007, April). *Contextualization and adaptation of Japanese Lesson Study in Singapore.* Paper presented at the American Education Research Association Annual Meeting, Chicago.

Fernandez, C. (2002). Learning from Japanese approaches to professional development: The case of lesson study. *Journal of Teacher Education, 53*(5), 393-405.

Gu, L. (2005, December). *Lesson study in China.* A Paper presented at the First Annual Conference on Learning Study. Hong Kong Institute of Education, Hong Kong.

Isoda, M., Shimizu, S. Miyakawa, T., Aoyama, K., & Chino, K. (Eds.). (2006). Special issue on the APEC-Tsukuba International Conference: Innovative teaching mathematics through lesson study. *Tsukuba Journal of Educational Study in Mathematics 25.* Tsukuba: Mathematics Education Division.

Lave, J., & Wenger, E. (1991). *Situated learning: Legitimate peripheral participation.* Cambridge: Cambridge University Press.

Leont'ev, A. N. (1977). *Activity, consciousness, personality.* Englewood Cliffs, NJ: Prentice Hall.

Lewis, C. (2002a). Does lesson study have a future in the U.S.? *Nagoya Journal of Education and Human Development, 1,* 1-23.

Lewis, C. (2002b). *Lesson study: A handbook for teacher-led improvement of instruction.* Philadelphia, PA: Research for Better schools.

Lewis, C. (2002c). What are the essential elements of lesson study? *CSP Connection.* 2(6), 1. Retrieved on February 22, 2005, from http://www.lessonresearch.net/newsletter11_2002.pdf

Lewis, C., Perry, R., Hurd, J., & O'Connell, M. P. (2006). Lesson Study comes of age in North America, *Phi Delta Kappan, 88*(4), 273-281.

Lewis, C., Perry, R., & Murata, A. (2006). How should research contribute to instructional improvement? A case of lesson study. *Educational Researcher, 35*(3), 3-14.

Lewis, C., & Tsuchida, I. (1998). A lesson is like a swiftly flowing river: Research lessons and the improvement of Japanese education. *American Educator, 22*(4), 12-17 & 50-52.

Lo, M. L., Pong, W. Y., & Pakey, C. P. Y. (2005). *For each and everyone: Catering for individual differences through learning studies.* Hong Kong University Press.

Lulce, A. (2005, June). *Making new pedagogies: Classroom interaction in Singapore classrooms.* Keynote address presented at an International Conference of the Centre for Research in Pedagogy and Practice: Redesigning, Pedagogy, Singapore.

Ministry of Education, Singapore. (2006). Education statistics digest. Retrieved 22 July 2008 from http://www.moe.gov.sg/education/education-statistics-digest/

Ministry of Education. (2007). *Mathematics syllabus primary.* Singapore: Author.

Sato, M. (2006, September). *Vision, strategies and philosophy for school innovation in Japan: Designing school as learning community.* Presidential address at the Korean Presidential Committee on Educational Innovation, Future, Innovation and Educational Strategies, South Korea.

Stigler, J.W., & Hiebert, J. (1999). *The teaching gap: Best ideas from the world's teachers for improving education in the classroom.* New York: Summit Books.

Wenger, E. (1998). *Community of practice.* Cambridge: Cambridge University Press.

Wertsch, J. (Ed.). (1981). *The concept of activity in Soviet psychology.* Armonk, NY: Sharpe.

Wilson, S.M., & Berne. J. (1999). Teacher learning and the acquisition of professional knowledge: An examination of research on contemporary professional development. *Review of Research in Education, 24*, 173-209.

Chapter 5

Teacher Change in an Informal Professional Development Programme: The 4-I Model

YEAP Ban Har **HO Siew Yin**

This chapter describes a study on teacher change within a large research project which investigated the effects of using word problems that require students to engage in sense-making. Case studies of several teachers who participated in the study were used to develop a model of teacher change. This model, referred to as the 4-I Model, describes teacher change when a new initiative is introduced and is exemplified by four types of teachers who: ignore the initiative; imitate practices recommended for the implementation of the initiative; integrate the principles of the new initiative into their instructional practice; internalise the principles of the initiative.

Key words: professional development, teacher change, teaching problem solving, sense-making in word problems

1 INTRODUCTION

The Ministry of Education of Singapore (2004) identifies the quality of teachers as a key success factor in improving the education system. It hopes to empower teachers to "share and explore new pedagogies as well as to model desired mindsets" (p. 14). This hope necessitates teacher change. Nelson and Hammerman (1996) argued that it is important to understand "what it is in innovative professional development projects that makes this process of change possible" (p. 4) and to describe the nature of teacher change.

Teacher change is slow and difficult as it requires opportunities for teachers to grapple with new ways of thinking and teaching (Darling-Hammond, 1990). One experience necessary for teacher change is for teachers to see that the new ways of thinking and teaching are translated into student learning and higher achievement (Guskey, 1989).

In this chapter, we focus on describing teacher change in the classroom context. Cooney (1994) called for a systematic inquiry into the processes of teacher education and suggested that description of such processes is important. Section 2 is a review of key literature on teacher change. Section 3 describes the professional development programme within which the study was conducted. Section 4 describes the research methodology employed. Section 5 presents three case studies which contributed to the development of a model to describe teacher change. Section 6 describes the 4-I Model of teacher change. The chapter concludes with suggested applications for the model of teacher change (Section 7).

2 WAYS OF DESCRIBING TEACHER CHANGE

In her review on teacher development, Sowder (2007) provided a comprehensive catalogue of ways of describing teacher change. One approach is to describe teacher change by describing their instructional practice. For example, teachers' questioning techniques can be used as a measure of change. In particular, teachers who ask questions that focus on procedures and answers are said to have a *calculational orientation* while teachers who ask questions that focus on situations, ideas, and meaning are said to have a *conceptual orientation* (Thompson, Philipp, Thompson & Boyd, 1994).

Another approach is to describe teacher change by describing stages in teacher growth. For example, Schifter (1995) used changes in teachers' conception of mathematics as a measure of change. In the first stage, teachers perceive mathematics as an accumulation of isolated facts, definitions, and procedures. In the second stage, teachers see mathematics as comprising activities, but these activities are without mathematical structure. In the third stage, teachers see mathematics as comprising activities with mathematical structures. In the fourth stage, teachers perceive mathematics as inquiry, organised around big mathematics ideas.

The model of teacher change described at the end of this chapter combines these two approaches. While the first approach looks at teachers' instructional practice and the second at stages in teacher growth, the 4-I model looks at stages in teacher growth with respect to their instructional practices.

3 THINK-THINGS-THROUGH (T^3) PROJECT

The present study on teacher change took place within a larger research project called Think-Things-Through (T^3) that investigated the effects of the use of word problems that require the consideration of context on students, teachers and classroom environment over a period of three years. In the first two years of the project, five schools in different parts of Singapore were selected for the project. The schools represented the different school types in Singapore. Five of the schools were mixed-gender schools. One was a single-gender school. Three of them were government schools while the other two were government-aided schools. All the students in one grade level in four schools participated in the project. In one school, only two classes of students were in the project. In the third year of the project, a sixth school was included in the project to investigate the effects of duration on the research variables. The sixth school is a mixed-gender, government school. All the students in the sixth school participated in the project.

Thus, in five schools, most of the students participated in the project for all the three years. At the start of the project, these students were in Primary 3. In one school, the students joined the project in the third year when they were in Primary 5. The mathematics teachers of all these students were included in the project. As teachers may or may not teach a particular class for all the three years, some teachers were in the project for one year, others for two years, and the rest for all the three years. A sub-set of these teachers took part in the project on teacher change. Table 1 shows the number of schools, classes, and teachers in the project each year.

Table 1
Number of Schools, Classes of Students and Teachers in the T^3 Research Project

	First Year	Second Year	Third Year
Number of schools	5	5	6
Number of classes	37	37	42
Number of teachers	37	35[1]	41[2]

In the study, teachers were given 12 sets of word problems each year. These sample word problems require the solvers to consider the context included in the problem and are referred to as *non-standard* word problems. *Standard* word problems can be solved by identifying the correct mathematical operation or procedure and performing the subsequent computation correctly. In solving *non-standard* word problems, it is insufficient to be able just to do that. Figure 1 shows an example of a *standard* word problem and a *non-standard* word problem. In the given *standard* word problem, one merely needs to identify that addition is the correct operation to use on the two given distances. In the *non-standard* word problem, one has to consider the relative positions of David's house, Enling's house, and the school in order to suggest a possible solution. The context, i.e., the relative positions of the three places, is important. More examples are given at the T^3 project website, http://math.nie.edu.sg/T3.

It is well documented that students from different parts of the world suspend their ability to make sense when they solve *non-standard* word problems (Verschaffel, Greer & De Corte, 2000). Consider this *non-standard* word problem: John's best time to run 100 metres is 17 seconds. How long will it take him to run 1 kilometre? Students who give the solution as 17 seconds × 10 are considered to have ignored the context of the problem and assumed constant rate for a non-constant rate situation. The answer is an unrealistic response. In various studies undertaken by Verschaffel, Greer, and De Corte (2000) using different pairs of *standard* and *non-standard* word problems, fewer than 20% of the students in each study were able to give realistic responses to many of the *non-standard* word problems. This phenomenon of students

[1] Two teachers taught two classes each.
[2] A teacher taught two classes.

suspending their ability to make sense when solving word problems is also observed amongst Singapore students (Chang, 2004; Foong & Koay, 1997; Koay & Foong, 1996; Yeap & Abdul Ghani, 2001; Yeap, Chang & Abdul Ghani, 2002).

Standard Word Problem

Mr Farhan drove 6 km this morning.
He drove another 18 km in the evening.
How far did Mr Farhan drive during the 2 trips?

Who could be correct? Choose one answer.
(1) Kim only (2) Jega only (3) Kim and Jega ( )

Show your method.

Non-standard Word Problem

David and Enling are in Singapura Primary School.
David lives 6 km from the school.
Enling lives 3 km from the school.
How far do David and Enling live from each other?

Who could be correct? Choose one answer.
(1) Kim only (2) Jega only (3) Larry only
(4) Kim and Jega only (5) Kim, Jega and Larry ( )

Show your method.

Figure 1. Standard and non-standard word problems.

Cochran-Smith and Lytle (1999) distinguished teacher knowledge into three types: knowledge-for-practice, knowledge-in-practice, and knowledge-of-practice. Knowledge-for-practice is acquired through formal professional development courses. Knowledge-in-practice is learnt from other teachers in a professional community. Knowledge-of-practice is acquired by investigating one's own teaching. In Singapore, formal out-of-school professional development courses is the most common form of in-service training. This is also the most common form of professional development for teachers internationally (Jaworski & Wood, 1999). However, action research has gained popularity and is practiced by many teachers (Wong & Chua, 2006; see also Chapter 3). Lesson Study has also been used by a number of schools as a form of professional development (see Chapter 4).

The present study investigated teacher change when teachers generally learnt from one another in a professional community. While formal workshop-style professional development courses are useful in raising teachers' awareness towards certain initiatives, professional development that provides opportunities to carry out the initiatives in actual classrooms provide better learning experience (Jaworski & Wood, 1999). An informal in-school professional development programme was designed for the present study. The teachers were given a set of word problems to use in their mathematics classes. Teaching notes were also given to the teachers to emphasise the need to encourage students to consider the context of the word problems during the solution process. The teachers were also encouraged to talk to each other about the problems and ways to teach these problems. They were encouraged to form informal groups to solve these problems collaboratively. There was no structured training in the form of formal in-service courses.

4 RESEARCH METHODOLOGY

Each participating school had two to four teachers who volunteered to participate in the study on teacher change. There were 10, 12, and 12 teachers in the first, second, and third year respectively. It was decided to allow the teachers to volunteer for the study so that they were comfortable and willing to talk about their change process.

Each of these teachers was observed at least once teaching a lesson that included word problems. The lessons were video recorded, transcribed, and coded. Field notes were taken during the lessons, focusing on tasks and teacher actions, in particular how the teachers set up and implemented the tasks during the lessons. The teachers were also interviewed on several occasions. The interviews were based on the tasks used and the questions they asked in their most recent mathematics lessons that included word problems. It was decided to focus on the tasks and their implementation, as they are external representations of teachers' views about teaching and learning. Rich tasks are necessary but not sufficient conditions to engage students in high-level thinking. Silver and Smith (1996) argued that starting with a cognitively demanding tasks allows teachers to engage students in sharing their thinking, comparing approaches, making conjectures, and generalising. Stein and Lane (1996) reported that the way tasks were set up and implemented in a lesson had significant effects on student learning.

The data analysis was based on the model described by Schoenfeld (2007). He described a six-phase process starting with a real-world situation. In this study, the real-world situation was classroom lessons and descriptions of these lessons. Important aspects of the real-world situation were the tasks used by the teachers and teacher actions associated with the setting up and implementation of the tasks.

The analysis comprises two parts. The first part involves the tasks used by the teachers. The word problems used by the teachers were classified as either standard or non-standard illustrated as in Figure 1. The non-standard word problems used by the teachers were analysed with respect to its context and mathematical structure, and they were compared to the context and structure of the tasks given to the teachers in the study.

The second part involves teacher actions. Each lesson was broken up into several segments. A segment is defined to be a part of a lesson that corresponds to a main stage in solving word problem: understanding the problem, planning for its solution, carrying out the solution, and reflecting on the solution or extending the problem (Polya, 1957). In each segment of the lesson, instances when the teacher was emphasising the importance of context were noted and described.

On the basis of the data collected from lesson observation and interviews, a case was constructed for each teacher to represent the nature of classroom tasks they used, and the way these tasks were set up and implemented during the lessons. These cases were used to develop a model to describe teacher change. The cases were compared and categorised according to the type of change each teacher went through.

5 THREE CASE STUDIES

In this section, three selected cases are used to illustrate a model of teacher change. In each case, a pseudonym is used to describe a particular teacher. However, it does not indicate the gender, age, or racial group of that teacher. These were not considered in the study. The masculine form he/him/his/himself is used for convenience, regardless of the gender of the teacher.

5.1 *The Case of Yeo*

Yeo was observed teaching a Primary 4 lesson that included the following word problems taken from the textbook.

Carl and Ben had $4686 altogether.
Carl's share was twice as much as Ben's.

a. *How much was Ben's share?*
b. *How much was Carl's share?*
c. *If Carl spent $500 on some books, how much money had he left?*

Lisa had 1750 stamps.
Minah had 480 fewer stamps than Lisa.
Lisa gave some stamps to Minah.
In the end, Minah had 3 times as many stamps as Lisa.

a. *How many stamps did Minah have in the beginning?*
b. *How many stamps did Lisa have in the end?*

Yeo read aloud the word problems and asked a few comprehension questions, such as "Who had more money, Carl or Ben?" and a few metacognitive questions, such as "Are you sure this is right?" In both word problems, Yeo used diagrams to represent the situations and asked students to suggest the computations required to solve the problems. During the interview, Yeo said that he did not feel that it was necessary to get the students to pay too much attention on the context of the word problems as the solution does not depend on the stories. However, Yeo admitted that the amount spent on the books was "kind of a lot for books" and the number of stamps that Lisa and Minah had "was plenty as kids do not collect stamps nowadays."

Yeo also reported that he used the non-standard word problems from the Think-Things-Through Project. One of the non-standard word problems used was the following word problem.

> *Mini sausages are sold in packs of 6.*
> *Buns are sold in bags of 8.*
> *We do not want to have either the sausages or buns leftover.*
>
> Part 1
> *We use a mini sausage with a bun to make a hotdog.*
> *What is the fewest number of packs of sausages and bags of buns we should buy?*
>
> Part 2
> *We use 2 mini sausages with a bun to make a hotdog.*
> *What is the fewest number of packs of sausages and bags of buns we should buy?*
>
> Part 3
> *We use 3 mini sausages with a bun to make a hotdog.*
> *What is the fewest number of packs of sausages and bags of buns we should buy?*
>
> Part 4
> *We use 1, 2 or 3 mini sausages with a bun to make a hotdog.*
> *With a pack of sausages and a bag of buns, find the number of each type of hotdog we can make.*

When asked if he had created his own non-standard word problems, Yeo showed a small collection of teacher-created word problems. Most of the word problems had the same mathematical structure as the non-standard word problems from the Think-Things-Through Project. An example of Yeo's own creation is shown below.

> *Mrs Lim gives a chocolate bar and a drink to each student in her class. The chocolate bars come in boxes of 4. The drinks come in packs of 6. Mrs Lim does not want to have any left-over chocolate bars or drinks.*
>
> (1) *What is the smallest number of boxes of chocolate bars she should buy?*
>
> (2) *What is the smallest number of packs of drinks she should buy?*

Yeo described how he used the hotdog problem and his teacher-created problem with his students and confirmed that he did not bring up the question of the number of students in Mrs Lim's class. Yeo was expecting a solution of 3 boxes of chocolate bars and 2 packs of drinks – a reasonable solution if Mrs Lim's class has 12 or fewer students. Yeo, however, conceded that in the teacher-created problem, the solution depends on the number of students in Mrs Lim's class but he "did not think of it before" as the main process he engaged in in creating the non-standard word problem was "using the hotdog problem as a sample."

Interviews with Yeo on several occasions throughout the year confirmed that he used the non-standard word problems from the Think-Things-Through Project without further emphasis on the context of word problems other than that included in the worksheets. He created a few non-standard word problems by changing the context of the word problems without further change to the mathematical structure, and he taught other word problems without encouraging students to consider the role of context on the solution.

5.2 *The Case of Boo*

Boo was observed teaching a Primary 3 lesson based on a problem that he had created. The problem was to use exactly 19 ice-cream sticks to form the letters I (using 3 sticks) and T (using 2 sticks). Boo had previously used a non-standard word problem from the Think-Things-Through Project. That problem was to find the number of two-eye and three-eye monsters where the total number of eyes was 19. It is clear that Boo was able to create new non-standard word problems by changing the context of the example. The mathematical structure of the word problem was not changed, however.

There were several instances which indicated that Boo was able to transform the mathematical structures of given problems into new ones. Boo was rather proud of a collection of non-standard word problems that he had created. During the interview when he chose to talk about these problems, Boo explained the inspiration of each problem. The following example is illustrative of Boo's explanation. One of the problems that Boo had tried to use from the Think-Things-Through Project is counting the number of whole numbers less than 1000 that has at least one digit 5. See http://math.nie.edu.sg/T3/downloads/P3_T1_01.pdf. Boo said that this problem was too difficult for his class. He created the problem below.

Today we will find "lucky" numbers.
A number is "lucky" if it is a product in the 9 times table or in the 12 times table. Let's find out how many lucky numbers are there.

Boo felt that the lucky number problem could be used instead of the digit five problem as "they are both the same actually" because "both focus on double counting" but "they are in different topics." The given problem was to be done earlier in the school year, whereas multiplication of 9 and of two-digit numbers was to be done later in the school year "when the kids should be more ready for this idea of double counting."

In the lesson on 19 ice cream sticks, it was clear that Boo tried to encourage students to go beyond computation. Boo got the students to explain why $19 \div 2$, $19 \div 3$ and $19 \div 5$ do not make sense in this problem. Students were provided with ice-cream sticks to model the situation.

This behaviour was not reported when Boo described the way the students were solving standard word problems. The way Boo taught word problems in the textbooks and other standard curriculum materials resembled that demonstrated by Yeo. The interview data on Boo about the way he had taught division-with-remainder problems is characteristic of Boo. The following discussion is based on this word problem.

> *Toogol, Noogol and Zoogol collected twigs and dried leaves at Farmer Joe's garden. They shared 63 twigs among themselves. How many twigs did each of them get?*

He emphasised understanding the problem and selecting the correct operation to use. Boo spent more than half the time on working out the division algorithm. When asked if he did anything to emphasise the role of the problem context on the solution, Boo reported that he tried to get the students to imagine the scenario. When pressed for further ways to emphasise the role of context, Boo explained that he would have done so if there were 64 twigs instead of 63. He would then get the students to think about the remaining twig. Boo later realised that students who did not assume that Toogol, Noogol, and Zoogol would share the twigs equally may obtain a different answer.

In another interview, Boo was asked to change a set of standard problems that he recently used into non-standard ones. The word problems were used in the topic of multiplication of 2-digit number with a 1-digit number. Boo had earlier described using these word problems in the class where he emphasised the understanding of the problem, the choice of operation, and the consequent computation. Boo did not place much emphasis on the context. Boo had little difficulty in converting standard word problems into non-standard ones. For example, he showed how he changed two word problems in a worksheet such that he could emphasise the role of context in the solution. Boo changed this word problem

> *There are 9 rows of chairs in the hall. There are 23 chairs in each row. How many chairs are there in the hall?*

into this one

> *There are 9 rows of chairs in a hall.*
> *There is the same number of chairs in each row.*
> *Ali sits on one of the chairs. There are 11 chairs to his right and 11 chairs to his left. How many chairs are there in the hall?*

Boo reasoned that the new problem requires students to think about the actual situation to use the correct numbers in the computation. His intention was to encourage students to explain why the solution is not 22×9.

In another example, Boo changed this problem

Mrs Tan bought 8 movie tickets.
The price of each movie ticket was $8.
How much should Mrs Tan pay for the movie tickets?

into one that allowed him to ask questions like, what if 7 tickets were needed instead of 8. The word problem created by Boo is shown below.

Mrs Tan bought 8 movie tickets.
There was a special discount. The price of a pair of movie tickets was $14. How much should Mrs Tan pay for the tickets? What if she bought 7, instead of 8, movie tickets?

Boo also explained that the teacher could raise questions such as, if the price already included taxes and if one also knows the price of a single ticket.

Interviews with Boo on several occasions throughout the year confirmed that he expected students to consider the context of non-standard word problems given in the Think-Things-Through Project as well as when he recognised such problems in the textbooks. Boo was also able to convert standard word problems into non-standard ones. However, Boo did not always recognise non-standard word problems from the textbooks and other standard curriculum materials. In particular, Boo often reverted to the "understand-the-problem-and-select-the-operation" model of solving word problems. In the final interview, Boo cited his "increased awareness about the importance of the story in word problems" as one of the ways the project had changed him. While Yeo created non-standard word problems by changing the context of sample word problems without further change to the mathematical structure, Boo adapted the sample word problems by changing more than just the context. Boo also made the sample problems easier or more difficult by changing the mathematical structure and/or content to suit the students and topics he was teaching. In teaching word problems in the textbook and standard curriculum materials, Boo shared more similarities than differences with Yeo. In particular, both rarely used the approach that they used for non-standard word problems when teaching textbook problems. Boo's change was seen primarily in situations that were explicitly calling for an emphasis on problem context and less in implicit situations.

5.3 *The Case of Hoe*

Hoe taught a Primary 3 class in the first year of the project and a Primary 5 class in the third year. His case was similar to Boo's when Hoe was teaching the Primary 3 class. In one of the interviews, Hoe made a comment that he had always found the idea of emphasising story context important "but, previously, I somehow seldom do it in class" and "using the sample problems made me not to take the assumptions for granted."

Here, we discuss Hoe's teaching of the Primary 5 class. In the observed lesson, He taught a set of word problems about rates. He reviewed the topic using a set of examination questions. In solving this problem

> *A photocopier can print 40 copies in 30 minutes.*
> *At this rate, how many copies can it print in 1 hour and 30 minutes?*

Hoe asked: "Is 40 copies per minute realistic? You have seen a photocopy machine before. Think back. Is 40 copies per minutes reasonable?" Then he showed the class a few pamphlets describing the capabilities of a few brands of photocopiers. Before Hoe asked the students to solve the problem, he asked "Why is this phrase [at this rate] important to your answer?" After the students had given a solution, Hoe asked for possible answers when the assumption was removed. He then posed a second problem.

> *A runner can run 4 km in 30 minutes.*
> *At this rate, how far can she run in 1 hour and 30 minutes?*

and asked the students to respond to this problem. He led a discussion on the degree of realism of the situation. He asked whether it is realistic for a person to run 4 km in 30 minutes, and whether the assumption of constant rate is reasonable. He asked students to suggest what data were needed to answer the question given that the assumption of constant rate is not reasonable. Hoe ended the lesson by asking students to pose word problems where multiplication can be used and where multiplication is not a suitable model.

In various interviews, Hoe was able to show how he could encourage students to make sense of the situations given in the word problems. Hoe said he had always done that but "it was usually to understand the given situations rather than to question them." At one interview, Hoe brought a few test items written by his colleagues to ask if other solutions were possible depending on the assumptions made. It became increasingly clear that Hoe was looking at word problems from traditional sources such as textbooks and examinations with a critical eye and that he was encouraging his students to question the assumptions, the information given in word problems, and their answers.

6 THE 4-I MODEL OF TEACHER CHANGE

The 4-I model of teacher change describes the changes in a teacher's pedagogical practice when there is an education reform. In this study, the "reform" was the use of non-standard word problems in mathematics lessons.

It is possible that some teachers do not feel the need to use these word problems, and in the 4-I model, these teachers are described as *ignoring the reform principles*. They may continue to use word problems from the textbook, workbook, worksheets, and other standard instructional materials. In the above study, none of the case studies illustrates the stereotypical characteristics of teachers in this category. This was expected because these teachers had volunteered themselves to participate in the study.

Among the case studies, there were teachers who used the word problems given to them in the Think-Things-Through Project without further emphasis on the role of story context other than the ones included in the worksheets. These teachers continue to teach word problems in standard curriculum materials without paying attention to the story context. When these teachers created their own non-standard word problems, they changed the context of the given sample problems without changing the mathematical structure. In the 4-I model, teachers demonstrating these characteristics are described as *imitating the reform principles*. The case of Yeo illustrates the stereotypical characteristics of teachers in this category. In an education reform, teachers in this category depend on the strategies given to them and are not able to generate their own strategies to implement the reform.

In this study, there were teachers who were able to go beyond imitating the examples of non-standard word problems given to them. They were able to create their own examples that were not mere superficial modifications of the given examples. These teachers were able to change the mathematical structure of the given word problems to make the problems easier or more difficult. They were also able to integrate what they saw in the sample word problems into those that they encountered in the textbooks and other standard curriculum materials. In the 4-I model, teachers demonstrating these characteristics are described as *integrating the reform principles*. These teachers, however, continue to teach standard curriculum materials without paying attention to the story context. They had not internalised the idea of emphasising story context in the solutions of word problems to the point when they could apply the idea implicitly. The case of Boo illustrates the stereotypical characteristics of teachers in this category. In an education reform, teachers in this category are able to generate their own strategies to implement the reform based on some examples. However, while their actions are consistent with the reform principles in situations that explicitly connected to the reform, this may not be so in other less explicit situations.

In the 4-I model, teachers who have internalised the importance of story context in solving word problems are described as able to *internalise the reform principles*. The case of Hoe provides evidence of a teacher who was able to emphasise the role of story context even in standard word problems. In particular, this teacher often converted standard word problems into non-standard ones. Such a teacher would re-conceptualise word problems, standard or non-standard, as opportunities to engage students in mathematical modelling. In an education reform, teachers in this category are able to generate their own strategies to implement the reform given some examples. Their actions are consistent with the reform principles in a range of situations including traditional ones. Boo's actions were consistent with the reform principles in situations that are explicitly connected to the reform. However, Hoe's actions were consistent with the reform principles not just in situations that are explicitly connected to the reform; this consistency was maintained even in other situations.

Table 2 provides a summary of these four types of teacher change under the 4-I model.

Table 2
Teacher Change in an Education Reform

Type of Teacher Change	Characteristics
Ignore	The teachers ignore the given examples to implement the reform. They may, at best, use the given examples, if required to do so.
Imitate	The teachers use the given examples to implement the reform. They may, at best, change superficial features of the examples to generate their own.
Integrate	The teachers use and adapt the given examples to implement the reform. They are able to change more than the superficial features of the examples to generate their own. In traditional situations, they may revert to actions that are not consistent with reform principles. At best, they recognise this inconsistency when it is made explicit to them.
Internalise	The teachers use and adapt the given examples to implement the reform. They are able to apply reform principles even in traditional situations. They do not revert to actions that are not consistent with the reform principles.

7 APPLICATIONS OF THE 4-I MODEL

This 4-I model can be used in a range of investigations on teacher change in any education reform, not necessarily limited to mathematics education. In this section, we describe two other examples in which this model is used to investigate teacher variables in mathematics education research.

In the Enhancing Pedagogy of Mathematics Teachers (EPMT) project, the main aim is to enhance primary and secondary teachers' pedagogy in teaching for understanding. Teachers were taught various principles and strategies of teaching for understanding through traditional workshop-style in-service courses. They were subsequently asked to try some of these strategies in their classroom with an emphasis on collaboration among teachers in the same school and sharing among teachers from all the project schools. The teachers were asked to video tape their lessons once prior to and a few times during the entire school year. The 4-I model is used to study teacher change in this project.

In the Lesson Study in Primary Mathematics Project, the main aim is to study teacher learning in this alternative mode of professional development. In this project, each lesson study team identifies the learning outcomes for the teachers in the team. The lesson study cycle is then conducted to achieve the learning outcomes. The 4-I model is used to measure the degree of achievement of the learning outcomes by examining the type of teacher change with respect to the stated learning outcomes.

The 4-I model can be used to develop a more impactful assessment framework to assess teachers in in-service courses. Typical assessment frameworks for in-services courses tend to focus on variables that may not result in an impact on teaching practice. Assessment framework that focuses on teacher change has arguably more impact on teaching practice as the model measures teacher change in the classroom context.

The 4-I model of teacher change can be used to benefit research and professional development efforts. Its use in these ways would generate data that can further refine the model.

Acknowledgement

This study on teacher change is part of the larger project *Think-Things-Through* (T^3) *Project* funded by the Educational Research Fund EP2/03 YBH, Ministry of Education, Singapore.

References

Chang, S. H. (2004). *Sense-making in solving arithmetic word problems among Singapore primary school students.* Unpublished Master thesis, National Institute of Education, Nanyang Technological University, Singapore.

Cochran-Smith, M., & Lytle, S. (1999). Relationships of knowledge and practice: Teacher learning in communities. In A. Iran-Nejad & P. D. Pearson (Eds.), *Review of research in education,* Vol. 24, (pp. 249 – 304). Washington, DC: American Educational Research Association.

Cooney, T. J. (1994). Research and teacher education: In search of a common ground. *Journal for Research in Mathematics Education, 25,* 608 – 636.

Darling-Hammond, L. (1990). Instructional policy into practice: The power of the bottom over the top. *Educational Evaluation and Policy Analysis, 12,* 233 – 241.

Foong, P. Y., & Koay, P. L. (1997). School word problems and stereotyped thinking. *Teaching and Learning, 18*(1), 73-82.

Guskey, T. R. (1989). Attitude and perceptual change in teachers. *International Journal of Educational Research, 13,* 439 – 453.

Jaworski, B., & Wood, T. (1999). Themes and issues in in-service programmes. In B. Jaworski, T. Wood & S. Lawson (Eds.), *Mathematics teacher education: Critical international perspectives* (pp. 125 – 147). London: Falmer Press.

Koay, P. L., & Foong, P. Y. (1996). *Do Singapore pupils apply common sense knowledge in solving realistic mathematics problems?* Paper presented at the Joint Conference of Educational Research Association, Singapore and Australian Association for Research in Education, Singapore.

Ministry of Education. (2004). *To light a fire: Enabling teachers, nurturing students.* Singapore: Author.

Nelson, B. S., & Hammerman, J. K. (1996). Reconceptualizing teaching: Moving towards the creation of intellectual communities of students, teachers, and teacher educators. In M. W. McLaughlin & I. Oberman (Eds.), *Teacher learning: New policies, new practices* (pp. 3 – 21). New York: Teachers College Press.

Polya, G. (1957). *How to solve it* (2nd ed.). New York: Doubleday Anchor Books.

Schifter, D. (1995). Teachers' changing conceptions of the nature of mathematics: Enactments in the classroom. In B. S. Nelson (Ed.), *Inquiry and the development of teaching: Issues in the transformation of mathematics teaching* (pp. 17 – 25). Newton, MA: Centre for the Development of Teaching, Educational Development Centre.

Schoenfeld, A. H. (2007). Method. In F. Lester (Ed.), *Second handbook of research on mathematics teaching and learning* (pp. 69 – 107). Reston, VA: National Council of Teachers of Mathematics.

Shulman, L. S. (1987). Knowledge and teaching: Foundations of the new reform. *Harvard Educational Review, 57*(1), 1 – 22.

Silver, E.A., & Smith, M. S. (1996). Building discourse communities in mathematics classrooms: A worthwhile but challenging journey. In P. Elliott (Ed.), *Communication in mathematics, K-12 and beyond* (pp. 20 – 28). Reston, VA: National Council of Teachers of Mathematics.

Sowder, J. T. (2007). The mathematical education and development of teachers. In F. Lester (Ed.), *Second handbook of research on mathematics teaching and learning* (pp. 157 – 223). Reston, VA: National Council of Teachers of Mathematics.

Stein, M. K., & Lane, S. (1996). Instructional tasks and the development of student capacity to think and reason: An analysis of the relationship between teaching and learning in a reform mathematics project. *Educational Research and Evaluation, 20,* 50 – 80.

Thompson A. G., Philipp, R. A., Thompson, P. W., & Boyd, B. A. (1994). Calculational and conceptual orientations in teaching mathematics. In D. B. Aichele & A. F. Coxford (Eds.), *Professional development for teachers of mathematics* (pp. 79 – 92). Reston, VA: National Council of Teachers of Mathematics.

Verschaffel, L., Greer, B., & De Corte, E. (2000). *Making sense of word problems.* Lisse, The Netherlands: Swets & Zeitlinger.

Wong, K. Y., & Chua, P. H. (2006, May). *Promoting action research through the Zonal Centre of Excellence (COE) for Mathematics in Singapore.* Paper presented at the International Conference: Preparing teachers for a changing context. London: University of London, Institute of Education. Electronic publication.

Yeap, B. H., & Abdul Ghani, M. (2001). *Facilitating sense-making in primary mathematics through word problems.* Paper presented at AARE 2001 International Education Research Conference, Fremantle, Australia.

Yeap, B. H., Chang, A. S. C., & Abdul Ghani, M. (2002). Promoting thinking and sense-making through mathematics word problems. In D. W. K. Chan & W. Y. Wu (Eds.), *Proceedings of Thinking Qualities Initiative Conference* (pp. 177-178). Hong Kong: Centre for Educational Development.

Chapter 6

Singapore Master Teachers in Mathematics

Juliana Donna NG Chye Huat **FOO Kum Fong**

Master Teachers are role models of teaching excellence and are appointed based on strong pedagogical knowledge they have demonstrated over the years. Their main role is to develop and enhance the capacity of teachers through mentoring and demonstrating good teaching practice. Though appointed to serve the cluster schools, Master Teachers are expected to work beyond the confines of their clusters. As Master Teachers are at the apex of the teaching track, potential Master Teachers have to go through a rigorous accreditation process before their appointments. This chapter documents the journey of two Mathematics Master Teachers in the Primary (PMTT) and Secondary (SMTT) schools and their voices in their mission to serve as catalysts for the growth of committed and passionate teachers.

Key words: Master Teacher, time-tabled time, curriculum leader, teacher leaders, Learning Support for Mathematics, action research

1 THE TEACHING TRACK

The Education Service Professional Development and Career Plan was implemented by the Ministry of Education (MOE) in Singapore in 2002. This plan caters to the diverse and varied talents, abilities, and aspirations of all education officers. To develop teams of quality professional teachers, capable leaders, and dedicated specialists, three career tracks are introduced. They are Teaching track, Leadership track, and Senior Specialist track.

Teachers who have the aptitude and aspiration can pursue one of the three mentioned tracks. Classroom teachers with good track records and preferring to remain in the Teaching Track can progress on to become Master Teachers. Generally, the progression of the Teaching Track is from Classroom Teachers to Senior Teachers to Master Teachers. However, there is fluidity among the three tracks. It is possible for movements across the three tracks if the officers decide to do so and at the same time satisfy the criteria needed for the preferred track.

2 THE ACCREDITATION FRAMEWORK

To add rigour and put in place a comprehensive process for the appointment of the Senior and Master Teachers, MOE introduced a formal accreditation framework in September 2001. The accreditation process took effect with the promotion of the first three educational officers to Master Teachers in 2002. The schools were subsequently notified of the accreditation framework, and annually principals nominate potential Master Teachers and guide them in the accreditation process. Up to 2007, there were only 19 Master Teachers with two Master Teachers in Mathematics: one in the Primary and the other in the Secondary Mathematics.

After more than five years of consistently good performance as a teacher, an education officer who aspires to make progress in the teaching track can be nominated by the principal to be considered for appointment as a Senior Teacher. The officer then goes through a rigorous accreditation process which culminates in an interview with a selection panel. However, the interview does not guarantee appointment as a Senior Teacher.

2.1 *The Accreditation Process for Master Teacher Appointment*

A senior education officer who has been on the job for at least three years, with a good track record of strong pedagogical expertise and the potential to be a Master Teacher, can be nominated by the principal to be considered for this position. The accreditation process involves the submission of the professional portfolio which is aligned with the five accreditation standards given in Table 1.

Table 1
Five Accreditation Standards

Self	Standard 1	Professional Development
Students	Standard 2	Quality learning of pupils
	Standard 3	Character development and well-being of pupils
School & Partners in Education	Standard 4	Contribution to school organisation and development
	Standard 5	Collaboration with parents and community groups

Under the focus area of "self", the aspiring Master Teacher must demonstrate evidence of leading in professional development. The potential candidate must have engaged in continual learning through sharing of knowledge and expertise at various platforms, adopted innovative teaching practices, and engaged in educational research. Standards 2 and 3 fall under the area of "students." In Standard 2, the aspiring Master Teacher must have made provisions for quality learning of pupils by creating positive classroom environments conducive for learning. At the same time, this teacher, through innovative and creative teaching strategies, has been a source of inspiration to pupils of different abilities. Besides the cognitive aspect, the aspiring Master Teacher looks after the affective well-being of pupils as reflected in Standard 3. This is achieved through the inculcation of desirable values and attitudes in pupils and the creation of an environment that enhances their morale and well-being. In Standard 4, under the umbrella of "School and partners in education," the aspiring Master Teacher must be instrumental in bringing about positive school effectiveness and development by leading in school initiatives. Under the same broad area, in Standard 5, the aspiring Master Teacher has worked with parents and community groups to maximise the learning objectives of pupils.

An aspiring Master Teacher is expected to produce a professional portfolio encapsulating his/her efforts in these five accreditation standards. The portfolio documents his/her professional work for the past three years. The accreditation standards are assessed through this portfolio. Upon a successful evaluation of the portfolio, the aspiring Master Teacher is then invited for an interview at the cluster and ministry levels.

The accreditation process is an insightful learning journey. At each step, the aspiring Master Teacher reviews and reflects on the processes and outcomes of her teaching and learning journey, as well as on emerging trends in pedagogy on the educational landscape.

3 ROLES OF THE MASTER TEACHER

The Master Teacher is a role model to all teachers and serves as a resource person at the cluster level. He or she guides and advises teachers within the cluster to help introduce new methodologies and skills in the learning and teaching of subjects, including the character development of students.

The Master Teacher reports directly to the cluster superintendent who, in consultation with principals of the cluster schools, decides on the deployment of the Master Teacher. Hence, the job scope of the Master Teacher is diverse, depending on the needs of the particular cluster he/she is serving.

The following sections describe our journey as Master Teachers and also our contributions as (a) teacher leaders, (b) curriculum leaders, and (c) builders of a community of professional practitioners.

4 PERSONAL JOURNEYS TO BECOME MASTER TEACHERS

4.1 *PMTT*

I have taught for more than 35 years in a primary school and was Head of the Mathematics Department for about nine years. Being with and learning with children have always been my joy. MOE recognises my passion, dedication, and expertise in teaching and learning when in 1999, I was made the sole recipient (Primary School) of the President's Award for Teachers, the highest honour given to teachers in Singapore.

"It is the responsibility of every experienced teacher to ensure that quality teaching be passed on to new colleagues." This is one remark made by a teacher trainer that has remained vivid and stays with me for life. I also remember these words, "Don't say 'No' and close doors until you have tried it. Always give yourself and others a chance." In 2002, I was invited to take on the challenge of being one of a pioneer group of Master Teachers.

As a Master Teacher, my role is multifarious. I impact people, programme development and am at the same time a resource person in the area of primary mathematics teaching and learning. I am predominantly engaged in being a pedagogy "expert" to help improve the teaching and learning scene in Singapore. I report directly to the cluster superintendent, who with the cluster school leaders, supports me in my work. I work collaboratively with school leaders, Heads of Department (HODs), Senior Teachers, teachers, and beginning teachers.

4.2 *SMTT*

This is a reflective account based on my one year stint as a Master Teacher (Foo, 2008). Currently, I am the only Mathematics Master Teacher in the secondary school, serving the East Zone[1] . Under the guidance of my cluster superintendent and school leaders, I am deployed to work with the secondary as well as primary teachers. I do, however, lend my expertise and work collaboratively with Heads of Department in the primary and secondary schools within the cluster and the zone.

As a beginning teacher, I remembered a conversation with a Senior Teacher who taught me an invaluable lesson. I told her I just wanted to teach and not to be involved in anything else. In return, I would dutifully ensure that all my charges achieved good grades. Instead of the expected nod and praise, I got a sharp rebuke from her. "Then you are just a subject specialist! You are not a teacher. You must see yourself as an educator and not merely as just a subject specialist, unconcerned about the other aspects of the child's development." That was the invaluable lesson I learnt on the meaning of education that looks into the holistic development of the child and not merely his academic achievement. Those words have paved the way for my current appointment as a Master Teacher.

---

[1] The schools are grouped into four zones: North, South, East and West. The East Zone comprises seven clusters. I am currently serving the East 5 Cluster which is made up of 15 schools.

Prior to my appointment as a Master Teacher, I was a Mathematics Curriculum Planning Officer in the Ministry of Education from 2000 to 2002. My experience as a curriculum officer in crafting the syllabus had enabled me to develop a broad overview of how educational policies and changes in the syllabus, disseminated from the Ministry, impacted the schools and the pupils. My four years of secondment to teach in the National Institute of Education (NIE), a teacher training institute, where I helped to mould young teachers and prepare them for their responsibility in the schools, has added another dimension to the development of my current role as a Master Teacher. With the knowledge and invaluable experience that I have acquired through my various portfolio in the span of my twenty years of teaching career, as a Mathematics teacher, a Head of Department (HOD) of Mathematics, a Curriculum Planning Officer in the MOE, and a Teaching Fellow conducting pre-service and in-service courses for primary and secondary school teachers, I have been implicitly groomed and shaped to take the responsibility of a Master Teacher. The experience in the school, MOE, and NIE has equipped me with the skills and knowledge, not only to build the capacity of secondary and primary school teachers but also to drive curriculum innovations and classroom research.

5 AS TEACHER LEARDERS

5.1 *PMTT*

Under my leadership, the HODs and other key personnel in mathematics teaching and learning form the Mathematics Interest Group (MIG). I organise and chair termly meetings with the MIG. The objective includes creating a platform for these instructional leaders to share their professional knowledge on the latest trends, methodologies, and changes in mathematics learning. Department management as well as sharing on pedagogical aspects of the learning of mathematics are also part of the agenda. These meetings aim to bring about networking opportunities among the MIG members to encourage collaborative work beyond their respective schools. Members also network to leverage on one another's resources and expertise. As some of the members are not as experienced, they tap on the expertise of their more experienced colleagues. MIG members have successfully collaborated in conducting workshops and focus group discussions among teachers of different cluster schools teaching the same level. With my guidance, they have also shared materials to hold workshops for parents of their students as part of an effort to partner stakeholders in the children's learning. MIG members welcome such meetings, both as an avenue for professional sharing in the teaching and learning of mathematics as well as for bonding to effect better sharing. Such structured and regular meetings and get-togethers have helped build confidence within the MIG members working directly with me. The team has commented on how these meetings have helped level up their capacity as instructional leaders in mathematics teaching and learning in school.

Under my guidance, MIG members have collaboratively organised mathematics seminars on assessment item writing and professional development activities. Feedback has been positive with both key personnel and teachers being more aware of how assessment items should be set. A newly-appointed HOD expressed how her participation in such activities has improved her belief in herself as an instructional leader in her school.

I have set up a Mathematics Resource Centre as a venue for the conduct of professional sharing. I source and procure useful teaching materials for the centre. Teachers can use or loan these materials. Schools share their resources here. The centre is a repository for printed and electronic copies of assessment papers of all the schools within the cluster. Teachers may choose to refer to these resources to guide them in designing their school assessment papers. The centre is also used for professional discussions as well as for reflection purposes. Teachers have initiated their own small group meetings at the centre for their professional development. I am much encouraged by such initiatives on the part of the teachers.

Besides conducting workshops and offering pedagogical advice, I organise professional sharings and create opportunities for mathematics HODs and teachers to interact with mathematics educators from higher institutions of learning. This helps HODs to keep abreast of global trends in the teaching and learning of mathematics.

As a teacher leader, I guide HODs in organising and facilitating mathematics problem-solving workshops for parents to encourage home support for their pupils. These workshops are met with overwhelming success as parents are equipped with skills to help their children in their mathematics learning. Some HODs have now made such parental workshops a regular feature of their schools. It is indeed an achievement for me to see HODs taking ownership of such work in their schools.

Knowing mathematics and teaching mathematics can be quite different. Responding to requests by teachers who find difficulty in delivering a particular lesson, I conduct demo-lessons to role model as an effective teacher of mathematics. I carry out such lessons in real classrooms with real pupils. No simulation. It is a very challenging task because I do not know the children prior to the lesson. However, it is rewarding. Once, when I re-visited a "participating" school, the children came to me to talk about a previous lesson and asked questions! Other teachers are also invited to sit in while I teach so that more teachers can benefit from the lesson and experience. Such demo-lessons are appreciated and teachers are grateful for the opportunity to learn from them. I remember one such lesson after which a teacher-observer confessed her inadequacies and related how she had found the session extremely helpful both in terms of pedagogy as well as classroom management! I also experimented with an interactive mass lecture for about 200 12-year-old primary school pupils on the use of heuristics. This was in response to a request by a school to help motivate pupils in learning mathematics. The session was interspersed with hands-on solving of mathematics problems using heuristics. Despite the rather large number of attending pupils, their positive response showed their enthusiasm and interest to extend their learning.

As a teacher leader, I initiate the Cluster Level Sharing (CLS) sessions. These sessions bring together teachers of the 12 cluster schools teaching the same level of mathematics to meet and share experiences, challenges, and success stories of their mathematics teaching in their respective schools. HODs share their pedagogical strengths too, thus developing themselves beyond their own schools. The CLS operates in an informal manner and teachers, who are placed in small groups, are free to speak their minds. A scribe documents the discussions and these discussion notes are then collated and made available to all the cluster schools. Feedback confirms that teachers enjoyed these sessions held in the Cluster Mathematics Resource Centre and have asked for them to be a regular feature.

Newly appointed primary mathematics HODs are also given support. Tea sessions are organised for them. Together with officers from the Curriculum Planning and Development Division, I guide and advise experienced HODs in sharing their department management of mathematics with the newly appointed primary mathematics HODs. The sessions are conducted with small groups for better interaction. They have proven to be invaluable to the new HODs as feedback has been positive and the sessions planned for them are very much appreciated. These sessions will continue into 2009.

Master Teachers should lead in educational research. Working collaboratively with the Centre for Research in Pedagogy and Practice (CRPP) and the NIE, I was a co-investigator of a large-scale research, *Mathematics Assessments Project (MAP)*, conducted by NIE. The research project addresses the possibilities of having new assessments in Mathematics besides the tried and tested traditional pen and paper assessment. This research spreads over a period of two years, involving 16 schools and over 60 school teachers. The findings are shared at an international conference. A module on Action Research in Mathematics has been organised at the cluster to expose teachers to the rudiments of action research so that they can become more reflective in their practice and more aware of their delivery effectiveness when teaching their pupils. So far, the feedback has been encouraging.

5.2 *SMTT*

Strong pedagogical skills and expertise in one's subject knowledge is a pre-requisite of a Master Teacher. To look into the holistic development of our pupils, I need to move beyond expertise in my subject area and be knowledgeable in other aspects, such as alternative modes of assessment and social emotional learning. This is all the more so because our influence as a Master Teacher spans beyond the school level to the cluster and zone, as I provide advice and guidance to the teachers within the East Zone. In fact, I work with some teachers in the Normal Technical (NT) stream on infusing social emotional learning in Mathematics to help the NT students dispel their fear and build their confidence in the learning of mathematics. Just like the Senior Teacher who had made such an impression on me, the influence of a Master Teacher is far-reaching, though the fruits of labour are not immediately visible.

To support the initiative of *Teach Less, Learn More* (TLLM) (MOE, 2007), which has spurred greater exploration of pedagogical strategies, I conducted more than 30 sessions of school-based sharing for teachers within the first year of my appointment. These were usually done during the "time-tabled time," a structured time-frame created by schools to promote sharing and interactions among teachers to facilitate better student engagement within and beyond the classroom. As a teacher, I firmly believe that learning becomes more meaningful when the students are able to relate the subject to real-life situations. Hence, in many of the sharing sessions, the historical development of the topic and the connections of the topic to daily life are often included. Hopefully, the effort would dispel the notion that mathematical problem solving is just about the application of algorithms and prescribed procedures acquired through rote learning. I am also working, based on my aspiration that my efforts at sharing would grow over time, to encourage teachers to go beyond the well-defined examination syllabus to ignite the passion of students in the learning of Mathematics.

As a beginning Master Teacher, when I first started sharing on pedagogical practice and skills during the "time-tabled time," attendance and punctuality were a problem. I understand the teachers' predicament because teachers have a very demanding role to fulfill and they have to wear many hats. Besides classroom teaching, teachers are often involved in school-based projects and events. Even though the "time-tabled time" is scheduled within the curriculum time, being present for the sharing would rank low in priority if they need to attend to the immediate needs of the pupils. However, after the initial sessions, I noticed that the attendance improved dramatically. It was the attentiveness and appreciation in the later sessions that, in many ways, have motivated me to improve the efficacy of my sharing.

Hence to influence any changes, we need to be patient and more importantly, never to lose heart. Just like planting a seed and ensuring its healthy development, the soil must be skillfully treated. Similarly, a process of "re-culturing" must precede any form of "restructuring." A good idea or pedagogical approach is like a seed; for healthy growth to take place, the soil needs to be "cultured" and this requires time.

6 AS A CURRICULUM LEADER

6.1 *PMTT*

Curriculum design is part and parcel of the job scope of the Master Teacher. To impact schools in their introduction and implementation of new pedagogies, I work closely with officers from the Curriculum Planning and Development Department (CPDD), MOE to better understand policies and curriculum development. With such knowledge and exposure, I can guide and help cascade relevant issues to the school personnel. The introduction of the use of calculator is one such issue that school personnel need more information. I help to provide ground support by guiding personnel on the use of suitable calculators and in understanding how assessment will be like with the introduction of the calculators.

I initiate meetings among primary mathematics HODs where I share and involve experienced mathematics HODs in interacting with newly-appointed HODs to discuss curriculum leadership in mathematics. As such, key personnel are more aware of issues like dyscalculia, age-appropriateness in learning, Singapore mathematics, and alternative assessments in mathematics.

I have conducted numerous pedagogical workshops to better equip teachers with good skills in making mathematics learning meaningful and fun. These include approaches to teaching topics like statistics, fractions, decimals, geometry, capacity, and others. I have done a webcast on the teaching of mensuration which was meant to be viewed on line in times of emergency closure of schools. Besides workshops on mathematics topics, I have also conducted workshops on action research and alternative assessment. These workshops are mostly hands-on and cater to teacher needs from Primary 1 to 6. Theoretical underpinnings are also shared during such interactions to help teachers understand the basis of how concepts could be better developed, learnt, and delivered. Feedback for all these workshops has been positive. Because of requests, I have made repeat workshops for other groups of participants.

With the changing focus in learning tools, I collaborated with the officers from the Educational Technology Department (ETD) in introducing new ICT tools to the cluster schools. We spearheaded the use of the Geometer's SketchPad (GSP) and Hot Potatoes to address the need for more interactive ICT programmes in mathematics teaching and learning. With numerous training sessions, the key personnel of the 12 cluster schools eventually adopt the use of the Open Tools to enhance their pupils' learning of mathematics.

As a consultant to the Learning Support for Mathematics (LSM), I work collaboratively with officers from the Psychological Services Branch (PSB), MOE to guide, advise, and train teacher trainers in training and preparing newly appointed LSM teachers in their teaching of mathematics through the 4-prong approach. LSM supports the learning of children at Primary 1 who are unable to cope with the "mainstream" mathematics learning. Together with PSB officers, I have guided and prepared LSM teachers to present their experience of this 4-prong approach at a conference. The approach encourages the explicit inclusion of meta cognition, cognition, environment, and motivation aspects in the teaching and learning of mathematics for this group of children.

Instruction drives assessment. Curriculum reviews affect assessment modes. With the changing educational landscape, schools want to introduce alternative assessment in mathematics education to cater to the new pedagogies used. As a curriculum leader, I hand-hold, train, and give support to HODs to encourage them in making that necessary mind-set change. I introduce Journal Writing in mathematics as a form of assessment and train teachers to try it out. HODs are supportive. Teacher participants have given positive feedback on how they find it meaningful and beneficial after going through their pupils' work. However, some are still apprehensive. The challenge then is for me to be patient and give them more time while offering "partners-in-action" hand-holding for such teachers. This is where I work extremely closely with the HODs to help the teachers.

A Master Teacher should engage in continual learning. Together with teachers, I learn about lesson study models. We hope to make small improvements in professionally developing ourselves through this authentic way of peer planning, observation, and discussion. We will embark on this in 2008 so that teachers can professionally develop themselves within their own schools or by collaborating with colleagues from other schools who have shown similar interests.

A Master Teacher rises to the occasion when her expertise is needed beyond subject boundaries. As a Master Teacher, I work closely with the Senior Teachers of the cluster to further develop them. I lent my expertise in being a trainer for the Structured Mentoring Programme to prepare school mentors for their roles in helping beginning teachers in their roles in schools.

6.2 *SMTT*

Schools have now moved into a new phase in education that focuses on curriculum customisation, choice, and flexibility (MOE, 2005(a), 2005(b), 2007), to nurture a new generation of pupils with diverse abilities and talents. In creating such diverse pathways, we provide our young charges with more choices in education and encourage them to grow with their passions. This also means that schools must experiment with new pedagogical methods, new curricula, and new systems of assessment. The implication would be an expansion in my job scope as a Master Teacher. Besides being entrusted with the responsibility of guiding and mentoring teachers, my role includes providing the depth and knowledge to teachers in the cluster schools and lead in the customisation of school-based curricula. Just one year into my master teachership, some schools have already approached me about the implementation of alternative modes of assessment and teaching approaches, such as Lesson Study and Understanding by Design (Wiggins & McTighe, 2005).

To ensure that teachers understand the curriculum changes, a Master Teacher is expected to work collaboratively with the Curriculum Planning Officers to hold sharing sessions and provide further support to teachers who need assistance. As part of MOE's effort to continuously improve the teaching and learning of mathematics, the use of calculators at Primary 5 is introduced in 2008. To equip teachers with the knowledge and skills in integrating calculators into the curriculum, workshops have been conducted by Curriculum Planning Officers for the teachers. In spite of the workshops, some teachers remain apprehensive about the use of this tool. To help the teachers acquire the initial confidence and at the same time to show how calculators can facilitate the exploratory approaches, I have planned several cluster workshops.

As curriculum leaders, it is essential to know the "whys" besides the "hows." A lack of awareness and understanding of the policy intentions is one other area in which the Master Teacher could help to bridge in our meetings with teachers and key personnel. Similarly in the secondary schools where the topic on geometric proofs is a new inclusion in the Additional Mathematics syllabus, cluster workshops were planned to expand teachers' knowledge on geometric reasoning and teaching strategies.

To serve as a resource person at the cluster level, the Master Teacher needs to be mindful of emerging trends. With the vision of building Singapore into a global schoolhouse and a knowledge and research hub, research is very much promoted at every level in Singapore. In the education arena, especially in schools, action research is presented as a reflective tool, a self-enquiry process to solve problems within and outside the classroom environment. To promote educational research within the cluster, I conducted many sessions to help schools with their classroom research and lend support in the interpretation of data, both qualitative and quantitative.

7 BUILDING A COMMUNITY OF PROFESSIONAL PRACTITIONERS

7.1 *PMTT*

Besides looking into mathematics teaching and learning, it is part of my job scope to guide the Senior Teachers in the cluster. Working in collaboration with some school leaders, I initiated the Cluster Senior Teacher Guild to bring together all Senior Teachers in the cluster for concerted professional development and networking. They share one thing in common: passion for teaching. They may not be teaching the same subjects or levels. The Guild creates a platform for Senior Teachers to operate beyond their school levels. Collaboratively, we organise regular professional sharing sessions where Senior Teachers are given space and training to help them realise their mentoring roles as well as their roles as curriculum leaders in their respective schools. I look into the capacity building of the Senior Teachers. They are taken on learning journeys. A memorable one was taking all of them to a pottery session in which they moulded and created their own masterpieces. Along with the activity, they were led into analogising their roles as "moulders" of beginning teachers in wanting to become passionate and quality teachers. It is memorable because a couple of years after the journey, these Senior Teachers still reminisce about the experience. To them, the experience is aptly related to their roles. Some new pedagogical approaches include Systems Thinking in the Classroom (STiC) as a Pedagogical Tool and Learning through Conversations, among others. Learning journeys outside the education realm are organised to expand the teachers' perspectives beyond that of the formal education scene. An annual retreat or worktreat is organised on a full-day basis to create time for Senior Teachers to reflect and revitalise their passion in teaching. Positive feedback indicates that the Senior Teachers have benefited from such attention and care given to them. Senior Teachers have presented their successful lessons using STiC at conferences. The Guild opens up opportunities where teachers can develop and present their new learning at conferences and seminars.

The teachers' well-being is also looked into. Motivating teachers enhances the role-modelling aspect of a Master Teacher. I have conducted team building workshops, retreats, and talks for schools upon invitations by school leaders. I give regular presentations to inspire teachers at the bi-annual Induction Programme for all beginning teachers organised by MOE.

I am also involved in the Strategies for Effective Engagement and Development (SEED) of pupils where new pedagogical approaches for better student engagement are given the necessary support. I guide teachers in the SEED movement in sharing with others their experiences in adopting various pedagogies and alternative assessments at Primary 1 and 2.

7.2 *SMTT*

At the heart of the Master Teachers' job is to build the capacity of teachers within the cluster by mentoring, coaching, and demonstrating good teaching practice and model lessons. So far, I have demonstrated several lessons on topics such as Algebra, Trigonometry, and Statistics in the East Cluster schools, and more have been scheduled. Besides showcasing good teaching practices, the demonstration also serves to convince teachers the use of ICT tools as a means to enhance the learning of mathematics. Some teachers may still be inhibited to infuse technology in the lessons to enhance the effectiveness. I remembered a series of lessons on trigonometry which I conducted in Term 4, with a class of Secondary 3 pupils, which was observed by the mathematics teacher. With the use of *Graphmatica* for investigative tasks on trigonometric functions and *Gnuplot* as well as *Geometer Sketchpad* for illustration of certain key concepts, the excitement and enthusiasm generated by the pupils was the driving force that convinced that teacher of the need to use the same tools. The same teacher, who earlier on had not been too keen to learn the software, came forward to insist on sharing the use of the same ICT tools, to be conducted during the time-tabled time. Hence, the observations during my demonstration lessons have, in a very powerful way, indirectly created the "want" to learn, a change of mindset from the initial stage.

To develop teachers within the cluster into learning communities, the Master Teacher must first seek to understand with an open heart and listen with an open mind for meaningful synergy to take place. Just like the teacher who had earlier expressed reluctance to infuse technology but was willing to overcome her inhibitions, after observing the exhilaration and the value-addedness of using the ICT tools, the Master Teacher needs to listen with an open heart, to impact and influence change. Many a times, I have to remind myself to be sensitive to the needs and feelings of the teachers, so that the best in every teacher can be drawn out to build a community of professional practitioners. In this way, the cluster schools can collaboratively forge ahead to build up teaching excellence and uplift the quality of the education service.

8 CONCLUSION

Master Teachers are role models. They play pivotal roles in helping to bring out the best in every teacher, so students, in turn, are nurtured to their full potential. With schools taking greater ownership of their curriculum and programmes to cater to the diversity of learners' interests and abilities, Master Teachers will definitely impact the teaching and learning scene. To ensure that teacher quality is never compromised, Master Teachers will be in the forefront of education and act as catalysts to the growth of committed and passionate teachers.

References

Foo, K. F. (2008). So you want to be a Master Teacher? *ASCD Review, 14,* 53-56.

Ministry of Education, Singapore. (2005a). *Greater flexibility and choice for learners.* Retrieved June 24, 2007, from http://www.moe.gov.sg/media/press/2005/pr20050922a.htm

Ministry of Education, Singapore. (2005b). *Greater support for teachers and school leaders.* Retrieved June 24, 2007, from http://www.moe.gov.sg/press/2005/pr20050922b.htm

Ministry of Education, Singapore. (2007). *Teach less learn more.* Retrieved June 24, 2007, from http://www.moe.gov.sg/bluesky/tllm.htm

Wiggins, G., & McTighe, J. (2005). *Understanding by design* (Expanded 2nd ed.). Alexandria, VA: Association for Supervision and Curriculum Development.

Part II

Teaching and Learning of Mathematics

Chapter 7

Model Method: A Visual Tool to Support Algebra Word Problem Solving at the Primary Level

NG Swee Fong **Kerry LEE**

This chapter sets out the research agenda of a series of funded studies with the specific focus to investigate how the model method is used to solve algebra word problems. The Perception study explored how 145 Secondary 2 students perceived the model method and its relationship with letter-symbolic algebra. Algebra 1, Algebra 2, and Algebra 3 are related studies with the second building on the findings of Algebra 1 and Algebra 3 on the first two studies in the series. While all three algebra studies are underpinned by a cognitive-psychological framework, this chapter reports on the pedagogical aspects of the research. Algebra 1 and Algebra 2 worked with Primary 5 students. Both samples answered the same set of algebra word problems. To delineate the knowledge, processes and metacognitive skills necessary to solve algebra word problems, an additional four sets of mathematics tasks were administered to the students participating in Algebra 2. Algebra 3, a 6-year longitudinal study, commencing with students in K2, explores the development of algebraic thinking in students from Kindergarten to Secondary 3.

Key words: model method, algebra word problem, representation

1 INTRODUCTION

Between 2001 and 2007, four funded research studies examined how students used the model method to solve algebra word problems. The first, (henceforth the Perception study) investigated how 145 Secondary 2 students (13+) viewed the model method and its relation to algebra (Ng, 2003). A series of three related studies, Algebra 1, Algebra 2 and Algebra 3, underpinned by a cognitive-psychological design, was set up to investigate the relations between students' cognitions and algebra word problem solving. Algebra 1 and Algebra 2 have been completed; writing up of findings is in progress. Data collection for Algebra 3, a 6-year longitudinal study commenced in January 2007. Details of this study are discussed in the penultimate section of this chapter. This chapter focuses on the pedagogical aspect of the model method while the next chapter by Lee and Ng discusses the cognitive-psychological findings from Algebra 1, Algebra 2, Experimental study, fMRI studies 1 and 2, and Model Method Analysis System (MMAS). The diagram in Figure 1 provides an overview of the studies discussed in this chapter.

Problem solving is the focus of the Singapore mathematics curriculum (Curriculum Planning & Development Division, 2000, 2006a; Curriculum Planning Division, 1990) with arithmetic and algebra word problem solving forming an integral part. There is a huge corpus of research documenting difficulties students have with arithmetic word problems (Fasotti, 1992; Hegarty, Mayer & Monk, 1995; Morales, Shute & Pellegrino, 1985; Pape, 2003; Swanson, Cooney & Brock, 1993; Verschaffel, Greer & De Corte, 2000) and algebra word problems (Brenner et al., 1997; Clement, 1982; Koedinger & Nathan, 2004; MacGregor, 1991; Stacey & MacGregor, 2000). There is by now an extensive body of literature (Charles & Silver, 1988; Krulik, 1980; Mason, Burton & Stacey, 1982; Noddings, 1988; Schoenfeld, 1985; Silver, 1985) indicating that strategies for problem solving can be taught effectively. For example, Lewis' (1989) work with 96 U.S. college students showed that when they were taught a "diagramming method" to represent information, their performance in "compare arithmetic word problems" improved. The performance of second graders of average and above average ability in addition and subtraction word problems improved when they were taught to use schematic drawing to represent information (Willis & Fuson, 1988).

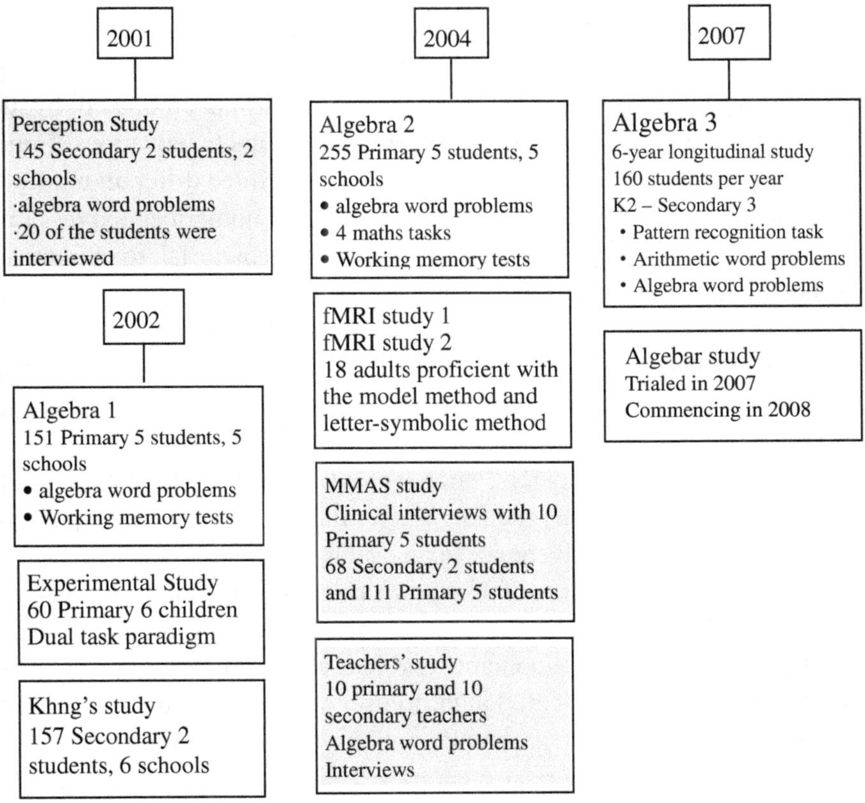

Note: No cognitive-psychological tests were used in studies in shaded boxes.

Figure 1. The various research studies.

Comprehension and representation of the information presented in a given problem constitute two important phases in the problem solving process (Kintsch & Greeno, 1985). The availability of different modes of representation (e.g., equations or schematics) makes it vital for students to learn how to represent information and to move from one mode to another (Kaput, 1987). Because representation is crucial, Singapore has adopted a unique approach to support students in word problem solving. Perusal of primary mathematics textbooks shows a focused and consistent use of the model method heuristic (henceforth model method) to solve arithmetic as well as algebra word problems, the type normally found in introductory algebra courses. A very simplistic description of the model method is that a series of rectangles is used as external and hence visual representation of the information presented in word problems. The quantity and qualitative relationship embodied by the rectangles are then represented by a series of arithmetic equations. Hence the model method requires students to work with three different modes of representation: (i) text, (ii) pictorial, and (iii) numerical expressions. Translations are between these modes, text to pictorial to numerical, although the order may not be in this sequence. For some problems, students may find it more useful to translate text to numerical expressions before representing this information pictorially. (An example of such a route is provided by the solution to the *Water problem* in Figure 9.)

In Singapore, although letter-symbolic algebra is introduced to Primary 6 (11+) students, the foci are on the construction, manipulation and evaluation of algebraic expressions in one unknown. The construction and solution of algebraic equations begin in Secondary1 (12+). The affordance of the model method has meant that primary students have access to a concrete and less abstract tool to solve algebra word problems. Students are taught to use the model method tool first to solve arithmetic word problems, where different length rectangles represent different numerical quantities. Later, when it is used to solve algebra word problems, the rectangles may signify a given quantity or the unknown value to be evaluated. The qualitative relationship of one unknown with another, for example "more than" or "less than", is suggested by the arrangement of the rectangles to one another, and also the type of lines used to draw rectangles. It is more common to use full lined rectangles to represent "more than" relationship as it is logical to represent what is there; while dotted lines are used to represent "less than" relationship, as "less than" signifies something missing. Also full lines are used to represent a quantity that is available, dotted lines for those that are removed. The form of, and the ways to, represent relationships are offered by teachers to their students. A model drawing then, is a pictorial representation of a word problem, the text form being its initial presentation.

Helping students make sense of complex word problems was the aim of the Ministry of Education when it introduced the model method in

1983 (Koh, 1987, 2006). Its introduction was premised on the belief that the solution of a word problem is made more accessible if the structure of the problem is articulated. The teaching of, and learning to use, the model method continue to play a significant part in the teaching and learning of primary mathematics. Teaching students the heuristic, however, does not determine how they choose to represent the information in a given problem. The model method is not an algorithm. The model method builds on students' knowledge of the part-whole relationship between numbers. When they use the model method, young students' impulsion to compute is effectively curtailed as they are required to process the text based information and think of how to represent pictorially, the quantitative and qualitative information. The model method is not a panacea to the problems discussed in word problem solving literature. It is generally accepted that successful solution of word problems is contingent upon an integrated and well-organised knowledge base, heuristic methods, metacognitive processes and a positive disposition towards problem solving. Also, successful word problem solving requires linguistic knowledge (Verschaffel, Greer & De Corte, 2000). These issues are still valid when using the model method.

The objective of this chapter is to provide the reader with information of the various research studies that investigated students' use of the model method as a problem solving heuristic. The following section begins with a brief description of how a local textbook introduces the model method. This is followed by a discussion of how students move between different representations. Discussion of the various completed studies follows. Next, illustrative examples of correct and wrong solutions from the seminal study with Primary 5 students, best serve the purpose to compare differences in how the model method is used to solve word problems.

2 EXAMPLES FROM A LOCAL TEXTBOOK

The model method builds upon two representational aspects of number: (i) size of a number and (ii) part-whole relationship of numbers. In the model method, rectangles are used to represent numbers, known or otherwise. Rectangles of differing lengths are used to represent numbers of differing sizes, the bigger the number, the longer the rectangle. Also a number can be represented by the sum of its parts. The examples in Figure 2 illustrate how length and the part-whole nature of the number could be used to solve arithmetic word problems presented in a Primary 2 text (Collars, Koay, Lee & Tan, 2007, p. 4). Figure 2(a) shows the model drawing for a combine word problem. Here two rectangles are drawn, a shorter rectangle to represent the number 10 and a proportionally longer rectangle, the number 20, the bigger of the two numbers. Since the objective of the question is to find the total number of stamps, then the numbers 10 and 20 are the two parts, and the sum of these two parts gives the whole 30, represented pictorially by the all embracing horizontal brace indicating the total length of the two rectangles. Using the same part-whole relationship, the model drawing in Figure 2(b) is a representation of a subtraction word problem where the whole is given and one of the missing parts has to be evaluated.

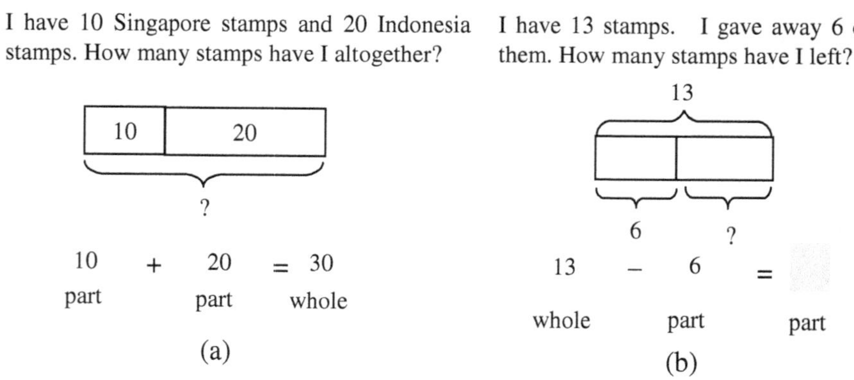

Figure 2. Part-whole relationship.

Similarly, the visual and concrete aspect of the model method makes it suitable to solve algebra word problems. The representation in Figure 3 from a Primary 5 text (Collars, Koay, Lee, Ong & Tan, 2005, p. 27) illustrates how the model method can be used to solve this algebra word problem. If letter-symbolic algebra were used, then letters represent unknowns and are to be operated on and evaluated. The model method does not require this level of abstraction. The rectangle representing the unknown value is identified as a unit and a unitary method is used to solve for the unknown unit. In the *Rahmad and Weng Ho* problem, the remaining amount of money held by Weng Ho is taken as the unit for comparison. The total amount of money held by Weng Ho is represented pictorially by the unit rectangle and the amount spent, represented by a rectangle constructed with broken lines. Five rectangles, each identical in length to Weng Ho's is used to represent the balance held by Rahmad. A "?" mark next to Rahmad's set of rectangles signifies the value to be evaluated. Once the unit upon which the rest of the model is based has been identified, and after the value of their purchases had been discounted, the relationship between the difference in money between the two friends can be articulated. Hence 4 *units* = \$117 - \$25. Since 4 *units* = \$92, therefore the value of one *unit* = \$23. Rahmad had \$140 at first. Letter-symbolic algebra solution may look like this. If x represents the amount of money Weng Ho had at the end of his purchase, then Weng Ho went shopping with $\$(x+117)$ and Rahmad went shopping with $\$(5x+25)$. Since Weng Ho and Rahmad started out with the same amount of money, then $5x+25=x+117$, which reduces to $4x=92$, a form which is equivalent to the unit representation of 4 *units* $=\$117-\$25$. Selection of the unit is important. The unknown unit is similar to the unknown x used to construct the algebraic equation, what Bednarz and Janvier (1996) described as the generator.

Rahmad and Weng Ho had the same amount of money. After Rahmad bought a shirt for \$25 and Weng Ho bought a pair of roller shoes for \$117, Rahmad had 5 times as much money as Weng Ho. How much money did Rahmad have at first?

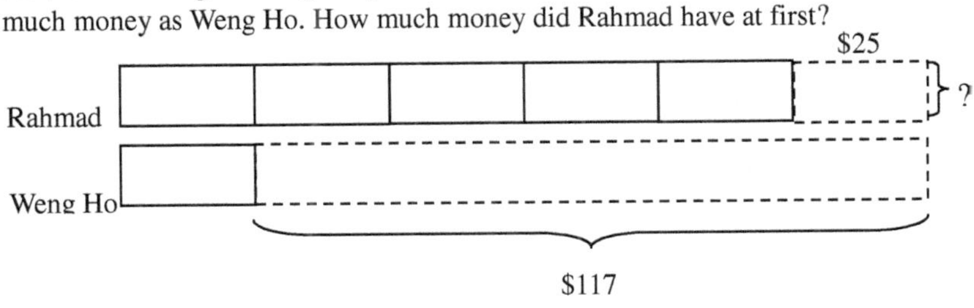

Figure 3. Rahmad and Weng Ho problem.

These three examples, by no means exhaustive, are meant to give the reader a sense of how the model method can be used to solve arithmetic and algebra word problems and how a local textbook scheme approved by the Ministry of Education presents it. The monograph *Singapore Model Method,* (Ministry of Education, in press) provides detailed descriptions of the model method and how it can be used to solve different categories of word problems.

3 MOVING BETWEEN DIFFERENT MODES OF REPRESENTATION

How do those who use the model method process the text information into the structure of the model? Three studies looked at this process. Using clinical interviews, 20 Secondary 2 students (Ng, 2003), 10 Primary 5 students (Ng & Lee, 2008) and 10 primary teachers (Ng, Lee, Ang & Khng, 2006) were asked to use the model method to solve algebra word problems. Although different questions were used in the three studies as each study had a different objective, nevertheless, there was a general structure to the translation of text-based information to the structure of the model drawing, to the arithmetic equations. Participants' solutions to the *Animal problem* are used to illustrate the translation process (see Figure 4). To facilitate the discussion, each sentence of the question is enumerated and the pictorial representation is provided. Participant here refers to teachers as well as students.

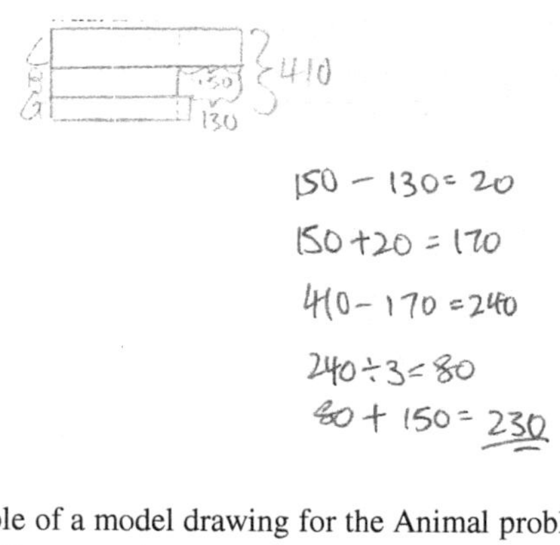

Figure 4. An example of a model drawing for the Animal problem where the mass of the dog was used as the generator.

(i) A cow weighs 150 kg more than a dog.
(ii) A goat weighs 130 kg less than the cow.
(iii) Altogether the three animals weigh 410 kg.
(iv)What is the mass of the cow?

Step 1: Participant first read the entire question.
Step 2: Participant reread the question to choose the unit.
Step 3: Participant drew the rectangle representing the unit. The rectangle was labelled, in this case "dog."
Step 4: Participant drew two rectangles, one for the mass of the dog and the difference rectangle which showed the difference in mass between dog and cow. This set of rectangles was labelled as cow.
Step 5: Participant read (ii).
Step 6: Participant drew two rectangles, one for the mass of the dog and a much shorter rectangle of 20 kg, which represented the difference in mass between dog and goat. The difference in mass between goat and cow was indicated by a horizontal brace carrying the mass 130 kg. Again this set of rectangles was labelled, in this case goat.
Step 7: Participant read (iii). A vertical brace was drawn to link all three rectangles and the total mass was indicated.
Step 8: Participant then evaluated the unknown generator by writing down a series of arithmetic equations.

Although any of the three masses could be chosen as the generator, the participants explained that the general practice was to choose the most obvious unknown, in this case, the dog. When the mass of the dog was the chosen generator, the three sets of rectangles represented the actual mass of the animals. When the mass of the cow was chosen instead (see Figure 7(b)), the sets of rectangles represented the relationships between the mass of the animal and that of the unit. Participants who chose the cow as the unit equalised the lengths of the different masses to that of the cow. First they drew the rectangle to represent the mass of the cow. The relationship between the mass of the dog and the cow was represented by a shorter rectangle and a horizontal brace with 150 kg was used to indicate the difference in mass between these two animals. Similarly the relationship between the goat and the cow was represented by attaching the "difference brace" of 130 next to a cow's rectangle. Participants knew that the mass of the cow could be evaluated directly when they chose it as the unit.

The numerical trace showed the participants used a series of undoing or unwinding of operations to find the value of the unknown generator/unit. For this problem, all relevant numerical information was indicated after construction of the model drawing. When the mass of the dog was the unit/generator, the numerical trace indicating the difference in mass between dog and goat was written after the drawing. This difference was calculated mentally during the drawing phase but made overt only during the numerical phase. The total in difference was recorded. If the dog was the generator, the unwinding process to solve for the unknown unit began by removing the total difference from the total mass of the three animals, 410 – 170. Since this difference represented the total of the three unknown units, then the value of one unit was found by dividing this total by 3. The mass of the cow was found by summing the 150 and 80.

The above description shows that participants who used the model method to solve word problems oscillated between any two of the three modes of representation as presented schematically in Figure 5. Oscillations between the text and the model drawing allowed for checking of the accuracy of the latter product. Checking the accuracy of the arithmetic expressions was achieved by oscillating between it and the model drawing.

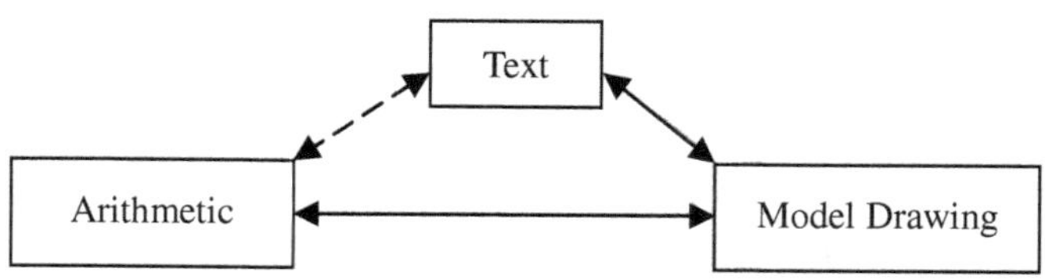

Figure 5. Schematic representation of how students work with different modes of representation when they use the heuristic model method to solve word problems.

Two products, the model drawing and the arithmetic expressions, are traces of participants' interpretation of the problem. If participants' interpretations are correct, then the model drawing is a logical representation of the text information and the arithmetic expression that of the model, and both these products are accurate representations of the problem. Because there are two products, any one wrong product would result in a wrong solution. For example, even if a set of correct arithmetic expressions were produced based on the model, construction of a wrong model would result in a wrong solution. Alternatively, a model which accurately represents the text information fails if the subsequent arithmetic expressions were wrong. If the arithmetic expressions are accurate, then its accuracy could be ascertained by checking it against the text. The accuracy of the solution could be tested by substituting its value into the pictorial representation.

The model drawing is a visual and concrete representation of the text-based information. The structure of the model drawing also shows the link between the arithmetic expressions and the text-based information. The very physical nature of the model drawing means that a specific text-based information can be pointed at and hence manipulated. For example, since specific sets of rectangles are used to represent masses of different animals, this means that the mass of a specific animal can be identified by pointing at that set of rectangles and should the information needs correcting, the specific rectangle can be identified and its length altered better to represent relationships between the masses of any two animals. Such identification and manipulation are not possible with algebraic representation.

4 FINDINGS FROM PERCEPTION STUDY: ALGEBRA 1 AND ALGEBRA 2

Although the Perception study, Algebra 1, and Algebra 2 were three different studies, grounded on different methodologies, there were, nonetheless, findings that were common across the three studies. In the Perception study, about 79% of the 145 Secondary 2 students (Ng, 2003) who took part in a pencil-paper test, found the model method a useful problem solving tool as it enabled them to visualise the information in the text. There was, however, a trade-off, as construction of the appropriate model was time consuming. Attention had to be given to the details necessary for the solution of a problem, and hence, what and how should be included in the model drawing. In comparison, algebra was more efficient and more importantly, there were problems that could be solved using algebra which were not possible with the model method.

In one item, students were asked to comment on the whether the accompanying model drawing logically represented that algebra word problem. Clinical interviews were conducted with 20 of the students. While all these students agreed that the model drawing was acceptable, they, nevertheless, were not satisfied with the proffered model drawing. Students highlighted how, and provided reasons to explain why the current drawing was confusing. This was indeed a serendipitous finding for their talk demonstrated how their attention oscillated between three modes of representation of the problem: (i) the information provided in the text of the question, (ii) how the different rectangles in the model drawing represented the quantitative values and qualitative relationships, (iii) how proportional reasoning was used to construct each rectangle, even when the exact numerical values were yet to be evaluated. The last activity showed that the students were estimating the value of the unknown even before calculating the value. Similarly, in another study (Ng & Lee, 2008), 10 Primary 5 students, in clinical interviews, were able to discuss whether they would accept two different but correct model drawings presented for the same problem. Again, the "mathematical talk" of these Primary 5 students showed how they oscillated between the different modes of representation to check the accuracy of the solutions. Because lengths of rectangles were used to signify size of numbers, it was possible for students to offer descriptions such as "Draw a longer rectangle for a bigger number" or "a shorter rectangle for a smaller number." These students were able to justify why the same model drawing could be used to solve different problems that had a similar structure but different numerical quantity. As one student succinctly put it:

> Numbers are put in the right places. Rectangles are drawn correctly. If you understand the question you don't have to change the size of the rectangles. (Student AT2) (Ng & Lee, 2008).

The visible and concrete nature of the rectangle means it can be manipulated. Its manipulability provides students with opportunities to talk about how they can use alternative representations for the same problem. More importantly, in both studies, the students showed that they were in control of the situation, possibly because the offered product was within their zone of proximal development (Vygotsky, 1978).

4.1 *Illustrative Examples of How Students Used the Model Method to Solve Word Problems – Algebra 1*

Algebra 1 (Lee, Ng, Ng & Lim, 2004; Ng & Lee, 2005,) explored the correlations between different measures of working memory and algebra word problem solving performance of 151 Primary 5 students who were specifically asked to use the model method as a problem solving tool. In this chapter illustrative examples focusing on students' responses to one arithmetic and four algebra word problems best serve the purpose of comparing how the model method is used to solve these problems.

Students' model drawings were analysed to code how they used the model method to solve word problems, the variation in model drawings, the kinds of errors they made when using the model method. Because students were not interviewed, the analyses are necessarily interpretive. Table 1 lists these questions and their facilities. Each problem taps on different knowledge. The arithmetic *Enrolment problem* and the algebra *Furniture problem* are homogenous word problems as the same comparison "more than" is used to relate between different objects. The *Animal problem* is a non-homogenous question because two comparisons "more than" and "less than" are used to describe the relations between different animals. While the *Book problem* requires multiplicative reasoning, the *Age problem* and the *Water problem* require knowledge of fractions as well as multiplicative reasoning. Furthermore, the *Age problem* requires a "before and after" drawing to represent temporal change.

Comparison of correct solutions against those that were wrong, highlights how correct solutions were evidence of an integrated and well-organised knowledge base of numbers and fractions, relationships between the operations (that subtraction is the inverse of addition, division the inverse of multiplication), knowledge that model method heuristic demands great attention to details, and clarity in drawing, metacognitive processes in the choice of generator and good linguistic knowledge, and confidence with the method. The correct model drawing solutions were evidence of good practices of problem solving. Students' model drawings indicated clearly, the selected generator, the difference rectangle and what was to be evaluated, paying attention to details when constructing the model. While some students constructed correct model drawings, the corresponding arithmetic equations were wrong. Such errors could be due to a lapse in attention or misinterpretation of the model drawing. Wrong models could be evidence how students misinterpreted the relationships presented in the text or a failure to check the model drawing against the text. Some of these errors may be evidence of other problems such as poor linguistic skills.

Table 1
A Selection of the Items Used in Algebra 1.
(Facilities are provided in whole numbers and per cent)

Enrolment problem 95 (63%)	Dunearn Primary school has 280 pupils. Sunshine Primary school has 89 pupils more than Dunearn Primary. Excellent Primary has 62 pupils more than Dunearn Primary. How many pupils are there altogether?
Furniture problem 67 (47%)	At a sale, Mrs Tan spent $530 on a table, a chair and an iron. The chair cost $60 more than the iron. The table cost $80 more than the chair. How much did the chair cost?
Animal problem 57 (38%)	A cow weighs 150 kg more than a dog. A goat weighs 130 kg less than the cow. Altogether the three animals weigh 410 kg. What is the mass of the cow?
Age problem 22 (15%)	Mr Raman is 45 years older than his son, Muthu. In 6 years time, Muthu will be $\frac{1}{4}$ his father's age. How old is Muthu now?
Water problem 30 (20%)	A tank of water with 171 litres of water is divided into three containers, A, B and C. Container B has three times as much water as container A. Container C has 1/4 as much water as container B. How much water is there in container B?
Book problem 22 (15%)	A school bought some mathematics books and four times as many science books. The cost of a mathematics book was $12 while a science book cost $8. Altogether the school spent $528. How many science books did the school buy?

4.1.1 Good Practices as Evidenced by Correct Model Drawings

A total of 95 (63%) students gave the correct response to the *Enrolment problem*. In Figure 6, both model drawings showed that each rectangle was clearly drawn with the generator value of 280 clearly indicated in each rectangle. Also the difference rectangles were clearly indicated, their lengths were proportionally drawn to show that 89 was bigger than the 62. While students constructed similar model drawings, they presented different arithmetic expressions. For example, while 25 students summed the product of the value of the common generator with the differences in enrolment (Figure 6(a)), 69 others summed the enrolment of each school (Figure 6(b)).

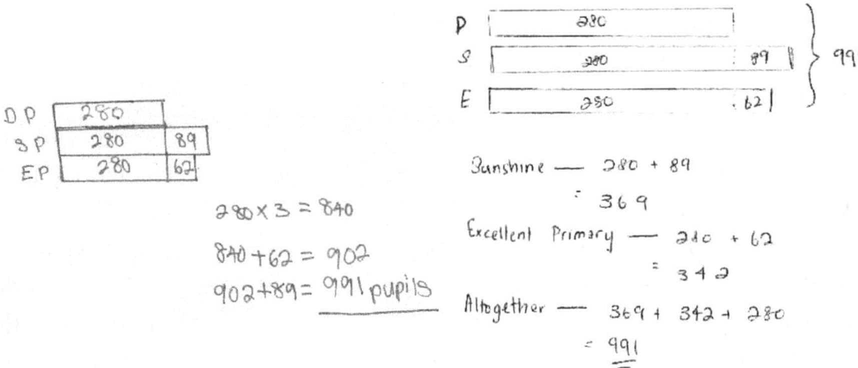

(a) (b)

Figure 6. Enrolment problem: (a) Sum product; (b) Sum individual enrolments.

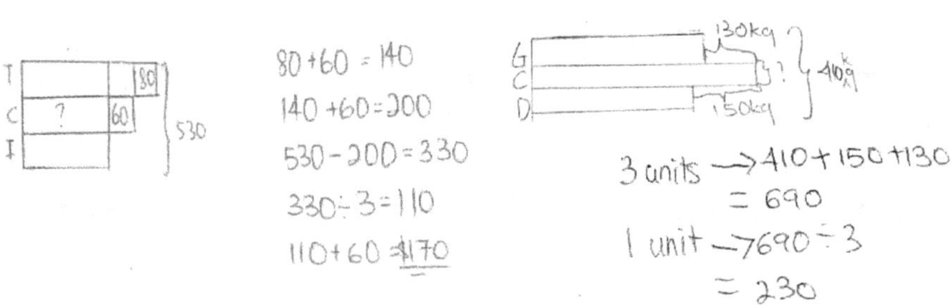

(a) (b)

Figure 7. (a) Furniture problem; (b) Animal problem.

The solution to the *Furniture problem* was direct as homogenous comparison "more than" related the different items. Sixty-seven (45%) students gave the correct model drawings in Figure 7(a). The model drawing was clear, with the difference in cost clearly indicated in each of the difference rectangle. A question mark "?" was used to indicate the variable to be evaluated. The total cost was indicated to the right of the rectangles. The undoing/unwinding process was used to evaluate the unknown generator. Compared to the *Furniture problem*, solution to the *Animal problem* was more complex as non-homogenous comparisons "more than" and "less than" were used in the latter. A total of 57 (38%) students successfully solved this item, 44 choosing the mass of the dog as the generator (see Figure 4) and 13 others using the mass of the cow (Figure 7(b)). Again all relevant information was indicated clearly in the model drawing.

Age problem: Twenty two (15%) students were successful with this problem. The model method solution to this problem is distinct from those discussed above as this requires the construction of a before and after model drawing representing the temporal change. The two correct solutions in Figure 8 illustrate how the visual nature of the model drawings makes it possible to represent, despite a change of six years, the constancy in age difference between father and son, knowledge of this fact is the key to the solution. The after model of both responses showed in detail, the multiplicative relationships between the ages of father and son.

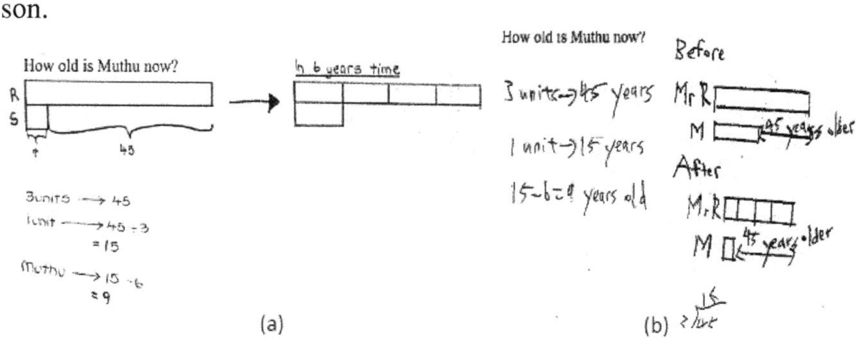

Figure 8. Age problem. (a) Age difference and multiplicative relationship; (b) Difference before and after.

Water problem: Multiplicative reasoning is necessary to solve this problem. For a more complete discussion of the solutions to this problem see Ng and Lee (2005). The examples in Figure 9 illustrate how solutions could be found, using *concepts* of, but not *operations* with, fractions. Concepts of fractions are necessary to transform the water problem to one where operations with whole numbers sufficed. By translating the relationship that "Container C has ¼ as much water as container B" to its equivalent "Container B has 4 times as much water as C," and then checking for the lowest common multiple for 4 and 3, leads to the correct solution (model drawings in Figure 9(a) and (b)).

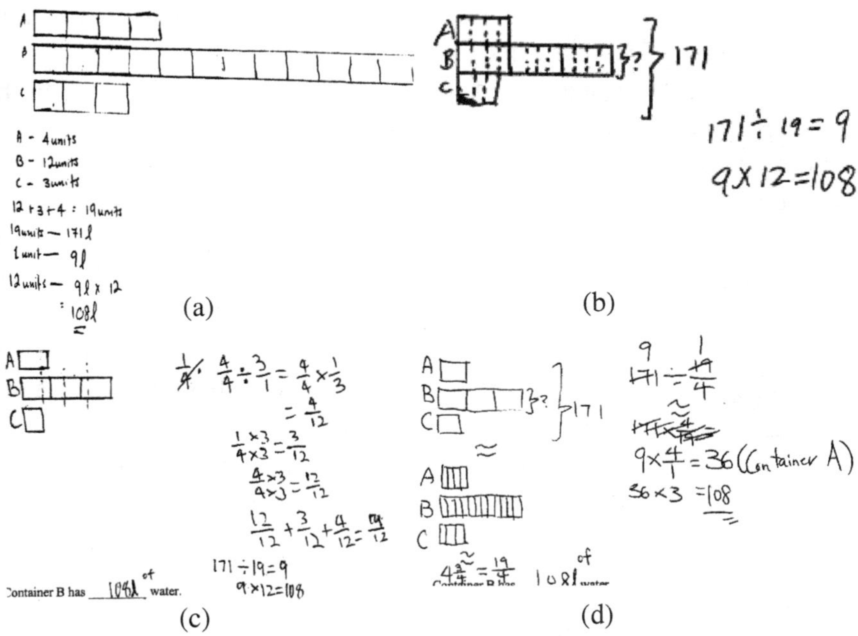

Figure 9. Water problem. (a) Lowest common multiple of 4 and 3; (b) Brief numerical solution; (c) Copious calculation; (d) From simple to detailed drawing.

Solution in Figure 9(c) and (d) demonstrates how the model drawings and fractions were used to solve the problem. All the model drawings illustrate how important it was to be meticulous in the representations. In particular, in comparison with the others, the numerical solutions in Figure 9(b) was spartan, the details in the model drawing suggested otherwise, as it showed the exact relationships between the three containers. Once those details were in, the solution could be found. Because the model drawing in Figure 9(c) had fewer details, the numerical expressions involving fractions were copious, reflecting student's operations with fractions. The solution in Figure 9(d) showed how a simple model drawing evolved into a detailed representation of the problem. The student's rather unorthodox work with whole numbers and fractions suggests a good command of the division by fraction algorithm, as the third arithmetic equation was correct. Also, this example shows that some students have knowledge beyond those taught in primary mathematics as division of whole numbers by fractions is taught only in secondary mathematics.

Book problem: Students' solutions showed they were flexible in the way they used the rectangles. In the above problems, while the rectangles could be used to signify a given number or an unknown number, different lengths indicating numbers of varying sizes, this was not the case for the rectangles in the *Book problem.* Here, the rectangles were used to represent the actual cost of the books, and the model drawing represented a group of mathematics and science books, where the proportional relationship between the books was delineated, four science books for every one mathematics book. The number of such groups could be found by determining how many groups of \$44 there were in \$528. Since there were 12 such groups, there must be 48 science books. Since the rectangles represented cost of the books, a correct drawing

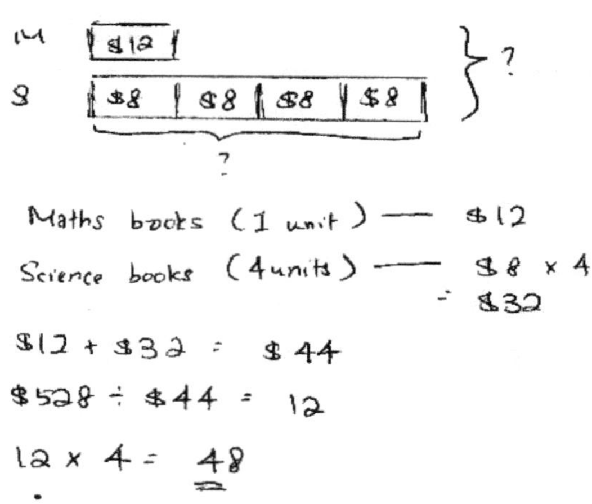

Figure 10. Book problem.

would have rectangles of different lengths representing the actual cost of each type of book. All the correct model method solutions, however, were similar to that in Figure 10, where the rectangles were of the same length, suggesting that students may be using the model drawing as an mnemonic aid. In cases such as this, meaningful relationships as exhibited by the number of rectangles are more important than the accuracy of the lengths of rectangles.

4.1.2 Types of Errors Evidenced by Model Drawings

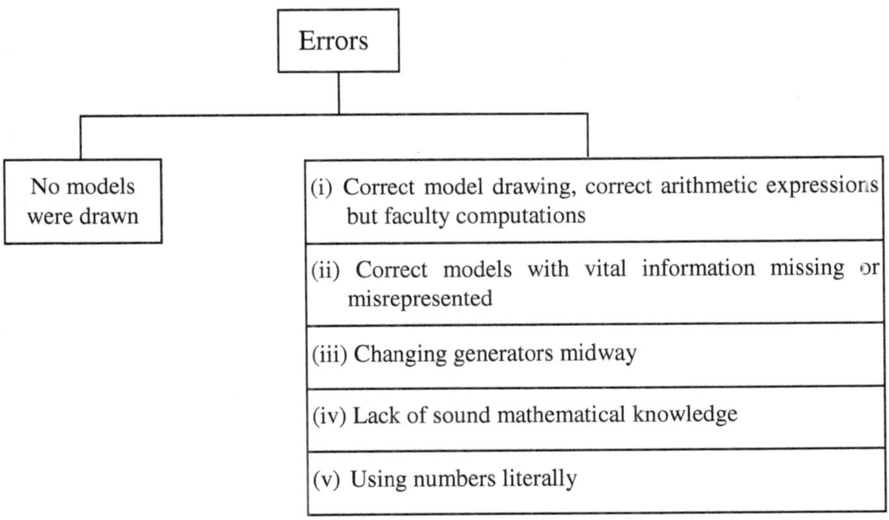

Figure 11. Five categories of errors.

Figure 11 lists the five categories of errors made by students. After discounting solutions which did not use the model method, partly correct solutions using the model method formed about 30% of the errors. There were three categories of such errors. As suggested by the title, "Correct model drawing, correct arithmetic expressions but faulty computations," students made errors in the arithmetic calculations. Only a small percent were errors of this type. Errors categorised as "Correct models with vital information missing or misrepresented" were examples of how important it was to exercise care in representing information and the meanings suggested by the representations. "Changing generators midway" could suggest a lack in monitoring activity of the students as attention was needed to identify the generator for the subsequent unknown. Examples from (ii) and (iii) show that model drawing that lacked precision as well as clarity in the choice of generators eventually resulted in a wrong set of arithmetic equations. Errors categorised as "Lack of sound mathematical knowledge" support the theory that successful problem solving is in part supported by integrated real world as well as mathematical knowledge. Examples from "Using numbers literally" show how these students' may have problems reading and understanding the questions. Here, every rectangle corresponded to exactly one number presented in the question. Relationships were not addressed.

Correct model drawing, correct arithmetic expressions, but faulty computations: Students who made such errors produced a correct model drawing and its related arithmetic equations. They, however, failed to answer the question. The example in Figure 12(a) shows that rather than proceeding to find the total enrolment, the student stopped after the partial sums were found. The example in Figure 12(b) illustrates a solution where, rather than finding the cost of the chair, the student stopped after finding the cost of the iron. Such solutions suggest that these students may not have understood the question, or did not re-read the question to check if they have fulfilled its expectation.

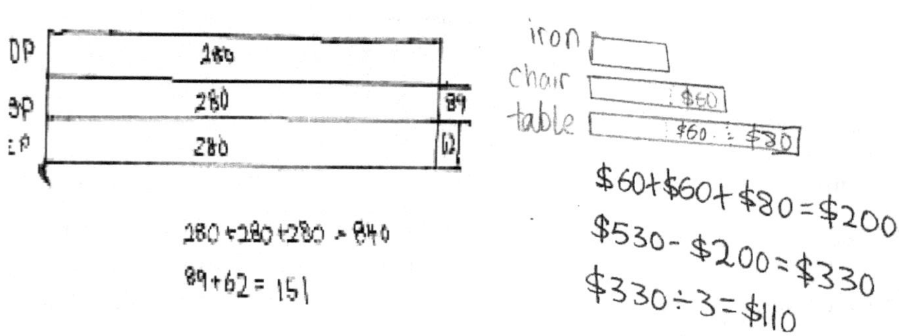

(a) (b)

Figure 12. (a) Enrolment problem, partial sum; (b) Furniture problem, wrong item.

Correct models where vital information is missing or misrepresented resulting in wrong arithmetic expressions. Errors of this type were common among algebra word problems. In Ng (2003, p. 13), 114 of the 145 students explained that:

> All information provided in the question had to be identified and included in the representation. Should a detail be missing, it would not be possible to solve the problem.

The effects of such missing information are reflected in the students' solutions. A total of 11 students (7%) constructed the correct model drawing of the *Furniture problem* (see Figure 13(a)) but the arithmetic expressions did not capture all the information expressed in the model drawing. In this example, the cost of the iron was the generator, and cost of the chair was built upon the cost of the iron, the cost of the table against the chair; and the differences in cost denoted by the marked rectangles. Although the model drawing was correct, detailed information, however, was missing from the rectangle representing the cost of the table. While the table cost \$80 more than the chair, it meant that it was (\$80 + \$60) more than the iron. One possible reason why this additional but vital detail was missing in the arithmetic equations was

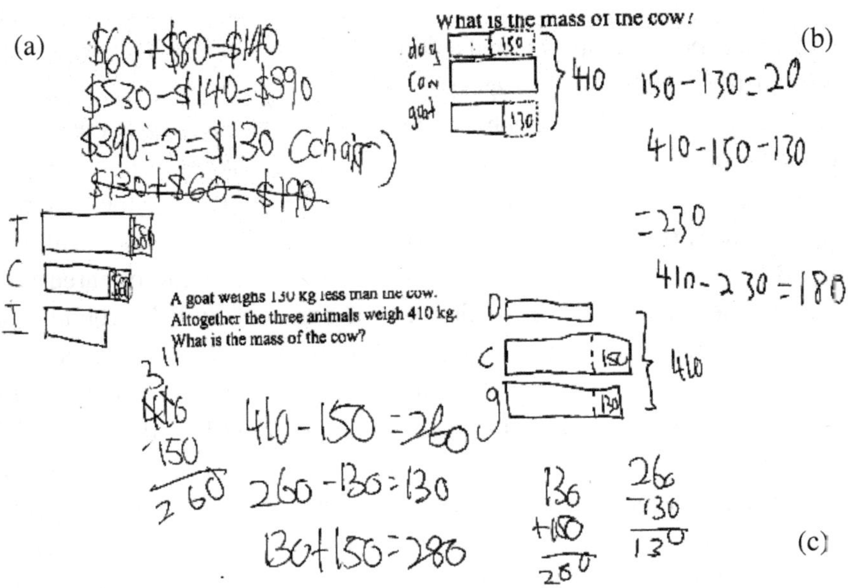

Figure 13. (a) Furniture problem, vital information not identified; (b) Animal problem, misinterpreted "less than" with "more than." (c) Animal problem, wrong relationship.

because it was not made obvious in the rectangle representing the cost of the table. Hence, while the actual value of the three units should be $\$530-\$60-\$60-\$80=\$330$, the wrong total of \$390 was found instead.

In the *Animal problem*, the animals were related by non-homogenous qualitative relationships such as "less than" and "more than", the animal of reference changed for each animal. These two changes could be reasons why the facility rate for the *Animal problem* was lower compared to the first two problems. The erroneous drawing in Figure 13(b) shows how 19 students misinterpreted the relationship that the goat was "130 kg less than the cow." Correct interpretation would result in two rectangles representing the mass of the goat, one representing the mass of the dog and a shorter rectangle of 20 kg, representing the difference in mass between the goat and the dog. The alternative interpretation of this error could be that these students may be attempting to equalise the lengths of the two shorter rectangles to that of the cow. If this was completed successfully, solution of the problem would lead directly to the mass of the cow. Instead, the drawing showed that the rectangles were used to represent literally the numerical difference between the animals without specific reference to the other pair in the comparison, suggesting that these students did not process with care such relationships between the different animals. If the mass of the cow was the chosen generator, hence the unknown unit, then the value of three unknown units could be found by equalising the rectangles for the mass of the goat and the dog to that of the cow. The arithmetic equation $410+130+150=690$ represents the total value of three units. Hence, the mass of the cow, the unit of one unit, will be found when $690\div3=230$. The set of arithmetic equations, however, has no particular objective as it does not give the value of any unit since the unknown unit is not clear. While another 17 students clearly identified the mass of the dog as the generator (Figure 13(c)), the drawings showed that they continued to represent the relationship "less than" as they did "more than" and, furthermore they used the wrong generator in the comparison. The rectangles suggest that the mass of the cow and goat were based on that of the dog and the goat is now 130 kg heavier than the dog.

Changing generators midway: Such errors were common to arithmetic as well as algebra word problems. The second set of rectangles was constructed correctly, the choice of generator was clear. For different reasons, the third set of rectangles was wrong. For example, in the *Enrolment problem* (Figure 14(a)), instead of using the enrolment of Dunearn to construct the enrolment of Excellent Primary, 37 (25%) of the students based the enrolment of Excellent Primary on that of Sunshine Primary. In the *Furniture problem*, 38 (26%) of the students compared, wrongly, the cost of the table against the chair, and not the iron (Figure 14(b)).

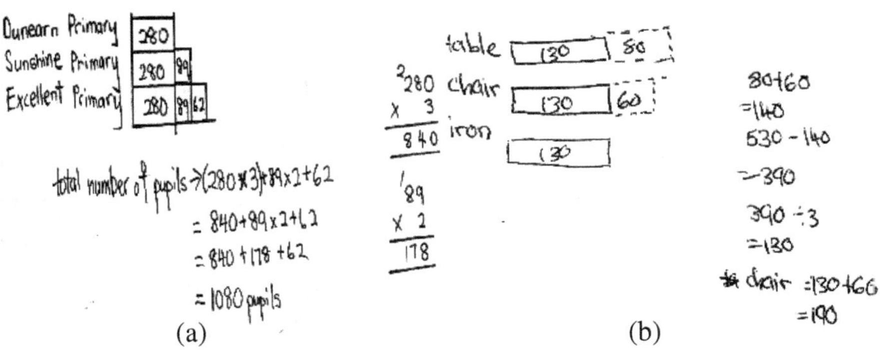

Figure 14. (a) Enrolment problem; (b) Furniture problem.

Lack of sound mathematical knowledge: Wrong solutions to the *Age problem* (Figure 15(a)) showed that students did not know how to use the fact that the difference in age between any two individuals remains constant despite temporal change. Consistent with the suspension of sense making literature in word problem solving (Vershaffel, Greer & De Corte, 2000), solutions showed students combining the different numbers to arrive at improbable solutions. Erroneous solutions for the *Water problem* showed an avoidance of fractions (Figure 15(b)). While the multiplicative relationship between A and B was correctly represented, that between B and C was converted to suggest that C had 4 times as much water as B, rather than the other way round. Eighty one (54%) students exhibited such errors, evidence of how concepts of fractions continue to challenge students.

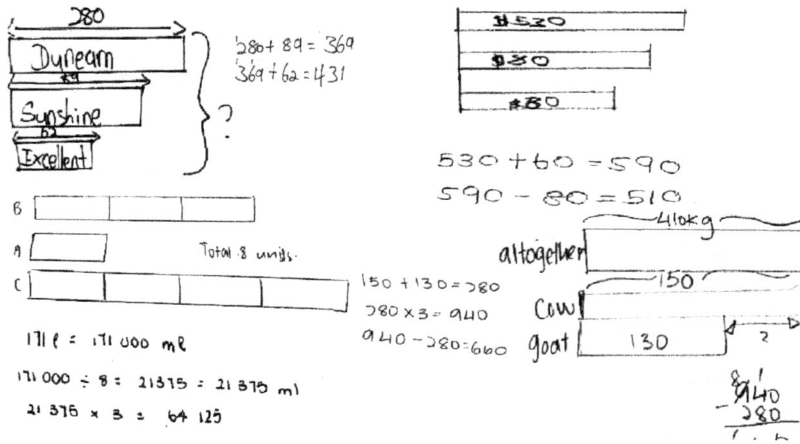

Figure 15. (a) Age problem. (b) Water problem, avoid fraction.

Using numbers literally: The examples in Figure 16 show how rectangles are used to represent numbers as they were presented in the word problems, a problem commonly found in 10 students (8%). In the *Enrolment problem* only the difference rectangles were drawn. By using rectangles to represent numbers as they were presented in the word problems meant that all algebra word problems were transformed into

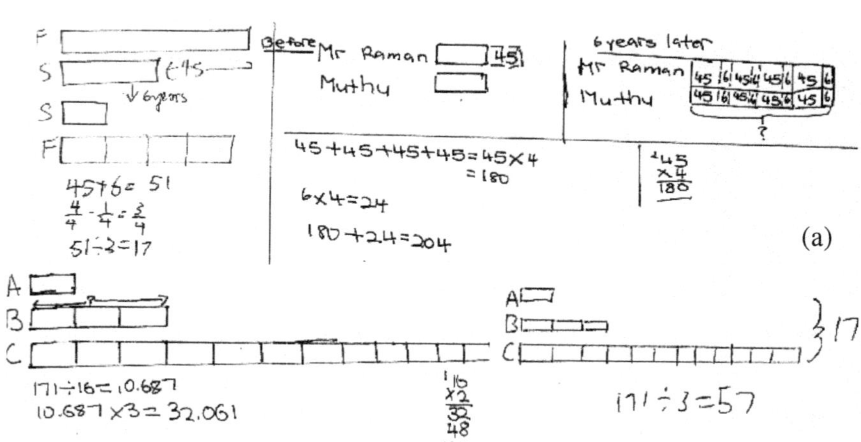

Figure 16. Using numbers literally.

arithmetic word problems. In such errors, the rectangles represented the difference relationships between the related items. No generator was identified. These responses suggest that the students may require help with arithmetic word problems, and work with algebra word problems should be delayed utill mastery of arithmetic word problems has been achieved.

4.2 *Algebra 2*

A suite of studies was designed to answer questions raised in Algebra 1. What particular skills, processes, content knowledge, metacognitive skills, and linguistic skills did successful students have when they were able to use the model method to answer the word problems? Four tasks were designed to answer this question. Table 2 provides a summary and objectives of the tasks.

A total of 255 Primary 5 students (132 boys) completed the above tasks and the set of algebra word problem solving task similar to that in Algebra 1. They were also required to complete the battery of cognitive and psychological tasks similar to those in Algebra 1. In Algebra 2, however, tasks testing the central executive component of working memory were further fractionated to identify what aspects of cognition correlated to algebra word problem solving. See Lee and Ng in Chapter 8 for a detailed discussion of the findings.

The Teachers' Perception Study explored primary and secondary school mathematics teachers' perceptions of the relationship between the model method and letter-symbolic algebra, their ability to translate between the two representations, and also primary students' epistemology of the model method (Ng, Lee, Ang & Khng, 2006). Primary teachers chose to use the model method, and secondary teachers, letter-symbolic algebra, to solve a set of algebra word problems. Primary teachers, unlike their secondary counterparts, were comfortable with the model method and translation of model method solutions to their equivalent algebraic forms was not an issue for them. Secondary teachers critiqued the teaching of the model method as students' knowledge of this heuristic meant that many secondary students, instead of using

Table 2

Summary of the Tasks Used in Algebra 2

Componential tasks and the expectations	Details of tasks
Integration tasks comprised three sets, A, B and C.	
Given questions, some using consistent language and others, inconsistent language, which relationships would students find more challenging?	12 items in Set A. 6 word problems using consistent language, and 6, inconsistent language Consistent language: May has \$78. June has \$25 less than May. How much money does June have? Inconsistent language: Li Peng watched 15 movies in a year. She watched 1/5 as many movies as Hui Leng. How many movies did Hui Ling watch in a year?
When provided with all the relevant information, were students able to construct an appropriate model	5 items in Set B. For each item, the givens and the unknowns were highlighted. While they were required to construct a possible model, students, however, were not asked to solve the problem.
When provided with items where relational terms "more than", "less than" and "n times as many", could students interpret these information to solve simple word problems.	10 items in Set C. 10 questions tested whether students would interpret relationships. This is one such example. John and Shiyang have some money. John has 1/5 as much money has Shiyang. (A) John has less money than Shiyang (B) John has more money than Shiyang.
Planning Tasks Given all the information, could students construct the appropriate arithmetic equations	5 items in this set. In each item, the givens, unknowns and a possible model drawing were provided. Students were asked to use the information to construct appropriate set of arithmetic equations to solve for the unknown.
Content knowledge To establish mathematical content knowledge of students	25 items in this set. Students solved mathematics problems which were not text based. Here are three examples. (i) $8911 \div 11 =$ (ii) $\frac{1}{3}$ of $6 =$ (iii) $\frac{1}{4}$ of $8 = \frac{1}{3}$ of ____
Translation task To test whether students were able to read a mathematics problem to identify what were the givens in a problem and what they were required to solve.	8 items in this set. Students had to identify what information was provided, what was missing and what they were required to find.

letter-symbolic algebra, continued to use the model method to solve algebra word problems. The examples in Figure 17 illustrate teachers' concerns that some secondary students, instead of letter-symbolic algebra, continued to use the model method to solve algebra word problems.

A total of 157 Secondary 2 students from six secondary schools took part in Khng's study which adopted a cognitive-psychological design. Despite her request that they used letter-symbolic algebra, many students continued to use the model method or a mixture of the model method and letter-symbolic algebra to solve ten algebra word problems, which had an adverse effect on solving accuracy (see Figure 17). Such examples are

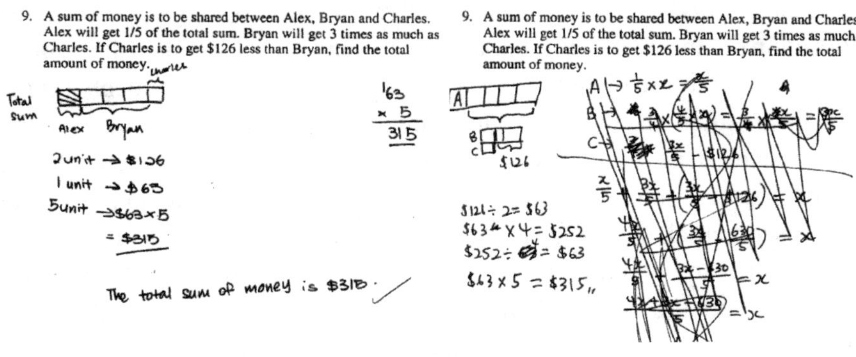

Figure 17. Evidence of how students' knowledge of the model method as a problem solving heuristic continued to intrude into their work with algebra word problems.

evidence of how students' knowledge of problem solving heuristics could work against their learning of letter-symbolic algebra. These students seemed to be pre-occupied with solving the problem rather than with the method for solving the problem, in this case letter-symbolic algebra, a predicament discussed by Lins, Rojano, Bell, and Surtherland (2001). Khng's findings, however, showed that

> domain-general processing abilities such as inhibitory and intellectual abilities also contributed significantly to individual differences in algebra word accuracy on top of the contribution from algebra knowledge. In particular, these results lend support to the view that successful problem solving depends not only upon the acquisition of domain-specific knowledge, but also the ability

to inhibit intrusions from previous knowledge or alternate competing schemes. (Ng, Lee, Ang & Khng, 2006, p. 232)

In the model method, rectangles may either be used to signify a specific quantity or an unknown. In learning mathematics, students may develop an intuitive understanding of a concept long before they can handle an abstract representation of that concept (Carpenter, Corbitt, Kepner, Lindquist & Reys, 1981). Does the experience using the model method help primary students develop an intuitive understanding that the rectangles may signify a variable, a precursor to letters as variables in letter-symbolic algebra? The intrusions of the model method into students' work with letter-symbolic algebra seem to suggest that students do connect the two representations. Ng and Lee (2008), using clinical interviews with ten Primary 5 students, found that these students seemed to have such an intuitive understanding, that rectangles and letters seemed to have a synonymous role – that is they are receptacles for numbers. In model drawing, the lengths of the rectangles were inconsequential. As long as the model drawing was logical, the same model drawing could be used to solve questions with a similar structure. Because of the small sample size, the conclusions could not be generalised. Hence, a bigger study was conducted with 111 Primary 5 students and 68 Secondary 2 students. Students used a custom-designed IT tool, the Model Method Analysis System (MMAS) to solve an identical set of word problems as those in Ng and Lee (2008). By using the pixel size to compare the length of the rectangles, we hope to ascertain whether (i) students used proportional reasoning to draw the rectangles, (ii) students had an intuitive understanding that the rectangles could be treated as variables. The manuscript for this study is in preparation. Figure 18 shows the screen shots of two students' responses to the same problem. The response in Figure 18(a) shows a one-to-one correspondence between the number of rectangles drawn and the number of marbles. Figure 18(b) shows another student perceiving the rectangle as a single unit representing a given number.

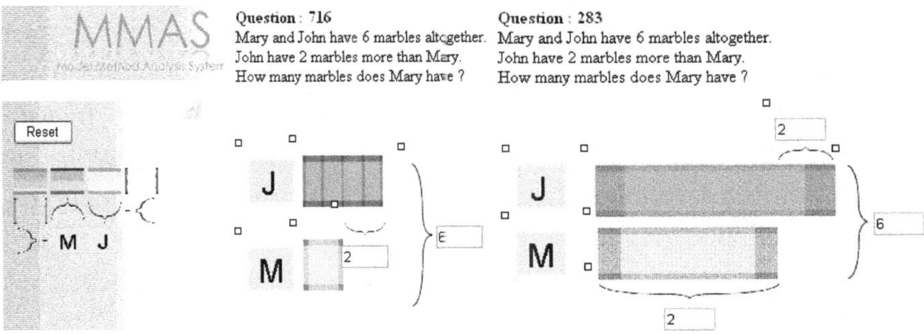

(a) (b)

Figure 18. (a) One-to-one correspondence between number of rectangles and number of marbles. (b) Rectangle as a unit.

Because of the concrete and visual nature of the model method, the secondary mathematics teachers (Ng & Lee, 2003) saw it as a primary school method, one very different from letter-symbolic algebra. How accurate is such a perception? Does this mean that these two methods recruit different areas of the brain? In Chapter 8, Lee and Ng discusses how Functional Magnetic Resonance Imaging (fMRI) technology was used to answer this and related questions.

4.3 *Ongoing Research: Algebra 3 and Algebar Study*

How students used the model method to solve algebra word problems and how different skills contributed to success in algebra word problem solving were the focus of Algebra 1 and Algebra 2 respectively. Ng (2004) analysed how algebraic thinking is developed in the Singapore primary mathematics curriculum. The study found that three approaches: (i) problem solving approach, (ii) generalisation approach, and (iii) functional approach, were used to develop algebraic thinking across the primary years. As students grow cognitively, how is students' algebra word problem solving supported by their ability to recognise patterns, identify functional relationships, and describe change? It is the aim of Algebra 3, a cross-sectional and a 6-year longitudinal study, to track and explore how K2 students (5+) develop across the years. In particular, this study focuses on and explores the interactions between the different approaches to developing algebraic thinking and algebra word problem solving. Besides sets of age-specific cognitive-psychological and linguistic tasks, a battery of grade-specific mathematics tasks, focusing on pattern recognition, identification of functional relationships, generalisation, describing change have been constructed for this purpose. Data collection started at the beginning of 2007, with about 160 students from these grade levels: K2, Primary 2, Primary 4, and Primary 6. These students are tracked each year till they reach Secondary 3.

The Algebar Study looks at how to support students make the transition between model method and letter-symbolic method. Empirical evidence from Secondary 2 students and secondary teachers (Ng, Lee, Ang & Khng, 2006) suggests that knowledge of the model method may complicate the learning of letter-symbolic algebra. Secondary school teachers often find that early learners of letter-symbolic algebra may not be that motivated to learn the skills needed to solve algebra word problems. With the affordance of a concrete and visual representation for numbers and unknowns, and arithmetic procedures to solve for the unknown, students can avoid engaging with the representational and transformational activities, generalising and justifying activities, activities which students find challenging (Kieran, 1989; Kilpatrick, Swafford & Findell, 2001). Given an algebra word problem, it is a common practice among many secondary students to construct a model drawing and then its equivalent algebraic equation and then reverting to arithmetic methods to evaluate the unknown unit, hence dispensing the need to transform equations (see Figure 17). Students asked whether it was necessary to use letter-symbolic algebra to solve word problems when the model method was just as effective. Such students were often confused with the ability to solve an algebra word problem with the necessity to learn letter-symbolic algebra, a more powerful problem-solving tool needed for higher mathematics and the sciences. Students' preference for the arithmetic based methods is consistent with empirical evidence (Koedinger & Nathan, 2004) which showed students continuing to use arithmetic methods they have learned in primary schools to solve context-based or story sums but had difficulties with representational and transformational activities related to letter-symbolic algebra. It must be understood that although students are not curtailed in their choice of methods in problem-solving situations, the learning of algebraic method is a key component of secondary mathematics curriculum (CPDD, 2006b).

To engage students in the algebraic method and its related representational and transformational activities, Algebar, a software tool, developed jointly by the Ministry of Education and the National Institute of Education, is being trialed in selected secondary schools. The specific objective of Algebar is to help students make the link between the model method and letter-symbolic algebra and its related representational and transformational activities. Given an algebra word problem, students who choose to use the model method, are prompted to construct the equivalent algebraic equations, beginning with definition of the variable. Appropriate prompts are provided to support students in the construction of the set of equivalent algebraic equations. The inbuilt self-checking system means that students can work independently of their peers and teachers. The screen shot in Figure 19 shows a positive response offered by the self-checking system. This software was trialed with two secondary schools and evaluation of this tool begins in 2008.

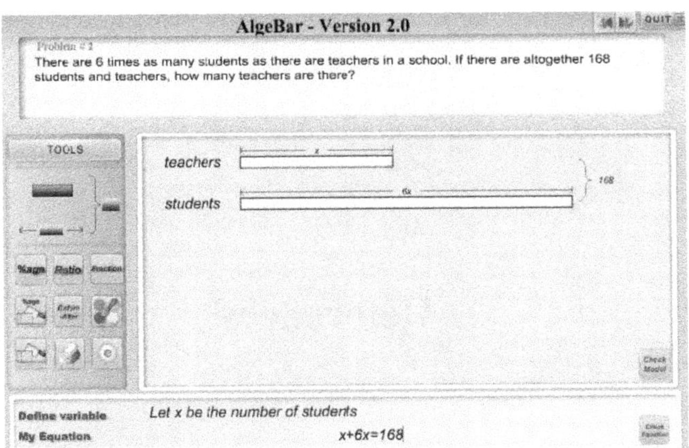

Figure 19. Screen shot of the Algebar tool used to help students make the link between the model method and letter-symbolic algebra.

5 CONCLUSIONS

The model method is an important part of the Singapore primary mathematics curriculum. Does the model method support students in word problem solving? Algebra 1 shows that students who took Mother Tongue (e.g., Mandarin, Tamil or Malay) at the highest level performed better than those who took mother tongue at the second level. Because students who took Mother Tongue at the highest level formed only one-fifth of the sample, the difference in performance cannot be generalised across all schools in Singapore. Nevertheless, it is an important finding and more research is needed to address this difference in performance. Also Algebra 1 found that students' performance with the one arithmetic word problem was not at mastery level. Perhaps more time should be spent helping students improve their performance with arithmetic word problems before the introduction of algebra word problems.

This chapter sets out the research agenda with the specific focus on investigating students' use of the model method to solve algebra word problems. The studies reported here include those that have been completed, are in progress and will come on stream. While a cognitive-psychological design underpins Algebra 1, Algebra 2, and Algebra 3, a pedagogical lens is used to look at the work of the students. Readers are directed to look at the work of Lee and Ng in Chapter 8 for findings from a cognitive-psychological perspective.

References

Bednarz, N., & Janvier, B. (1996). A problem-solving perspective on the introduction of algebra. In N. Bednarz, C. Kieran & L. Lee (Eds.), *Approaches to algebra: Perspectives for learning and teaching* (pp. 115–136). Rotterdam, The Netherlands: Kluwer Academic Publishers.

Brenner, M. E., Mayer, R. E., Moseley, B., Brar, T., Durán, R., Reed, B. S., & Webb, D. (1997). Learning by understanding: The role of multiple representations in learning algebra. *American Educational Research Journal, 34*(4), 663–689.

Carpenter, T. P., Corbitt, M. K., Kepner, H. S. Jr., Lindquist, M. M., & Reys, R. E. (1981). *Results from the Second Mathematics Assessment of the National Assessment of Educational Progress.* Reston, VA: National Council of Teachers of Mathematics.

Charles, R. I., & Silver, E. (Eds.) (1988). *The teaching and assessing of mathematical problem solving.* Reston, VA: National Council of Teachers of Mathematics.

Clement, J. (1982). Algebra word problem solutions: Thought processes underlying a common misconception. *Journal for Research in Mathematics Education, 13,* 16 – 30.

Collars, C., Koay, P. L., Lee, N. H., & Tan, T. S. (2005). *Shaping maths: Coursebook 5A.* Singapore: Federal Publications.

Collars, C., Koay, P. L., Lee, N. H., & Tan, T. S. (2007). *Shaping maths: Coursebook 2A* (2nd ed.). Singapore: Federal Publications.

Curriculum Planning & Development Division (2000). *2001 Mathematics syllabus (primary).* Singapore: Ministry of Education.

Curriculum Planning & Development Division (2006a). *Mathematics syllabus (primary) 2007.* Singapore: Ministry of Education.

Curriculum Planning & Development Division (2006b). *A guide to teaching and learning of O-level mathematics 2007.* Singapore: Ministry of Education.

Curriculum Planning Division (1990). *Mathematics syllabus (primary).* Singapore: Ministry of Education.

Fasotti, L. (1992). *Arithmetical word problem solving after frontal lobe damage: A cognitive neuropsychological approach.* Lisse, The Netherlands: Swets & Zeitlinger.

Hegarty, M., Mayer, R. E., & Monk, C. A. (1995). Comprehension of arithmetic word problems: A comparison of successful and unsuccessful problem solvers. *Journal of Educational Psychology, 87*(1), 18 – 32.

Kaput, J. J. (1987). Representation systems and mathematics. In C. Janvier (Ed.), *Problems of representation in the teaching and learning of mathematics* (pp. 19–26). Hillsdale, NJ: Lawrence Erlbaum Associates.

Kho, T. H. (1987). Mathematical models for solving arithmetic problems. In *Mathematical Education in the 1990's.* Proceedings of the Fourth Southeast Asian Conference on Mathematical Education (pp. 345–351). Singapore: Institute of Education.

Kho, T. H. (2006). Mathematical model for solving arithmetic problems. In P. Y. Lee (Ed.), *Teaching secondary school mathematics: A resource book* (pp. 367–377). Singapore: McGraw Hill.

Kieran, C. (1989). The early learning of algebra: A structural perspective. In S. Wagner & C. Kieran (Eds.), *Research issues in the learning and teaching of algebra* (pp. 33 – 53). Reston, VA: National Council of Teachers of Mathematics.

Kilpatrick, J., Swafford, J., & Findell, B. (Eds.). (2001). *Adding it up: Helping children learn mathematics.* Washington, DC: National Academy Press.

Kintsch, W., & Greeno, J. G. (1985). Understanding and solving word arithmetic word problems. *Psychological Review, 92,* 109 – 129.

Koedinger, K., & Nathan, M. J. (2004). The real story behind story problems: Effects of representations on quantitative reasoning. *The Journal of the Learning Sciences, 13*(2), 129 – 164.

Krulik, S. (Ed.) (1980). *Problem solving in school mathematics: 1980 NCTM Yearbook.* Reston, VA: National Council of Teachers of Mathematics.

Lee, K., Ng, S. F., Ng, E. L., & Lim, Z. Y. (2004). Working memory and literacy as predictors of performance on algebraic word problems. *Journal of Experimental Child Psychology, 89,* 140 – 158.

Lesh, R., Post, T., & Behr, M. (1987). Representations and translations among representations in mathematics learning and problem solving. In C. Janvier (Ed.), *Problems of representation in the teaching and learning of mathematics* (pp. 33 – 40). Hillsdale, NJ: Lawrence Erlbaum Associates.

Lewis, A. B. (1989). Training students to represent arithmetic word problems. *Journal of Educational Psychology, 81*(4), 521 – 531.

Lewis, A. B., & Mayer, R. E. (1987). Students' miscomprehension of relational statements in arithmetic word problems. *Journal of Educational Psychology, 79*(4), 363 – 371.

Lins, R. C., Rojano, T., Bell, A., & Sutherland, R. (2001). Approaches to algebra. In R. Sutherland, T. Rojano & A. Bell (Eds.), *Perspectives in school algebra* (pp. 1- 11). Dordrecht, The Netherlands: Kluwer Academic Press.

MacGregor, M. E. (1991). *Making sense of algebra: Cognitive processes influencing comprehension.* Geelong, Australia: Deakin University.

Mason, J., Burton, L., & Stacey, K. (1982). *Thinking mathematically.* Reading, MA: Addison-Wesley.

Mayer, R. E., Lewis, A. B., & Hegarty, M. (1992). Mathematical misunderstandings: Qualitative reasoning about quantitative problems. In J. I. D. Campbell (Ed.), *The nature and origins of mathematical skills* (pp. 137–154). Amsterdam: North-Holland.

Ministry of Education (in press). *Singapore Model Method.* Singapore: Author.

Morales, R. V., Shute, V. J., & Pellegrino, J. W. (1985). Developmental differences in understanding and solving simple mathematics word problems. *Cognition and Instruction, 2*(1), 41 – 57.

Ng, S. F., & Lee, K. (2005). How primary five pupils use the model method to solve word problems. *The Mathematics Educator, 9*(1), 60 – 83.

Ng, S. F., & Lee, K. (2008). As long as the drawing is logical, size does not matter. *The Korean Journal of Thinking & Problem Solving, 18*(1).

Ng, S. F. (2003). How secondary two express stream students used algebra and the model method to solve problems. *The Mathematics Educator, 7*(1), 1 – 17.

Ng, S. F. (2004). Developing algebraic thinking in early grades: Case study of the Singapore primary mathematics curriculum. *The Mathematics Educator, 8*(1), 39 – 59.

Ng, S. F., Lee, K., Ang, S. Y., & Khng, F. (2006). Model Method: Obstacle or bridge to learning symbolic algebra. In W. Bokhorst - Heng, M. Osborne & K. Lee (Eds.), *Redesigning pedagogies: Reflections from theory and praxis* (pp. 227–242). Rotterdam: Sense Publishers.

Noddings, N. (1988). Preparing teachers to teach mathematical problem solving. In R. I. Charles & E. A. Silver (Eds.), *The teaching and assessing of mathematical problem solving* (pp. 244-258). Reston, VA: National Council of Teachers of Mathematics.

Pape, S. J. (2003). Compare word problems: Consistency hypothesis revisited. *Contemporary Educational Psychology, 28,* 396 – 421.

Polya, G. (1971). *How to solve it: A new aspect of mathematical method* (2nd ed.). Princeton, NJ: Princeton University Press.

Schoenfeld, A. H. (1985). *Mathematical problem solving.* New York: Academic Press.

Silver, E. A. (1985). *Teaching and learning mathematical problem solving: Multiple research perspectives.* Hillsdale, NJ: Lawrence Erlbaum Associates.

Stacey, K., & MacGregor, M. (2000). Learning the algebraic method of solving problems. *Journal of Mathematical Behaviour, 18*(2), 149 – 167.

Swanson, H. L., Cooney, J. B., & Brock, S. (1993). The influence of working memory and clsssoification ability on children's word problem solution. *Journal of Experimental Child Psychology, 55,* 374 – 395.

Verschaffel, L., Greer, B., & De Corte, E. (2000). *Making sense of word problems.* Lisse, The Netherlands: Swets & Zeitlinger.

Vygotsky, L. S. (1978). *Mind in society: The development of higher psychological processes.* Cambridge, MA: Harvard University Press.

Willis, G. B., & Fuson, K. C. (1988). Teaching children to use schematic drawings to solve addition and subtraction word problems. *Journal of Educational Psychology, 80*(2), 192 – 201.

Chapter 8

Solving Algebra Word Problems: The Roles of Working Memory and the Model Method

Kerry LEE   NG Swee Fong

Although students in Singapore compare favourably with their international peers in mathematical performance, like elsewhere, there are significant individual differences in performance. In this chapter, we describe findings from a programme of research that focused on the role of working memory and higher cognitive capabilities on mathematical performance. Findings from two large scale behavioural studies suggest that the role of working memory may be domain-general and affects performance in both literacy and algebraic word problem solving. Amongst the various components of problem solving, identifying and understanding the quantitative relationships amongst protagonists in word problems are particularly resource intensive. Aspects of these findings are supported by findings from a functional neuroimaging study, which also shows that symbolic algebra is more demanding of working memory resources than is the "model method".

Key words: working memory, individual differences, model method, algebraic word problem, IQ

1 INTRODUCTION

In the latest Trends in International Mathematics and Science Study (Martin et al., 2004; Mullis et al., 2004), students in Singapore outperformed their international peers in both mathematics and science.

Findings from previous surveys suggest parental expectation, pedagogical practices, and teachers' input as factors likely to have contributed to this achievement. However, like in other countries, there are significant individual differences in students' mathematics performance. In the past, when students were streamed based on their academic performance, 10 – 15% of students were relegated to the lower ability grouping. In our programme of research, we focused on identifying some of the key cognitive capabilities that contribute to individual differences in mathematics performance.

In the past few years, we have conducted a series of studies to examine the relationship between higher cognitive functioning and performance in algebra word problems. In particular, we have focused on the role of working memory. Working memory is involved in short-term memory storage, reasoning, problem solving, and other higher cognitive tasks that require simultaneous representation and manipulation of information. It contains information that is current or active in our mind. In the context of problem solving, it serves as an interface between information contained in the problem and established knowledge in long-term memory. There are a number of working memory models (Miyake & Shah, 1999); our work was guided by Baddeley and Hitch (1974). The latest version of their model consisted of four components: central executive, phonological loop, visual spatial sketchpad, and an episodic buffer (Baddeley, 2000). Roles attributed to the central executive have changed over the years: from a manager of attentional resources (Baddeley & Hitch, 1974) to a system involved in other higher cognitive or executive functions, such as inhibition or resisting interference from inappropriate responses, planning, managing switches between problem solving strategies, and making available information stored in long-term memory (Baddeley, 1996). Both the phonological loop and the visual spatial sketchpad are short-term storage systems. The former is responsible for short-term maintenance of auditory and articulatory information. The latter performs similar functions for information of a visual or spatial nature. The episodic buffer was the newest addition to the model. It was postulated as a structure that facilitated exchange of information between the central executive and long-term memory (for a detailed introduction to Baddeley's working memory model, see Baddeley & Logie, 1999; Baddeley, 2000).

Reliable relationships between working memory and performance in arithmetic tasks have been documented. Bull and Johnston (1997), for example, found processing speed, capacity for remembering phonological information, and long-term memory measures predicted performance on the Group Mathematics Test (Young, 1980). Other studies emphasised the difference between short-term storage and executive resources. In contrast to short-term storage measures, which index the number of bits of information that can be acquired and remembered within a short period, central executive measures index the amount of such information that can be remembered while engaging in a reasoning or comprehension task at the same time. Gathercole and Pickering (2000a) replicated Bull and Johnston's finding but found the contribution of phonological short-term storage to be overshadowed by central executive capacity. The contribution of phonological storage was more prominent only when mental arithmetic tasks were used. Such tasks may be particularly taxing on the phonological system because both the problem and interim solutions have to be kept in mind. In an experimental demonstration, Fürst and Hitch (2000) showed that suppression of access to phonological storage reduced ability to perform mental addition. However, such suppression had little effect when numbers to be added remained visible during the task. In contrast, suppression of access to the central executive had a detrimental effect regardless of problem visibility. Both of these studies show that access to executive resources is critical to arithmetic problem solving.

We focused on algebra word problems for several reasons. First, as noted by Kintsch (1998), it offers a richer context than arithmetic. Nonetheless, computation in such problems is closely connected and is quintessentially arithmetic in nature. Word problems also offer an opportunity to study the interplay between students' mathematical and linguistic competencies. On a more applied level, mathematical problem solving is the central focus of the elementary curriculum in Singapore (Figure 1). Algebraic thinking, often in the form of word problems, is introduced early in Primary 4 as a vehicle for students to acquire skills in mathematical pattern recognition, quantitative comparison, and operation reversal. These foundation skills serve as a bridge to symbolic algebra, which is taught in the secondary or high school years.

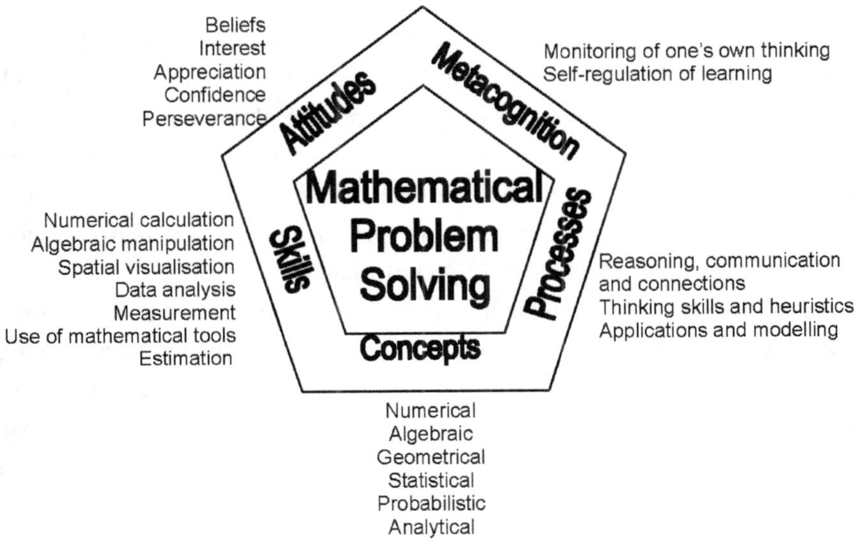

Figure 1. Singapore Primary school mathematics curriculum.

We were interested in algebra for another reason. Students, and many adults, find algebra difficult. Anecdotal observations suggest young students enjoy mathematics. However, their enjoyment and success with mathematics are often reduced when they begin to encounter algebra. One reason for this decline is that students have a misconception that arithmetic and algebra are disjointed subjects and that algebra involves solving for x and y. From the mathematics education perspective, algebra is often perceived as the cornerstone of mathematics as well as a gateway to learning higher mathematics. However, algebra is more than just symbolic manipulation and solving for x and y: it is a way of thinking that forms the bedrock of mathematics. Hence, it is important that all students be given the opportunity to learn algebra that allows them to see its relevance in relation to the world they live in. Given the role of working memory in arithmetic computation, we built on that work and examined the role of working memory in algebraic problem solving.

To design effective and well-founded pedagogy that caters for diversity in abilities requires imagination as well as a sound understanding of the cognitive prerequisites of algebra. Some of these prerequisites and difficulties include the surprisingly problematic and multi-faceted relationship with prior arithmetic learning (Stacey & Chick, 2004). Previous studies showed that success in algebra is dependent on foundation knowledge that stems from early childhood, including competency in arithmetic. Yet, one of our studies suggests that early exposure to algebraic problems is perceived by secondary school teachers as counter-productive (Ng et al., 2006).

In this chapter, we provide a summary of our studies that examined the relationship between working memory and algebraic problem solving. We also present several studies that examined the interconnection between problem solving strategies taught in the primary versus secondary school years. In this chapter, we focused on the cognitive psychological bases of problem solving. Readers who are more interested in the mathematical implications of these studies should refer to Chapter 7 of this volume.

2 WORKING MEMORY AND ALGEBRAIC PROBLEM SOLVING

Previous studies showed that working memory predicts mathematical performance. Most of these studies focused on foundation skills in numeracy and basic arithmetic operations (e.g., Bull & Scerif, 2001; Fayol et al., 1987; Gathercole & Pickering, 2000b; Passolunghi & Siegel, 2001; Swanson, 1994). In our first study, Algebra 1, we extended the literature by examining the relationship between working memory capacity and performances on algebra word problems (see Lee et al., 2004, for a full report of this study).

In this study, one of our main concerns was to clarify the relationship between working memory, literacy, and IQ. Previous studies showed that working memory and IQ measures are closely related (e.g., Kane & Hambrick, 2005). Working memory also strongly predicts performance on reading comprehension tasks (Daneman & Merikle,

1996). When measures of literacy and IQ are entered into the same regression models as working memory, the contribution of the latter is often greatly attenuated (e.g., Bull & Scerif, 2001). Given the central role of working memory in information processing models, this reduction in contribution is somewhat surprising. In this study, we examined whether the relationship between working memory and algebraic performance was indirect, as mediated by literacy and IQ.

Students from Primary 5 were administered a 10-item mathematical problem solving task. They also completed the Working Memory Test Battery for Children (WMTB-C, Pickering & Gathercole, 2001), an abbreviated IQ test from the Wechsler Intelligence Scale for Children (Wechsler, 1991), and tests of comprehension from the Wechsler Objective Reading and Language Dimensions, Singapore (Rust, 2000). Because work on the episodic buffer was still in its infancy, the WMTB–C assessed capacity in three domains of working memory: the central executive, phonological loop, and visual spatial sketchpad.

Results showed reliable and moderate to strong correlations between all working memory measures and mathematical performance. Students with greater working memory capacity performed better in both the literacy task and in the word problems. The various predictors (central executive, performance IQ, and literacy performance) accounted for 49% of variation in mathematical problems. Based on previous findings showing that the ability to read and comprehend problems is critical for problem solvers (e.g., Blessing & Ross, 1996; Bobrow, 1968; Cummins Dellarosa Denise et al., 1988; Mayer & Hegarty, 1996; Nathan, 1992; Kintsch & Greeno, 1985; Reusser, 1990), we examined whether performance on the word problems was mediated by literacy level. In our path analysis, the best fitting model showed working memory to have both direct and indirect effects on problem solving performance. The indirect effects were mediated by literacy and IQ (see Figure 2). These findings show that while better literacy skills contribute to better mathematical performance, higher central executive capacity also contributes to better performance in both literacy and mathematics word problems.

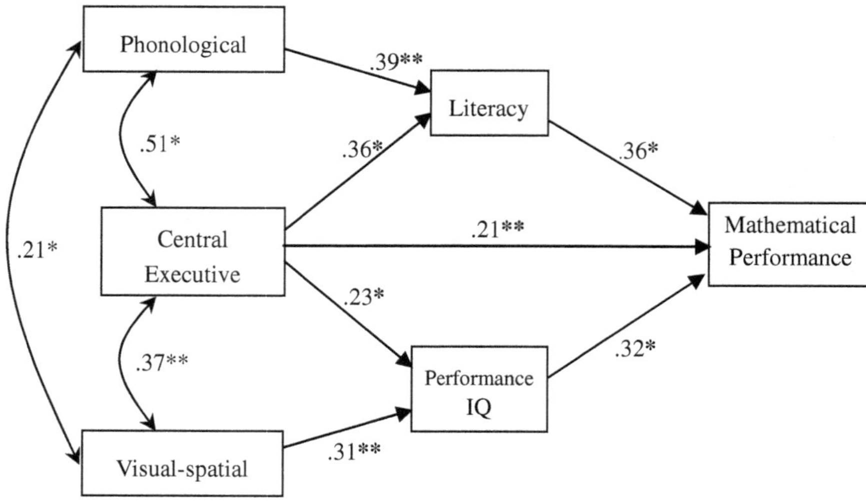

Note. $^*p < .05$, $^{**}p < .01$, $R^2 = .46$

Figure 2. A path diagram showing standardised path coefficients and correlation ratios between measures of working memory, literacy, IQ, and algebraic performance.

In a follow-up study, Algebra 2 (Lee, Ng, & Ng, in press), we expanded our search to include executive functions. Three types of executive functions have been studied in the literature: updating, shifting, and inhibition. Updating efficiency is closely related to working memory capacity (Miyake et al., 2000; St Clair-Thompson & Gathercole, 2006) and refers to ability to monitor and refresh contents of the mind. Shifting is synonymous with mental flexibility and refers to ability to switch from one cognitive process to another. Inhibition is multifaceted (Friedman & Miyake, 2004; Nigg, 2000) and refers to ability to resist interference from unwanted or irrelevant stimuli. Aspects of executive functioning have been found predictive of mathematical performance (Bull et al., 1999; Bull & Scerif, 2001; Clair-Thompson & Gathercole, 2006; Passolunghi et al., 1999; Passolunghi & Siegel, 2001). In this study, we examined whether they provided additional explanatory power for individual differences in algebraic performance.

In addition to executive functions, this study also differs from Algebra 1 in that we used a number of measures to index different components of problem solving. Most prior studies suggest that there are at least two components in problem solving: problem representation and solution formation (Briars & Larkin, 1984; Kintsch & Greeno, 1985; Mayer & Hegarty, 1996; Nathan, 1992; Reusser, 1990). The former refers to abilities to understand the problem, retrieve relevant information, and integrate that information with task requirement. Solution formation refers to processes involved in formulating a problem solving procedure and processes involved in computation. In this study, we used separate measures of problem representation and solution formation to index students' competency. In addition, we included tasks targeting areas of known difficulties. One of which was deciphering the quantitative relationship amongst protagonists depicted in a problem, e.g., the well known students/professor problem that require transformation of text statements such as "there are six times more students than professors" to equations (see Chapter 7).

The data showed that measures of working memory predicted 25% of variance in problem solving. They also accounted for similar amount of variation in tasks designed to index problem representation and solution formation. Of particular interest was the finding that deciphering the quantitative relationship amongst protagonists was particularly working memory intensive. Although performance on this component of problem solving was correlated with measures of shifting and inhibition, it was mostly strongly predicted by participants' updating capacity. These findings reinforce findings from Algebra 1 showing that working memory plays a significant role in algebraic problem solving. It adds to the literature by suggesting that particular components of problem solving are more resource intensive.

3 AN EXPERIMENTAL STUDY

Although findings from both Algebra 1 and 2 showed a close relationship between working memory capacity and algebraic problem solving, the evidence is correlational in nature. In a subsequent study, we examined their causal dependency using an experimental design. Because this study has not been reported elsewhere, we include more details here.

If ability to solve algebraic problems is truly dependent on working memory resources, suppression or reducing access to such resources should have a detrimental effect on performance. In this study, we used a random number generation (RNG) task to achieve this effect. Using a dual task paradigm, participants were asked to perform the RNG while they were solving algebraic problems. The RNG task is known to require executive resources as it requires participants to monitor their responses and to avoid producing well learned numeric sequences (Baddeley, 1996).

In the first experiment, 60 Primary 6 students participated; each child was administered one of three tasks: complex span, literacy, or algebraic problems. The complex span task, commonly used to measure working memory capacity, was used as a manipulation check to ensure that RNG has its desired effect. The one we used was adapted from the listening span test (Working Memory Test Battery for Children, Pickering & Gathercole, 2001). In each trial, participants were given five seconds to read a sentence (e.g., clocks eat apples) and were asked to decide whether the sentence was true or false. After all the sentences in a trial were presented (the number of sentences varied from two to six), participants were asked to recall the last words of each sentence, in the sequence in which they were presented.

In Lee et al. (2004), it was argued that working memory contributes to performance in both literacy and algebraic tasks. In this experiment, the literacy task was included to examine the relative effect of working memory resource suppression on both literacy and algebraic tasks. The literacy task was adapted from the vocabulary subtest of the Wechsler Intelligence Scale for Children (Wechsler, 1991). It differs from the often-used vocabulary test in that students were given short definitions of words and were asked to come up with the most difficult words that matched those descriptions. For example, "spectacles" or "glasses" were considered appropriate responses for "something that helps people with bad vision see more clearly." This procedure was used to approximate more closely the kind and the amount of written response used in the literacy and algebraic tasks. The algebraic task consisted of seven word problems that were drawn from the curriculum. Participants were instructed to solve the questions using a graphical heuristics called the model method that is taught in all schools.

Each participant was administered their assigned task on its own followed by a parallel version of the same task – counter balanced – in a dual task condition on a subsequent day. In the dual task condition, participants were instructed to generate a random number, ranging from one to ten at one-second intervals. Examples of both random and non-random sequences were presented by the experimenter and participants were instructed to avoid familiar sequences, like birthdays or telephone numbers. The actual experiment only proceeded when the experimenters were satisfied that the students understood what they had to do in the RNG task.

The results showed poorer performance in the suppression than in the no suppression condition for all three tasks (see Figure 3), $F(1, 57) = 90.46$, $p < .01$, partial $\eta^2 = .61$. Of interest was that the magnitude of this effect differed amongst tasks, $F(2, 57) = 6.21$, $p < .01$, partial $\eta^2 = .18$. It had the greatest effect on the algebraic task in which performance decreased in accuracy from 70% correct in the no suppression condition to 40% in the suppression condition. There was a similar drop in performance for the Listening Span task (from 50% to 25%) but a smaller decrease for the literacy task (from 43% to 32%). These differences do not seem to be an artefact of any compensatory behaviour. There was no evidence of students favouring the RNG task to the detriment of the algebraic, literacy, or Listening Span task. Analyses of the randomness of numbers showed that, similar to the main performance tasks, students performed more poorly in the dual task than in the baseline condition. They produced more redundancy (R score, Towse & Neil, 1998), $F(1, 45) = 6.39$, $p = .02$, partial $\eta^2 = .12$, and stereotyped responses (Adjacency score, Towse & Neil, 1998), $F(1, 45) = 7.55$, $p < .01$, partial $\eta^2 = .14$, in all three tasks.

These findings are consistent with the view that working memory has a causal role in problem solving. However, it is possible that the suppression effect is merely due to the imposition of a second task that needed to be performed simultaneously as the complex span, literacy, or algebraic task. If this is the case, the poorer performance could be

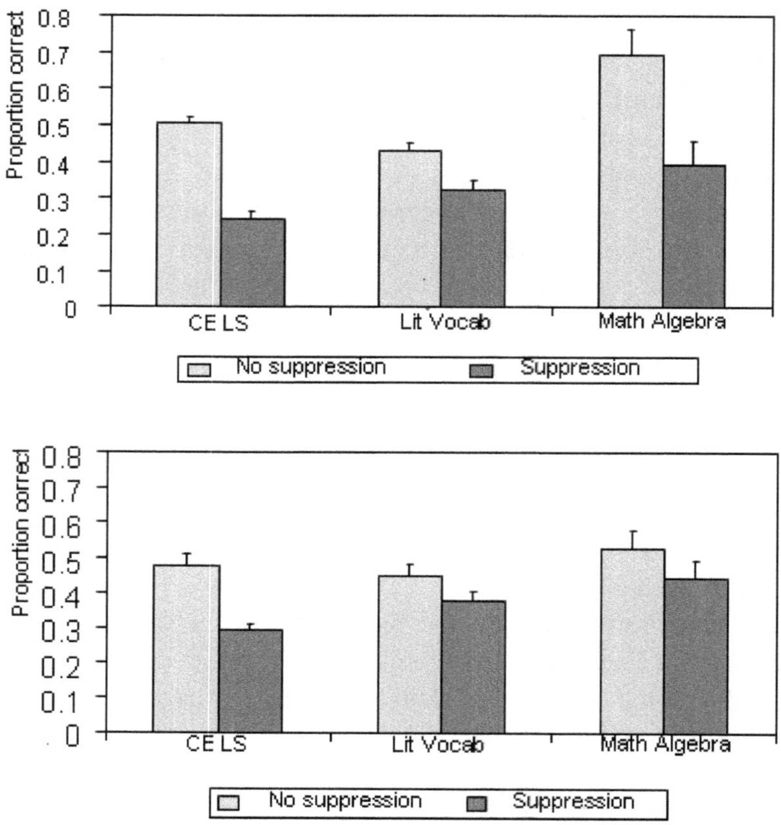

Note. Upper panel shows results from executive suppression. Lower panel shows results from articulatory suppression. CELS = Listening Span task, Lit Vocab = vocabulary task.

Figure 3. Proportion correct by task and suppression conditions.

attributed to general resource limitation rather than reduced access to working memory resources per se.

To address this concern, we conducted a second study in which the RNG task was replaced by an articulatory suppression task in which participants were asked to repeat the numbers 2-4-7 during the whole session. Because this task has minimal mnemonic load and executive monitoring requirement, it was expected to impose lower working memory demands. On account of the large suppression effect found in the first experiment, we tested only 30 students in this experiment.

Similar procedure and material were used, the only difference being the suppression task.

The results showed that articulatory suppression had a detrimental effect on all three tasks (see Figure 3, lower panel). Although the interaction effect between task and suppression failed to attain significance, on account of the smaller sample size, we examined the suppression effect for each task. The finding suggests that articulatory suppression produced poorer performance in the Listening Span task only. We also compared findings across the two experiments. Though both RNG and articulatory suppression resulted in poorer performance, RNG produced a stronger suppression effect.

An aspect of the findings from Experiment 2 bears additional comments. Performance on the algebraic task in the no suppression condition was noticeably lower than that found in Experiment 1. The performances of participants in the suppression conditions were similar. Because different students participated in the two experiments, it is important to look at suppression performance relative to performance in the no suppression performance. If articulatory suppression had the same effect as RNG, it should have rendered suppression performance in Experiment 2 even lower. As performance does not seem affected by a floor effect, the finding is consistent with a weaker suppression effect when articulatory suppression is used.

Findings from the two experiments suggest that performances on both the literacy and algebraic tasks are causally dependent on the availability of working memory resources. Findings from Experiment 2 suggest that the suppression effect found in Experiment 1 is due not to just having to perform a second task but is due to additional constraints in working memory resources imposed by the RNG task.

4 THE MODEL METHOD

The studies we described above show that working memory plays a significant and causal role in algebraic problem solving. A related issue that has occupied our attention is the difficulty students have in learning symbolic algebra. As is described in detail elsewhere (see Chapter 7) students in Singapore are exposed to algebraic problems earlier than students in many countries. In primary school, students are taught a variety of non-symbolic heuristics to give them early access to these complex questions. The most popular of which is the model method. Students are taught to represent known and unknown quantities using rectangular bars that specify known quantitative relationships amongst protagonists in word problems (see Figure 4). Typically, students use the model method until they enter secondary or high school, at which stage symbolic algebra is taught.

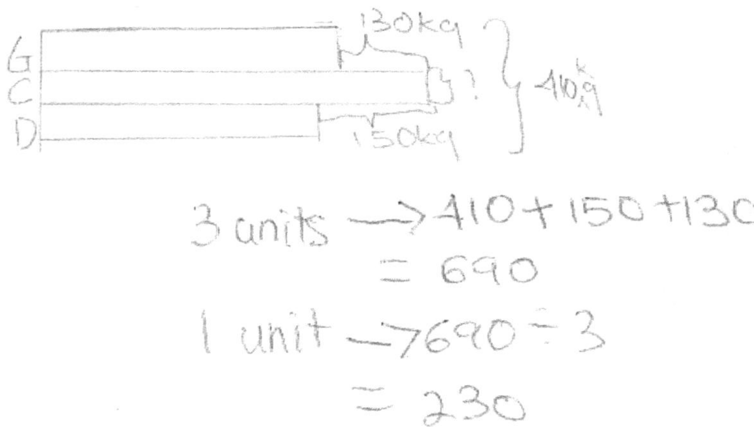

Note. A child's response to the question "A cow weights 150 kg more than a dog, a goat weighs 130 kg than the cow. Altogether the three animals weight 410 kg. What is the mass of the cow?"

Figure 4. The model method.

Considerable time and effort are expanded to teach the model method in the primary years. One concern we have is whether exposure to the model method helps students learn symbolic algebra. Proponents argue that the model method gives students earlier access to complex mathematical problems (Kho, 1987). Opponents argue that students are taught to do the same thing twice and the use of multiple methods simply confuses them. Indeed, interviews conducted with a small number of secondary teachers found they perceived the method as child-like, non-algebraic, and thought it a hindrance to the learning of symbolic algebra (Ng et al., 2006).

We addressed these issues from two different perspectives. First, we examined these issues indirectly by looking at students' use of models. Second, we examined whether the model method engaged different cognitive processes as compared to symbolic algebra.

5 STUDENTS' USAGE OF MODELS

In learning algebra, one of the more difficult concepts is the notion of variable. Students often find it difficult to appreciate how a single letter, *x*, can be used to represent vastly different numbers. Our investigation focused on students' perception of the rectangles used in the model method. Specifically, we examined whether they see them as serving the same functions as do variables in symbolic equations.

Although pictorial in appearance, these rectangles have more in common with the *x* and *y* in symbolic equations than with pictorial representation of objects. From a curricular perspective, students in Singapore are exposed to model-like representation from an early age. In the early primary years, teddy bears and other familiar objects are used to depict quantities in teaching numeracy and arithmetic. Once students are familiarised with the basic arithmetic operations, teddy bears are replaced with rectangles that represent particular quantities (Collars et al., 2007a, 2007b). By the time algebraic problems are introduced in the latter parts of Primary 4, such rectangles are used not so much to represent specific quantities but to represent quantitative relationship amongst protagonists in word problems. In demonstrating model drawing, teachers tend to emphasise the importance of relative length rather than absolute length. For the question depicted in Figure 4, for example, what is emphasised is that the rectangle representing the cow should be longer than that representing the goat, which in turn should be longer than that representing the dog. Our study examined whether this fundamental change in the role of representational agents is understood and is implemented by students.

We designed a computer programme called Model Method Analysis System (MMAS) that allows students to draw models on screen. Students were presented with a number of simple algebra word problems that were identical in mathematical structure. Across questions, the size of operands varied: from ones, tens, hundreds, to thousands. Number size was also manipulated within each range: from small, medium, to large. We were interested in the way students depicted operands of increasing magnitude. Students with well-developed knowledge of the representational properties of models were expected to realise that the absolute lengths of models were unimportant. Instead, they were expected to emphasise the importance of relative length. Students with less developed knowledge were expected to draw longer models for larger operands. The computer programme provided a more easily quantifiable and a standardised method for eliciting and measuring responses.

We tested 111 11-year-olds and 68 14-year-olds. The older students were expected to have better understanding of the abstract nature of model representation and to have more stable model drawings across operand size. Because using models for the depiction of algebraic question was still relatively new for the younger students, we expected them to reason in a more concrete manner and to draw longer rectangles for larger operands.

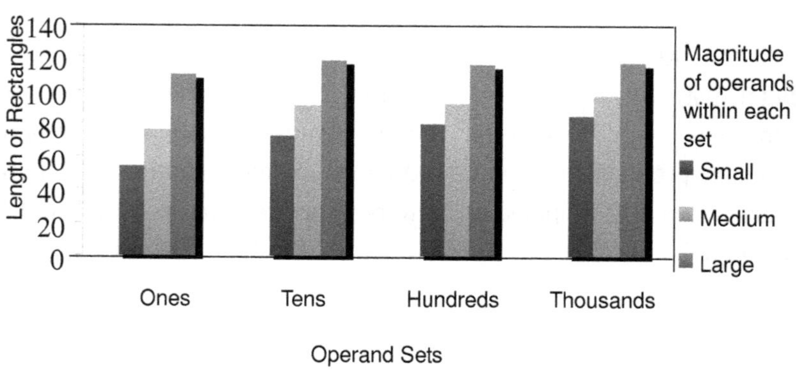

Note. Length of rectangles was measured using pixel count and was relative to screen resolution.

Figure 5. Students' model drawing by size of operands.

Preliminary analyses showed no reliable age related differences (Lee, Ng, Khng, & Guerrero, in preparation). Both groups drew longer rectangles for problems with larger operands (see Figure 5). This increase was particularly pronounced for the ones versus the tens. Length of rectangles also increased when number size increased within each number range. These findings suggest that even by the early secondary years, the average child still did not have a firm understanding of the abstract nature of representational models. Although these findings say little about the efficacy of pedagogies involving models, they suggest that the notion of variable is challenging, even for those who were taught with the model method in their primary years.

6 THE MODEL AND SYMBOLIC METHODS COMPARED

Our next set of studies examined whether there are differences in cognitive processes engaged by the model versus symbolic methods. On the surface, the two methods appear different. The former utilises pictorial representations that are demarcated by alphanumeric values. The symbolic approach uses only alphanumeric values. Whether these surface characteristics result in cognitive differences in processing is unclear. In an effort to provide a multifaceted examination of this issue, we conducted two experiments using functional magnetic resonance imaging. Performance of cognitive tasks results in hemodynamic activities in different neural systems. Functional neuroimaging uses differences in the electromagnetic properties of oxygenated versus deoxygenated blood to provide measures of brain activation. When combined with suitable experimental designs, functional neuroimaging can identify brain regions that are jointly or preferentially activated by specified cognitive tasks (for an introduction to fMRI, see Hernandex-García et al., 2004; Huettel et al., 2004).

In our first experiment, we focused on the initial stages of algebraic problem solving (Lee et al., 2007). As mentioned earlier, most information processing theories of mathematical problem solving specify two components: problem representation and problem solution. In our first experiment, we focused on processes associated with problem representation.

Eighteen adult participants, pre-tested and matched for competency in both the model and symbolic method (over 90% accuracy on a task

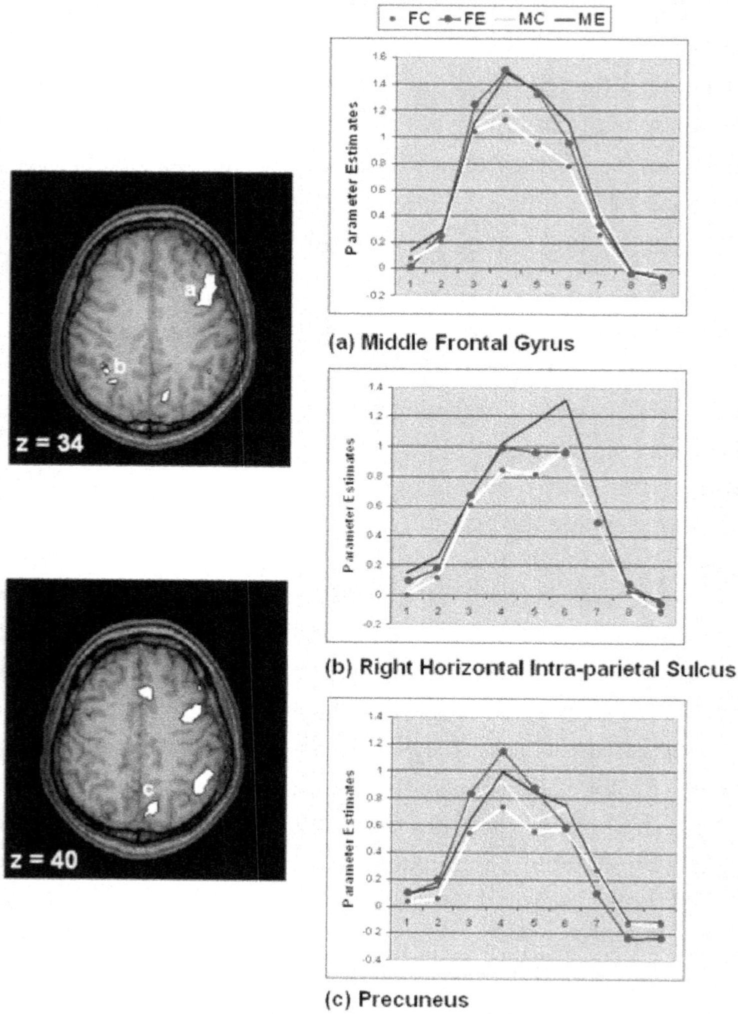

Figure 6. Activation map for similarity between the model and symbolic methods. A threshold of $p < .001$, uncorrected, was used to determine whether a voxel was activated. The left side of each transverse slice represents the right side of the brain. ME = Method experimental, MC = Model control, FE = Symbolic experimental, FC = Symbolic control.

similar to those used in the experiment and with less than 5% difference in accuracy across the two methods), were presented with simple word problems, e.g., James has 50 fewer watches than Mike; how many watches does James have? In the scanner, they were asked to transform mentally these questions into either model representations or equations. Each question was followed by the presentation of a solution, in either a model or an equation format. Participants were asked to compare the presented solution and the one they had in mind, and to identify whether it was correct.

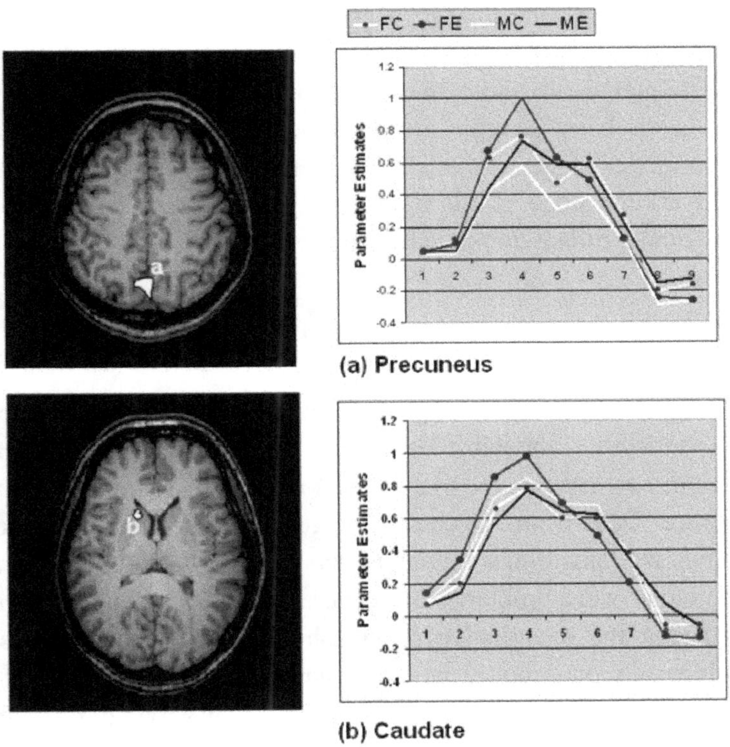

Figure 7. Activation map for difference conjunction (FE > ME) and (FE > FC) and (FE > MC). A threshold of $p < .001$, uncorrected, was used to determine whether a voxel was activated. The left side of each transverse slice represents the right side of the brain. ME = Model experimental, MC = Model control, FE = Symbolic experimental, FC = Symbolic control.

Transforming word problems into either models or equations activated areas associated with working memory processes (see Figure 6). Of particular interest was that the symbolic approach activated more strongly the basal ganglia, the posterior superior parietal lobule, and the precuneus (see Figure 7). These areas have been associated with attentional selection and orientation: both of which are associated with the executive component of working memory.

Our second experiment focused on solution formation (Lee, Yeong & Ng, in preparation). Participants were presented with either model representations or equations and were asked to compute the solutions. The findings were similar to that of the first experiment. The symbolic approach activated working memory areas more so than did the model method. Of interest was that in both experiments, there was greater activation in the basal ganglia in the symbolic condition. According to the ACT-R model (Anderson et al., 2003), this may reflect the retrieval of procedural knowledge. If this were the case, it suggests that the use of equations is more reliant on procedural retrieval.

7 CONCLUSIONS

Our findings show that working memory plays a significant role in algebraic problem solving. Evidence came from correlational, experimental, as well as real time measures using functional neuroimaging. Data from Algebra 1 suggest that the role of working memory may be domain-general and affects performance in both literacy and mathematics word problem solving tasks. Data from Algebra 2 show that identifying the quantitative relationships amongst protagonists in word problems is particularly working memory intensive. Further, the evidence suggests that this component of problem solving is resource intensive not just because it is generally more difficult. In contrast, it seems that identifying such relationships is particularly reliant on updating capacity. Although performance on this task was also predicted by inhibitory and switching efficiency, these relationships were rendered non-reliable when updating was included in the regression model. Perhaps what is driving this relationship is having to entertain the quantities possessed by each protagonist and having to evaluate different interpretations of relational arrangements as specified in the word problem.

One promising implication of these findings is that it may be possible to improve both problem solving performance and literacy by improving working memory capacity or by reducing the working memory load of problem tasks. Research on working memory improvement is still at an early stage. However, there are already a number of demonstrations using computer training programmes that show promise (e.g., Klingberg et al., 2005). Our finding that quantitative evaluation is particularly resource intensive may also assist teachers in developing more focal pedagogical intervention. A remaining question is whether there are windows in students' development at which their working memory systems are particularly sensitive to intervention. We are currently in the midst of a longitudinal project that examines these issues (see Chapter 7).

References

Anderson, J. R., Qin, Y., Sohn, M. H., Stenger, V. A., & Carter, C. S. (2003). An information-processing model of the BOLD response in symbol manipulation tasks. *Psychonomic Bulletin & Review, 10,* 241-261.

Baddeley, A. (2000). The episodic buffer: a new component of working memory? *Trends in Cognitive Sciences, 4,* 417-423.

Baddeley, A., & Hitch, G. J. (1974). Working Memory. In G. A. Bower (Ed.), *Recent advances in learning and motivation* (pp. 47-90). New York: Academic Press.

Baddeley, A. (1996). Exploring the central executive. *The Quarterly Journal of Experimental Psychology, 49A,* 5-28.

Baddeley, A. D., & Logie, R. H. (1999). Working memory: The multiple-component model. In A. Miyake & P. Shah (Eds.), *Models of working memory* (pp. 28-61). Cambridge: Cambridge University Press.

Blessing, S. B., & Ross, B. H. (1996). Content effects in problem categorization and problem solving. *Journal of Experimental Psychology: Learning, Memory, and Cognition, 22,* 792-810.

Bobrow, D. G. (1968). Natural language input for a computer problem solving system. In M. Minsky (Ed.), *Semantic information processing* (pp. 146-226). Cambridge, MA: MIT Press.

Briars, D. J., & Larkin, J. H. (1984). An integrated model of skill in solving elementary word problems. *Cognition and Instruction, 1,* 245-296.

Bull, R., & Johnston, R. S. (1997). Children's arithmetical difficulties: Contributions from processing speed, item identification, and short-term memory. *Journal of Experimental Child Psychology, 65,* 1-24.

Bull, R., Johnston, R. S., & Roy, J. A. (1999). Exploring the roles of the visual-spatial sketch pad and central executive in children's arithmetical skills: Views from cognition and developmental neuropsychology. *Developmental Neuropsychology, 15,* 421-442.

Bull, R., & Scerif, G. (2001). Executive functioning as a predictor of children's mathematics ability: Inhibition, switching, and working memory. *Developmental Neuropsychology, 19,* 273-293.

Clair-Thompson, H. L., & Gathercole, S. E. (2006). Executive functions and achievements in school: Shifting, updating, inhibition, and working memory. *Quarterly Journal of Experimental Psychology, 59,* 745-759.

Collars, C., Koay, P. L., Lee, N. H., & Tan, T. S. (2007a). *Shaping maths: Coursebook 4A.* Singapore: Federal Publications.

Collars, C., Koay, P. L., Lee, N. H., & Tan, T. S. (2007b). *Shaping maths: Coursebook 4B.* Singapore: Federal Publications.

Cummins, D. D., Kintsch, W., Reusser, K., & Weimer, R. (1988). The role of understanding in solving word problems. *Cognitive Psychology, 20,* 405-438.

Daneman, M., & Merikle, P. M. (1996). Working memory and language comprehension: A meta-analysis. *Psychonomic Bulletin and Review, 3,* 422-433.

Fayol, M., Abdi, H., & Gombert, J. E. (1987). Arithmetic problems formulation and working memory load. *Cognition and Instruction, 4,* 187-202.

Friedman, N. P., & Miyake, A. (2004). The relations among inhibition and interference control functions: A latent-variable analysis. *Journal of Experimental Psychology: General, 133,* 101-135.

Fürst, A. J., & Hitch, G. J. (2000). Separate roles for executive and phonological components of working memory in mental arithmetic. *Memory and Cognition, 28,* 774-782.

Gathercole, S. E., & Pickering, S. J. (2000a). Assessment of working memory in six- and seven-year-old children. *Journal of Educational Psychology, 92,* 377-390.

Gathercole, S. E., & Pickering, S. J. (2000b). Working memory deficits in children with low achievements in the national curriculum at 7 years of age. *British Journal of Educational Psychology, 70,* 177-194.

Hernandex-García, L., Wager, T., & Jonides, J. (2004). Functional brain imaging. In H. Pashler & J. Wixted (Eds.), *Stevens' handbook of experimental psychology, Volume 4, Methodology in experimental psychology* (3rd ed.) (pp. 175-221). New York: Wiley.

Huettel, S. A., Song, A. W., & McCarthy, G. (2004). *Functional magnetic resonance imaging*. Sunderland, MA: Sinauer Associates.

Kane, M. J., & Hambrick, D. Z. (2005). Working memory capacity and fluid intelligence are strongly related constructs: Comment on Ackerman, Beier, and Boyle (2005). *Psychological Bulletin, 131,* 66-71.

Kho, T. H. (1987). Mathematical models for solving arithmetic problems. In *Mathematical Education in the 1990's*. Proceedings of the Fourth Southeast Asian Conference on Mathematical Education (pp. 345–351). Singapore: Institute of Education.

Kintsch, W. (1998). *Comprehension: A paradigm for cognition.* New York: Cambridge University Press.

Kintsch, W., & Greeno, J. G. (1985). Understanding and solving word arithmetic problems. *Psychological Review, 92,* 109-129.

Klingberg, T., Fernell, E., Olesen, P. J., Johnson, M., Gustafsson, P., Dahlström, K. et al. (2005). Computerized training of working memory in children with ADHD: A randomized, controlled trial. *Journal of the American Academy of Child & Adolescent Psychiatry, 44,* 177-186.

Lee, K., Lim, Z. Y., Yeong, S. H. M., Ng, S. F., Venkatraman, V., & Chee, M. W. L. (2007). Strategic differences in algebraic problem solving: Neuroanatomical correlates. *Brain Research, 1155,* 163-171.

Lee, K., Ng, S. F., Ng, E. L., & Lim, Z. Y. (2004). Working memory and literacy as predictors of performance on algebraic word problems. *Journal of Experimental Child Psychology, 89,* 140-158.

Martin, M. O., Mullis, I. V. S., Gonzalez, E. J., & Chrostowski, S. J. (2004). *TIMSS 2003 international science report: Findings from IEA's trends in international mathematics and science study at the fourth and eighth grades.* Chestnut Hill, MA: TIMSS & PIRLS International Study Center, Boston College.

Mayer, R. E., & Hegarty, M. (1996). The process of understanding mathematical problems. In R. J. Sternberg & T. Ben-Zeev (Eds.). *The nature of mathematical thinking* (pp. 29-53). Hillsdale, NJ: Lawrence Erlbaum Associates.

Miyake, A., Friedman, N. P., Emerson, M. J., Witzki, A. H., Howerter, A., & Wager, T. D. (2000). The unity and diversity of executive functions and their contributions to complex "Frontal Lobe" tasks: A latent variable analysis. *Cognitive Psychology, 41,* 49-100.

Miyake, A., & Shah, P. (1999). *Models of working memory: Mechanisms of active maintenance and executive control.* Cambridge: Cambridge University Press.

Mullis, I. V. S., Martin, M. O., Gonzalez, E. J., & Chrostowski, S. J. (2004). *TIMSS 2003 international mathematics report: Findings from IEA's trends in international mathematics and science study at the fourth and eighth grades.* Chestnut Hill, MA: TIMSS & PIRLS International Study Center, Boston College.

Nathan, M. J. K. (1992). A theory of algebra-word-problem comprehension and its implications for the design of learning environments. *Cognition & Instruction, 9,* 329-389.

Ng, S. F., Lee, K., Ang, S. Y., & Khng, F. (2006). Model Method: Obstacle or bridge to learning symbolic algebra. In W. Bokhorst - Heng, M. Osborne & K. Lee (Eds.), *Redesigning pedagogies: Reflections from theory and praxis* (pp. 227 – 242). Rotterdam: Sense Publishers.

Nigg, J. T. (2000). On inhibition/disinhibition in developmental psychopathology: Views from cognitive and personality psychology and a working inhibition taxonomy. *Psychological Bulletin, 126,* 220-246.

Passolunghi, M. C., Cornoldi, C., & De Liberto, S. (1999). Working memory and intrusions of irrelevant information in a group of specific poor problem solvers. *Memory and Cognition, 27,* 779-790.

Passolunghi, M. C., & Siegel, L. S. (2001). Short-term memory, working memory, and inhibitory control in children with difficulties in arithmetic problem solving. *Journal of Experimental Child Psychology, 80,* 44-57.

Pickering, S. J., & Gathercole, S. E. (2001). *Working memory test battery for children.* Kent: Psychological Corporation.

Reusser, K. (1990). From text to situation to equation: Cognitive simulation of understanding and solving mathematical word problems. In H. Mandl, E. De Corte, S. N. Bennett & H. F. Friedrich (Eds.), *Learning & instruction: European research in an international context* (pp. 477-498). Oxford: Pergamon.

Rust, J. (2000). *Wechsler objective reading and language dimensions (Singapore).* London: Psychological Corporation.

St Clair-Thompson, H. L., & Gathercole, S. E. (2006). Executive functions and achievements in school: Shifting, updating, inhibition, and working memory. *Quarterly Journal of Experimental Psychology, 59,* 745-759.

Stacey, K., & Chick, H. (2004). Solving the problem with algebra. In K. Stacey, H. Chick & M. Kendal (Eds.), *The future of teaching and learning of algebra: The 12th ICMI study* (pp.1 - 20). Boston, MA: Kluwer.

Swanson, H. L. (1994). Short-term memory and working memory: Do both contribute to our understanding of academic achievement in children and adults with learning disabilities? *Journal of Learning Disabilities, 27 (1),* 34-50.

Towse, J. N., & Neil, D. (1998). Analyzing human random generation behavior: A review of methods used and a computer program for describing performance. *Behavior Research Methods Instruments & Computers, 30,* 583-591.

Wechsler, D. (1991). *Wechsler intelligence scale for children* (3rd ed.). San Antonio, TX: Psychological Corporation.

Young, D. (1980). *Group mathematics test.* Kent: Hodder and Stoughton.

Chapter 9

Understanding of Statistical Graphs among Singapore Secondary Students

WU Yingkang **WONG Khoon Yoong**

This paper provides a review of four studies about what Singapore primary and secondary school students understand about standard statistical graphs. On the basis of international studies, a framework about Understanding of Statistical Graphs (USG) is proposed to cover four types of operation on graphs: graph reading (GR), graph interpretation (GI), graph construction (GC), and graph evaluation (GE). A major part of this review is about the large scale survey of secondary school students (n = 907) conducted by Wu in April 2003. This study found that secondary school students performed better in graph construction and graph reading than in graph interpretation and graph evaluation. Many students could not explain their statistical reasoning with graph items. Statistically significant differences were found by grade level and stream. The findings suggest that a more balanced coverage of USG is required at both the primary and secondary levels.

Key words: understanding of statistical graph, graph reading, graph interpretation, graph construction, graph evaluation, pictogram

1 INTRODUCTION

In the age of information, the ability to manage and make sense of the large amount of data encountered in everyday life is an important skill to acquire. Graphical representations are frequently used to display such data, and software such as *Excel* is a handy tool used to construct these graphs. Some graphs are poorly designed, and others are unscrupulously abused to deceive the readers as found in advertisements or to support a particular point (e.g., Wainer, 1984/1992). Hence, the school mathematics curriculum has to help students become competent and critical users of statistical graphs. Research on students' understanding of statistical graphs (USG) has become an important research theme internationally (Konold & Higgins, 2003; Shaughnessy, Garfield, & Greer, 1996). Some studies deal with students' understanding of graphs in general (Friel & Bright, 1995, 1996; Li & Shen, 1992), while others concentrate on understanding of particular types of graphs (Berg & Philips, 1994; Padilla, McKenzie, & Shaw, 1986). Further information about students' USG can be gleaned from graph items in large scale studies such as the Third International Mathematics and Science Study (TIMSS) (Pereira-Mendoza & Kaur, 1999).

In Singapore, there are few studies about USG, and the rest of this paper provides a review of four such studies.

2 THE FOUR STUDIES AND SINGAPORE STATISTICAL GRAPH TOPICS

These four studies are summarised in Table 1. Two studies are small-scale interviews of P3 (Bay, 2004) and P5 pupils (Goh, 2004), one

Table 1
Summary of Four Singapore Studies on Statistical Graphs

Researcher	Title	Participants	Topics
Bay Wee Fon (2004)	Students' interpretation of statistical graphs	20 P3 girls of mixed abilities; interview	Pictograph (about SARS), bar graph (about unemployment)
Goh Sock Lai (2004)	Primary 5 students interpretation of graphs in the real world	24 P5 pupils of mixed abilities; interview	Bar graph, line graph
Lim Chien Chong & Berinderjeet Kaur (1992)	Graphs – What do they say?	160 S3 girls.	Scatter plot (cost versus weight, height of son versus height of father)
Wu Yingkang (2005)	Statistical graphs: Understanding and attitude of Singapore secondary school students and the impact of a spreadsheet exploration	907 S1 to S3 students (different streams); statistics test and interview	Bar graph, pie chart, dot diagram, line graph, pictogram, histogram

is small survey of S3 girls (Lim & Kaur, 1992), and one a comprehensive survey of a representative sample of S1 to S3 students from the four streams (Special, Express, Normal Academic, Normal Technical) (Wu, 2005).

These studies cover standard statistical graph topics found in the Singapore mathematics curriculum (Ministry of Education, 2006). The statistical graph topics in the curriculum have been changed slightly over the years, but the main topics remain the same as shown in Table 2.

Table 2
Statistical Graph Topics in the Singapore Mathematics Curriculum (2006 version)

Level	Statistical Graph Topics
P1 (Primary 1)	Picture graphs: (a) collect and organise data; (b) make picture graphs; (c) use a symbol/picture to represent one object; (d) read and interpret picture graphs in both horizontal and vertical forms.
P2	Picture graphs: (a) make picture graphs with scales; (b) read and interpret picture graphs with scales; (c) solve problems using information presented in picture graphs.
P3	Bar graphs: (a) read and interpret bar graphs in both horizontal and vertical forms; (b) read scales; (c) complete a bar graph from given data; (d) solve problems using information presented in bar graphs.
P4	Line graphs: (a) read and interpret line graphs; (b) solve problems using information presented in line graphs.
P6	Pie charts: (a) read and interpret pie charts; (b) solve 1-step problems using information presented in pie charts.
S1 (Secondary 1)	Construction and interpretation of: tables; bar graphs; pictograms; line graphs; pie charts; histograms (exclude unequal intervals). Purposes and use, advantages and disadvantages of the different forms of statistical representations.
S2	Drawing simple inference from statistical diagrams.
S3/4	Interpretation and analysis of: dot diagrams; stem-and-leaf diagrams.
	Interpretation and analysis of: cumulative frequency diagrams; box-and-whisker plots.

3 UNDERSTANDING OF STATISTICAL GRAPHS (USG): A FRAMEWORK

Over the years, researchers and statistics educators have developed different formulations about statistical thinking and USG at different

Understanding of Statistical Graphs (USG)

Graph Reading (GR)

Extract data directly from one or more graphs and generate information by operating on explicit data shown in one or more graphs.

- R1 Extract one piece of data shown in a graph
  R1a: identify its value.
  R1r: describe its meaning.
- R2 Compare or integrate several pieces of data shown in a graph.
  R2a: give final answers.
  R2r: show workings or reasoning to the final answer.
- R3 Compare or integrate all the data shown in one or more graphs.
  R3a: give final answers.
  R3r: show workings or reasoning to the final answer.

Graph Interpretation (GI)

Form opinions from one or more graphs.

- I1 Infer from the scale used or the data shown in one or two graphs
  I1a: give final answers.
  I1r: describe the effect of scale or justify how to arrive at the final answer.
- I2 Infer from the data shown in a graph based on given conditions.
  I2a: give final answers.
  I2r: show workings or reasoning to the final answer.
- I3 Infer from the data shown in two graphs based on one's own criterion
  I3a: give final answers.
  I3r: justify how to arrive at the final answer.

Graph Construction (GC)

Display data in a graphical form.

- C1 Construct a specifier.
- C2 Construct a designated statistical graph: type, specifiers, scale, label, title.

Note: Specifiers convey values about the data shown in a graph.

Graph Evaluation (GE)

Evaluate a graph on its correctness or effectiveness.

- E1 Correctness: whether or not a graph is correctly constructed.
  E1a: identify right or wrong.
  E1r: explain what is wrong with a graph.
- E2 Effectiveness: whether or not a graph is effective to support a viewpoint.
  E2a: identify which one is effective.
  E2r: explain why a graph is effective.

Figure 1. The framework of Understanding of Statistical Graphs (USG) (Wu, 2005).

grade levels (Groth, 2003; Jones, Thornton, Langrall, Mooney, Perry, & Putt, 2000; Mooney, 1999). These formulations take as a starting point the collection of data followed by various processes applied to the data, such as describing them in words and graphs, organising the data to identify patterns and interpret salient features, and solving problems related to the data. On the basis of these studies, Wu (2005) developed a detailed framework for USG to encompass four key aspects: Graph Reading (GR), Graph Interpretation (GI), Graph Construction (GC), and Graph Evaluation (GE). Each aspect can encompass several components as shown in Figure 1.

Wu (2005) used this framework to construct a test of USG consisting of 10 questions with 53 items. These items are labeled (a) if only a final answer or value is required, whereas those labeled (r) require descriptions, explanations, justifications, or reasoning to support the final answers. Type (r) items are included for two reasons. First, the ability to use mathematical language to communicate mathematical ideas and arguments precisely, concisely, and logically is an important aim of the Singapore mathematics curriculum. Secondly, these items allow the researchers to gain useful information about how students solve graphical problems. As an illustration, item 3 of the test of USG is given

3. The dot diagram below shows the heights in centimeters of all the pupils in a Primary 4 class.

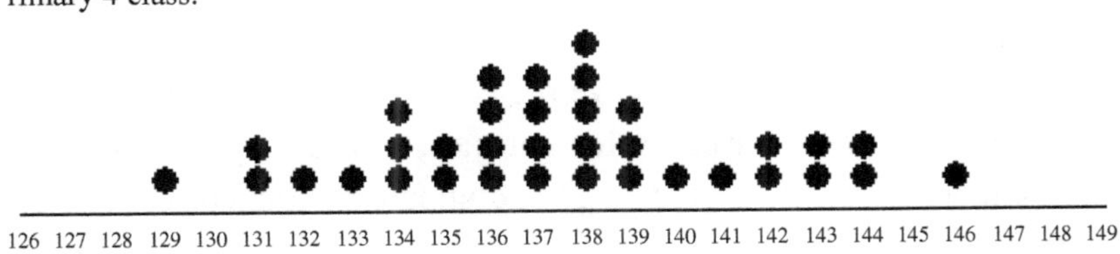

Heights of Primary 4 pupils in centimeters

(a) What is the *difference* in *heights* between the tallest pupil and the shortest pupil?

Answer____________

Working:

(b) If a pupil whose height is 165 cm claimed that he is in Primary 4, would you believe him?

Tick one only: Yes ☐ No ☐ Not sure ☐

Explanation

Figure 2. An item on dot diagram (Wu, 2005).

in Figure 2. For Part (a), giving an answer is a type (a) question and the working is a type (r) item. Part (b) is an item on graph interpretation.

4 GRAPH READING

In order to solve graph-related problems, one must be able to read off data values from a graph. This may involve extracting one piece of data in a graph (R1 in Figure 1), comparing or integrating several pieces of data shown in a graph (R2), and comparing or integrating all the data shown in a graph (R3). The change from R1 to R3 represents an increase in the amount of data used and in the sophistication of the operation involved. This division parallels the classification produced by Curcio (1987): "read the data" corresponds to R1; "read between the data" is divided into R2 and R3 in terms of whether several pieces of data or all the data shown in one or more graphs are used; "read beyond the data" relates to graph interpretation to be discussed in the later section.

Goh's study (2004) was based on Curcio's classification. The P5 pupils who were interviewed had little difficulty reading data from the given bar graph and the line graph but had some difficulty with reading between the data and reading beyond the data. Lim and Kaur (1992) reported that all the 160 girls in their study could read off values of one variable from a scatter plot, and about two thirds could compare two points in a scatter plot. With respect to the dot diagram in Figure 2, 73% of the 907 secondary students gave the correct answer to Part (a) (a) and 77% gave the correct reasoning for Part (a) (r). The surprising result that fewer students gave the correct answer compared to the correct reasoning was due to the computation errors made in calculating the difference. In general, the secondary students could read data from a pie chart or a dot diagram; they had more difficulty with reading multiple data from a cumulative line graph (around 50% correct) and a histogram (around 30% correct).

5 GRAPH INTERPRETATION

Graph interpretation is a desirable outcome of the Singapore mathematics curriculum, as shown in Table 2 above. According to Gal

(1998), the interpretation process should end in opinions about a given situation. Hence, GI is defined as "forming opinions from one or more graphs" in three different ways: form opinions by inferring from the scale used or the data shown in a graph (I1), from the data shown in a graph based on given conditions (I2), and from the data shown in two graphs based on one's own comparison criterion. Students could employ different strategies to form their opinions about data in graphical forms (Groth, 2003; Watson & Moritz, 1999).

In the TIMSS study (Pereira-Mendoza & Kaur, 1999), 44% of the S1 and 48% of the S2 Singapore students were able to make prediction from a line graph.

For Part (b) in Figure 2, there is no absolutely right answer, but opinions could be formed by combining knowledge about people's heights and the given data. About 17% did not answer this item. Only 16 students (2%) could make a valid judgment in this sense; for example, a S1 girl argued thus: "It is possible that the pupil whose height is 165 cm is in Primary 4 because some people have tall genes from their parents. It is not possible because the tallest pupil in the dot diagram is 146 cm and the average height is 138 cm." About 16% gave partially correct responses, arguing from the maximum height given in the dot diagram, the modal height, or that the given dot diagram only displayed the height of Primary 4 pupils from one class but not from the whole population. The majority of the responses (65%), however, did not refer to the data and cited only personal observations, for instance, "Some people are just born to be big and tall," or "I had a classmate who was 170 cm tall in Primary 4." From a statistical viewpoint, these responses are irrelevant to the given data and it shows that many students have difficulties in making reasonable predictions from a statistical graph (Watson & Pereira-Mendoza, 1996). Similarly, Lim and Kaur (1992) noted that some students may resort to intuition rather than graphical features to interpret scatter plot.

6 GRAPH CONSTRUCTION

Four strands of graph construction are apparent from published studies. The first strand examines the basic skills to complete a partially constructed graph or to construct a complete graph from a given data set (e.g., Berg & Philips, 1994). The second strand refers to the ability to construct a graph from verbal descriptions (e.g., Lim & Kaur, 1992). The third strand examines the ability to construct an effective graph which satisfies a given message or conclusion conveyed in a given data set (e.g., Chick, 2003). The fourth strand is about the effects of data-related factors, such as the size of a data set (e.g., Nisbet, 2002) and the type of the data (categorical versus numerical) (e.g., Nisbet, 2001), on students' performance in graph construction. Wu's study was restricted to the first strand only and it covered two components: construct a specifier[1] to complete a partially constructed graph (C1), and construct a designated statistical graph, including plotting all the required specifiers, scaling axes, giving labels to axes, and providing a title for the graph (C2).

Overseas studies show that students generally could construct a specifier to complete a partially constructed graph (e.g., Lenton, Stevens, & Illes, 2000), but some may not be able to draw a complete graph to include relevant labels for axes and scales (e.g., Berg & Philips, 1994; Padilla, McKenzie, & Shaw, 1986). In the TIMSS study (Pereira-Mendoza & Kaur, 1999), 92% of the S1 and 94% of the S2 Singapore students could draw a pictograph based on a given legend. One GC item in Wu's study asked students to construct a bar graph to represent some self-generated data. While about three quarters of the students could draw a correct bar graph, only 42% added a relevant title and 27% labeled the axes. It appears that these students did not take these graphical features seriously. Since titles and labels of axes are important constituents to make the graphs meaningful (Konold & Higgins, 2003; Kosslyn, 1989), more attention should be paid to helping students realise this function.

7 GRAPH EVALUATION

Unusual or misleading graphs often appear in the mass media. Hence, students should develop the ability to critically evaluate a graph. Graph

[1] Specifiers are used to represent data values. Specifiers could be the lines on a line graph, and the bars on a bar graph, or other marks that convey data values in a graph.

evaluation can cover the correctness (E1) and effectiveness (E2) of a graph against intended purposes, including the ability to spot errors in the given graph. This is an extension of graph interpretation.

Pictogram is often used in the mass media. There are two types of pictogram: a *number* pictogram uses different numbers of a same-sized picture to display different data values, and a *size* pictogram uses different sizes of a picture to represent different data values. Number pictograms are often taught to lower primary pupils without much difficulty. On the other hand, size pictogram is often misused in the mass media, in particular, using pictures that convey impressions of area or volume when the data values are only linear. The ability to evaluate this aspect of size pictogram was assessed in Wu's study. The size pictogram in Figure 3 below was given to illustrate that the number of private houses had doubled between 1980 and 2000. Students were asked whether it shows the information correctly or not. The misuse in this case is to suggest *volumes* that are not in the 1 : 2 ratio.

Figure 3. An item on size pictogram (Wu, 2005).

Although 78% of the students ticked the correct answer (No), only 29 students (3.2%) could give a convincing explanation, for example, that the "size" of the year 2000 house was more than twice that of the year 1980 house. About 35% of the students thought that the size of the house has nothing to do with the number of houses, as some of them still referred to number pictogram, as illustrated by the following exchange with a S3 Express boy:

R: Can you explain to me what this question means?

S: *It means like, number of houses is double, that means one, right, then two houses, not the size.*

R: So you mean there should be two houses rather than a big house?
S: Right.

About 34% treated the pictogram as "literal pictures," giving comments such as:

- It was because the Singapore last time houses were very small and this time they have more money to make a big house.
- House style should be different.
- The house in Singapore does not have chimney.

The inability of students to evaluate the correctness of a size pictogram has also been reported by Harper (2004) about a small sample of Grade 7 American students.

8 OVERALL PERFORMANCE ON GR, GI, GC, AND GE

Table 3 summarises students' scores in the test of USG, arranged in descending order by the percent scores: GC, GR, GE and GI. Graph constructions and graph reading items were found to be much easier compared to graph evaluation and graph interpretation. This trend is consistent with the results reported by Bay (2004) and Goh (2004) for primary pupils and other international studies. GI and GE items require students to give explanations or reasoning to justify their answers. Many students in this study were not familiar with these justification questions and sometimes they did not have the language skills to express themselves clearly. The whole test, GC, GR, and GI had acceptable internal consistency based on Cronbach's alpha, but GE had only moderate alpha. This last aspect requires further research.

The students also performed much better in type (a) items (answers only) than type (r) items (reasoning, descriptions, explanations, etc.). Among the 16 pairs of type (a) and type (r) items, the percentage of the correct responses to type (r) items, when the answers to the corresponding type (a) items were correct, ranged from 4.1% to 98.6%, with a median of 66.5% and a mean of 25.2%. This shows that the students had difficulties in expressing their thinking process about how they arrived at the correct answers, at least in a written form. Some students, who failed to express their thinking process in a written form, were able to do so orally during the interview.

Table 3
Summary of Students' Performance in the Test of USG

Aspect	Number of items	Cronbach's alpha	Total possible score	Raw score		Percent score	
				Mean	SD	Mean	SD
Graph construction (GC)	7	.72	14	9.1	3.6	64.9	25.8
Graph reading (GR)	22	.88	44	26.0	10.5	59.2	23.9
Graph evaluation (GE)	10	.54	20	9.6	3.1	48.0	15.7
Graph interpretation (GI)	14	.80	28	11.1	6.2	39.7	22.0
Overall	53	.93	106	55.8	20.4	52.7	19.2

Table 4 shows the correlation coefficients among the four aspects of USG. All the correlation coefficients are positive and statistically significant at the 0.001 level. This more stringent significant level was used by computing 0.05/6 to reduce the possible inflation of the Type I error according to the Bonferroni technique (Huck, 2000, p. 223). The result in Table 4 provides some empirical evidence that these four aspects describe the common underlying concept about students' understanding of statistical graphs. From a content analysis perspective, GR and GC deal with basic skills like reading data values from a graph and displaying a given data set in a graphical form. These skills are developed from an understanding of the structure and the uses of a graph. GR is also the basis for GI. In order to form opinions from the information shown in one or more graphs, one must be able to extract relevant information from a graph. Finally, GE is related to GR, GI, and GC because the ability to construct a correct graph helps one to evaluate whether or not a given graph is correctly constructed and the ability to read and interpret a graph helps one to evaluate whether or not a given graph is effective to support a certain point of view. Hence, both empirical data and content analyses support the inter-relationship among these four aspects of USG.

Table 4

Correlations among the Four Aspects of USG

	GR	GI	GC
Graph Interpretation (GI)	0.75*		
Graph Construction (GC)	0.56*	0.57*	
Graph Evaluation (GE)	0.64*	0.64*	0.49*

* $p < 0.001$.

9 RELATIONSHIPS BETWEEN STUDENT CHARACTERISTICS AND THEIR USG

Five three-way ANOVA tests were conducted to examine any statistically significant differences in students' scores in overall USG and in GR, GI, GC, and GE with respect to grade level, stream, and gender. The significance level of the ANOVA tests was set to be 0.01 (= 0.05/5) following the Bonferroni technique. The results are summarised in Table 5.

Table 5

Summary of the Five ANOVAs on Relationships between Student Characteristics and USG

Source	Overall USG	GR	GI	GC	GE
Grade level	√	√	√	√	√
Stream	√	√	√	√	√
Gender	×	×	×	×	×
Grade level × Stream	√	×	√	√	×
Grade level × Gender	×	×	×	×	×
Stream × Gender	×	×	×	×	×
Grade level × Stream × Gender	×	×	×	×	×

Note: √ statistically significant at 0.01 level; × not statistically significant at 0.01 level

There was no gender difference in the students' overall USG and in the four aspects. This means that boys and girls did equally well in the whole test as well as in the four aspects of GR, GI, GC, and GE. Therefore, it is not necessary for teachers to give any special attention to boys or girls in their teaching of statistical graphs. Older students generally scored higher in the whole test as well as the four aspects than younger ones, as expected, because of more opportunity to learn and maturity. There were significant differences in scores by stream in the order: Special (best), Express, Normal (Academic), and Normal (Technical) (worst). Indeed, Normal (Technical) students had such a poor understanding of statistical graphs that there were no significant differences in the scores across grade level for this group of students.

10 IMPLICATIONS AND CONCLUSION

The above review, in particular the findings from Wu's study (2005), suggests some implications for the teaching of statistical graphs and research in this area.

First, the teaching of statistical graphs should provide balance in the four aspects of graph reading, graph construction, graph interpretation, and graph evaluation. The first two aspects have been dealt with adequately in the current curriculum, whereas the last two are relatively new and should be given more attention. Teachers should design relevant tasks for their students to engage in graph interpretation and graph evaluation. Use of real life examples taken from the mass media will provide additional motivation so that the students understand the importance of these skills in handling everyday graphical information. Some teachers may need in-service training in this area.

Second, many students in Wu's study (2005) were handicapped in not being able to explain their reasoning in written form when they dealt with type (r) items, although some students could explain orally during interviews. Communication skill in mathematics is one component of the *Process* aspect of the Singapore Mathematics Framework. To inculcate this skill, the teacher should provide opportunity for students to talk more about different features and functions of statistical graphs and to write about them after class discussion. Similar proposal has been made by other researchers in statistics education (e.g., Friel & Bright, 1995, 1996).

Third, many students were found to be less than serious or knowledgeable about the use of labels in titles and axes on statistical graphs. Teachers should pay more attention to students' lack of concern about graphical conventions. More importantly, teachers need to help students understand why they should follow these conventions rather than simply show them what to do. For example, some students started the vertical axis from zero in the bar graph, but they did not know the reason for this rule.

Fourth, since statistical graphs are about everyday situations, it is to be expected that students will call on their knowledge of the world to interpret the graphs. This may be problematic if the students rely solely on this knowledge without considering the given data, as was reported above. Students should understand the need to employ data to accept or reject a hypothesis. This point is raised by Shaughnessy and Pfannkuch (2002) who considered that data are a prime requirement for judgments and decision making. Hence, classroom teaching should pay attention to help students develop this mindset.

Fifth, the wide variation of students' understanding of statistical graphs among different streams within the same grade level deserves careful consideration. *All* students, irrespective of the stream they are in, must develop a strong level of this understanding.

Sixth, nowadays information and communication technology (ICT) has been strongly recommended as an important tool to help students learn better about mathematics and statistics (e.g., Ben-Zvi, 2000; Shaughnessy, Garfield, & Greer, 1996). For example, ICT can facilitate the construction of a graph (the users have only to select the data and the type of graph required), allow the use of real data, and enable an easy translation between numerical and graphical representations of data. Elsewhere we have reported that *Excel* templates can help some students overcome misconceptions about certain statistical graphs but not others (Wu & Wong, 2007). For example, after working with an *Excel* template on size pictogram, all the 18 students involved in this exploratory activity still could not give a valid reason about the size of the picture against the value given. We suspect that the more realistic picture as shown in Figure 3 might require higher level of processing than the geometric cube used in the *Excel* template exploration. This hypothesis certainly requires further investigation.

Finally, but not the least, the framework in Figure 1 has been found to be useful in designing items that test related aspects of USG. However, it can be further refined by including more components, for example, how different data items in a graph are related. The four aspects may also form a hierarchy similar to the SOLO taxonomy (Biggs & Collis, 1991).

To conclude, research in statistics education in Singapore is still a largely unexplored area. This review has highlighted the limited number of studies in statistical graphs, but the findings can have important implications for classroom teaching and further research, as outlined above.

References

Bay, W. F. (2004). *Students' interpretation of statistical graphs.* Unpublished master's dissertation. National Institute of Education, Nanyang Technological University, Singapore.

Ben-Zvi, D. (2000). Toward understanding the role of technological tools in statistics learning. *Mathematical Thinking and Learning, 2*(1/2), 127-155.

Berg, C. A., & Philips, D. G. (1994). An investigation of the relationship between logical thinking structure and the ability to construct and interpret line graphs. *Journal of Research in Science Teaching, 31*(4), 323-344.

Biggs, J. B., & Collis, K. F. (1991). Mulitmodal learning and intelligent behavior. In H. Rowe (Ed.), *Intelligence: Reconceptualization and measurement* (pp. 57-76). Hillsdale, NJ: Lawrence Erlbaum Associates, Inc.

Chick, H. (2003). Transnumeration and the art of data representation. In L. Bragg, C. Campbell, G. Herbert & J. Mousley (Eds.), *Mathematics education research: Innovation, networking, opportunity: Proceedings of 26th Annual Conference of the Mathematics Education Research Group of Australia* (pp. 207-214). Sydney: MERGA.

Curcio, F. R. (1987). Comprehension of mathematical relationships expressed in graphs. *Journal for Research in Mathematics Education, 18*, 382-393.

Friel, S. N., & Bright, G. W. (1995, October). *Graph knowledge: Understanding how students interpret data using graphs.* Paper presented at the Annual Meeting of the North American Chapter of the International Group for the Psychology of Mathematics Education, Columbus, OH.

Friel, S. N., & Bright, G. W. (1996). *Building a theory of graphicacy: How do students read graphs?* Paper presented at the Annual Meeting of American Educational Research Association, New York. (ERIC Document Reproduction Service No. ED 395227).

Gal, I. (1998). Assessing statistical knowledge as it relates to students' interpretation of data. In S. P. Lajoie (Ed.). *Reflection on statistics: Learning, teaching and assessment in grades K-12* (pp. 276-295). Mahwah, NJ: Lawrence Erlbaum Associates.

Goh, S. L. (2004). *Primary 5 students interpretation of graphs in the real world.* Unpublished master's dissertation. National Institute of Education, Nanyang Technological University, Singapore.

Groth, R. E. (2003). *Development of a high school statistical thinking framework.* UMI dissertation services.

Harper, S. R. (2004). Students' interpretation of misleading graphs. *Mathematics Teaching in the Middle School, 9*(6), 340-343.

Huck, S. W. (2000). *Reading statistics and research* (3rd ed.). London: Addison Wesley Longman.

Jones, G. A., Thornton, C. A., Langrall, C. W., Mooney, E. S., Perry, B., & Putt, I. J. (2000). A framework for characterizing children's statistical thinking. *Mathematical Thinking and Learning, 2*(4), 269-307.

Konold, C., & Higgins, T. L. (2003). Reasoning about data. In J. Kilpatrick, W. G. Martin & D. Schifter (Eds.), *A research companion to Principles and Standards for School Mathematics* (pp. 193-215). Reston, VA: National Council of Teachers of Mathematics.

Kosslyn, S. M. (1989). Understanding charts and graphs. *Applied Cognitive Psychology, 3*, 185-226.

Lenton, G., Stevens, B., & Illes, R. (2000). Numeracy in science: Pupils' understanding of graphs. *School Science Review, 82*(299), 15-23.

Li, K. Y., & Shen, S. M. (1992). Students' weakness in statistical projects. *Teaching Statistics, 14*(1), 2-8.

Lim, C. C., & Kaur, B. (1992, November). *Graphs – What do they say?* Paper presented at the Sixth Annual Conference of the Educational Research Association, Singapore.

Ministry of Education, Singapore. (2006). *Mathematics syllabus: Primary.* Available from http://www.moe.gov.sg/cpdd/1_Primary%20maths%20syllabus%202007_for%20uploading%2024%20Jul.pdf

Mooney, E. S. (1999). *Development of a middle school statistical thinking framework.* UMI Dissertation Services.

Nisbet, S. (2001). Representing categorical and numerical data. In J. Bobis, B. Perry & M. Michelmore (Eds.), *Numeracy and beyond: Proceedings of the 24th Annual Conference of the Mathematics Education Research Group of Australasia* (pp. 378-385). Sydney: MERGA.

Nisbet, S. (2002). Representing numerical data: The influence of sample size. In A. Cockburn & E. Nardi (Eds.), *Proceedings of the 26th Annual Conference of the International Group for the Psychology of Mathematics Education,* PME 26 (Vol. 3, pp. 417-424). Norwich, UK: University of East Anglia.

Padilla, M. J., McKenzie, D. J., & Shaw, E. J. (1986). An examination of the line graphing: Ability of students in Grades seven through twelve. *School Science and Mathematics, 86*(1), 20-26.

Pereira-Mendoza, L., & Kaur, B. (1999). TIMSS Performance of Singapore secondary students (part A): Algebra, geometry and data representation, analysis and probability. *Journal of Science and Mathematics Education in Southeast Asia, 22*(2), 38-55.

Shaughnessy, J. M., Garfield, J., & Greer, B. (1996). Data handling. In A. J. Bishop, K. Clements, C. Keitel, J. Kilpatrick & C. Laborde (Eds.), *International handbook of mathematics education* (pp. 205-237). Dordrecht, The Netherlands: Kluwer Academic Publishers.

Shaughnessy, J. M., & Pfannkuch, M. (2002). How faithful is Old Faithful? – Statistical thinking: A story of variation and predication. *The Mathematics Teacher, 95*(4), 252-259.

Wainer, H. (1992). How to display data badly. Reprinted in R. L. Weber (Ed.), *Science with a smile* (pp. 252-278). Bristol: Institute of Physics Publishing. (Original work published in 1984)

Watson, J. M., & Moritz, J. B. (1999). The beginning of statistical inference: Comparing two data sets. *Educational Studies in Mathematics, 37*, 145-168.

Watson, J. M., & Pereira-Mendoza, L. (1996). Reading and predicting from bar graphs. *Australian Journal of Language and Literacy, 19*(3), 244-258.

Wu, Y. K. (2005). *Statistical graphs: Understanding and attitude of Singapore secondary school students and the impact of a spreadsheet exploration.* Unpublished PhD dissertation. National Institute of Education, Nanyang Technological University, Singapore.

Wu, Y. K., & Wong, K. Y. (2007). Impact of a spreadsheet exploration on secondary school students' understanding of statistical graphs. *Journal of Computers in Mathematics and Science Teaching, 26*(4), 355-385.

Chapter 10

Word Problems on Speed: Students' Strategies and Errors

JIANG Chunlian

This paper presents problem-solving strategies that Singapore students (n = 1002) used to solve three word problems on speed and the errors they made. It is surprising to find that the Primary 6 students performed significantly better than the Secondary 1 and Secondary 2 students, while the performance differences between the last two groups were not significant. The strategy analyses reveal that the Primary 6 students could use the arithmetic strategies, model method, and guess-and-check method more successfully than the secondary students. More effort needs to be made to bridge the gap between primary and secondary mathematics, and more word problems on speed need to be included in textbooks used in secondary levels.

Key words: problem-solving strategies, model method, word problems, speed, average speed

1 INTRODUCTION

Mathematical problem solving is central to mathematics learning and to enable pupils to develop their abilities in mathematical problem solving is the primary aim of the Singapore mathematics curriculum (Ministry of Education, Singapore, 2001a, 2001b). About a dozen of heuristics (strategies) such as "act it out" and "draw a model/diagram" are suggested in the mathematics syllabi. Heuristics and strategies are used interchangably in this study to cover the methods or problem-solving procedures that direct the search for a solution (Krulik & Rudnick, 1988). Being taught heuristics for solving problems, can Singapore students use them effectively? This becomes the question to be addressed in this study.

Word problems are often used for developing students' problem-solving abilities. The topic of speed was selected because it has a variety of applications in daily life (Cross & Mehegan, 1988; Cross & Pitkethly, 1988, 1989) and in mathematics from primary to university levels. The mathematics of change and variation, and in particular, the study of motion, is fundamental in elementary algebra and calculus instruction (Bowers & Nickerson, 2000). Several studies have included word problems on speed as a specific rate model of multiplication and division (Bell, Fischbein, & Greer, 1984; Fischbein, Deri, Nello, & Marino, 1985; Greer, 1992). However, these word problems on speed belong to the simplest category of the 13 motion problem categories identified by Mayer (1981) through an analysis of algebraic word problems in secondary school mathematics textbooks, but he did not examine how students actually solve the problems. This study seeks in part to fill this gap. It also considers the misconceptions about speed held by students at upper primary and lower secondary levels.

Therefore, the study was conducted to answer the research question "what strategies do Singapore students use to solve word problems on speed?" This paper first lists the speed topics included in mathematics curriculum, and then presents the strategies Singapore students used for solving three word problems on speed.

2 WORD PROBLEMS ON SPEED IN THE MATHEMATICS CURRICULUM

The main topics on speed in the Singapore curriculum are shown in Table 1. Word problems on speed appear as an independent unit in Primary 6 mathematics textbooks (Minstry of Educcation, Singapore, 2000a, 2000b) and under the units of rate, ratio, and proportion in Secondary 1 and Secondary 2 textbooks (Teh & Looi, 2002a, 2002b). There are also several word problems on speed under the unit of simultaneous linear equations. Therefore, this study dealt with students at Primary 6 (P6), Secondary 1 (S1), and Secondary 2 (S2) levels.

Table 1
Speed Topics in the Singapore Mathematics Curriculum (2006 version)

Level	Speed Topics
Primary 6	Speed and average speed: (a) Calculation of speed, distance, or time given the other two quantiites; (b) Writing speed in different units such as km/h, m/min, m/s, and cm/s; (c) Solving up to 3-step word problems involving speed and average speed.
Secondary 1	Speed, uniform speed and average speed: (a) Conversion of units (e.g., km/h to m/s); (b) Solving problems involving speed, uniform speed, and average speed.

3 THE THREE WORD PROBLEMS ON SPEED

The three word problems (Table 2) presented in this chapter belong to Mayer's (1981) Round Trip 1 category. Problem 1 is taken as a representative of problems that can be solved using arithmetic strategies. An arithmetic strategy is used when the problem solver writes down a mathematical statement involving one or more operations on the numbers given in the problem (Fong & Hsui, 1999). Problems 2 and 3 represent problems that cannot be easily solved using arithmetic strategies. Furthermore, knowledge of inverse proportion is needed for solving Problem 2, but it is not appropriate for solving Problem 3.

Table 2
The Three Speed Problems

No.	Text
1	Yesterday afternoon Mike drove to the kindergarten to take back his son. It was raining on the way to the kindergarten, so he had to drive at a slow speed of 39 km/h, and it took him $1\frac{1}{3}$ hours to get there. He returned soon at a faster speed along the same way. When he came back home, his wife told him that his average speed for the round trip is 52 km/h (Ignoring time spent in the kindergarten). Find his speed on the way back.
2	Sunday morning, Rebecca and her parents went out to enjoy the natural scenery. On the way to the destination, they travelled at a slow speed of 40 km/h. On the way back, they drove at a faster speed of 120 km/h. When they came back home, they found that they had been out for 2 hours. Find the average speed for this round trip (Ignoring time around the destination).
3	On the first day of this new term, Teacher Lee went to the bookshop to take back the ordered textbooks. On the way to the bookshop, his speed was as slow as 24 km/h because of the heavy traffic. On the way back, the traffic was light so he took only one hour. If the average speed for the round trip was 36 km/h, find the speed on the way back (Ignoring time spent in the bookshop).

4 STRATEGIES THAT CAN BE USED FOR SOLVING THE THREE PROBLEMS: A FRAMEWORK

Several strategies can be used to solve the three problems (Table 3). The first strategy is widely used by students for solving mathematical problems (Fong & Hsui, 1999). The following five strategies are listed in mathematics syllabi and are considered appropriate for solving word problems on speed. Proportion method and unitary method are added by referring to the textbooks used in Singapore and Hart's (1981) project for

Table 3
Strategies that Can be Used for Solving the Three Problems

No.	Strategy category	Definition
1	Arithmetic strategies	It is used where the subject writes down a mathematical statement involving one or more operations on the numbers given in the problem (Fong & Hsui, 1999).
2	Algebraic strategies	It is used when one or more unknowns are chosen as variables and equation(s) is set up.
3	Model method	It is used when the solution is suggested by or follows a model or a diagram (Kho, 1987).
4	Guess-and-check	It involves the following two steps: (a) Make a guess of a certain answer or the unknown in the problem based on an estimation. (b) Check if the constraints given in the question or implied from some of the question statements are satisfied. If all the constraints are satisfied, the guess is correct; the answer has been obtained or can be worked out. All processes will end at this point. If the constraints are not satisfied, the guess will be refined or adjusted, and another guess will be made, then another round of guess-and-check will begin.
5	Logical reasoning	Logical reasoning strategy is used when some forms of "if-then" reasoning are used (Van De Walle, 1993).
6	Unitary method	Unitary method involves finding the value equivalent to one unit of a quantity from an equating statement and obtaining the value equivalent to more units of the quantities using the value for one unit just found (Fong, 1999; Fong & Hsui, 1999; Yuen, 1995).
7	Proportion method	A proportion method is used when proportional properties (direct and inverse proportions) are used.
8	No strategy	It refers to the absence of a written response and where only pieces of information taken from the question are written down but without further working (Fong & Hsui, 1999).

solving ratio and proportion problems. "No strategy" is also included because students were often found to leave the question blank (Fong & Hsui, 1999; Gallagher et al., 2000). The rest of this section illustrates how to use these strategies to solve Problem 2 in the study. One important reason for choosing Problem 2 as the example is that all the strategies listed in the table can be used.

Arithmetic strategies. As aforementioned, this strategy is not appropriate for Problem 2. However, if a student knows the formula for finding the average speed (AS) of a round trip when speeds on the way to (S_1) and fro (S_2) are given, i.e., $AS = \frac{2 \times S_1 \times S_2}{S_1 + S_2}$, he/she may use it. One possible solution of using this strategy could be "$AS = \frac{2 \times 40 \times 120}{40 + 120} = 60$ km/h." In the study, solutions where a student makes manipulations on numbers taken from the problem were also classified under this category. For example, a solution "120 + 40 = 160, 160 ÷ 2 = 80 km/h" is said to use the arithmetic strategies.

Algebraic strategies. This is adopted from the heuristic "Use equations." Two typical solutions using this strategy are shown below.

Solution 1:

Let time on the way to the destination be x hours,

Then: $40x = 120 \times (2-x)$.

Solving it, $x = 1\frac{1}{2}$ h.

$1\frac{1}{2} \times 40 = 60$ km, $60 \times 2 = 120$ km,

$120 \div 2 = 60$ km/h.

Solution 2:

Let distance of one way be x km, then:

$\frac{x}{40} + \frac{x}{120} = 2$.

Solving it, $x = 60$ km.

$60 \times 2 = 120$ km,

$120 \div 2 = 60$ km/h.

In Solution 1, the time on the leaving trip (T_1) is chosen as the variable and the equation is set up based on the equality of distances on the two ways. Then the distance of the leaving way (D_1), the total distance (TD), and the average speed (AS) for the whole journey could be found. After T_1 has been found, the time and the distance on the way back (coded as T_2 and D_2 respectively) can be found, TD can also be found by putting D_1 and D_2 together. In Solution 2, the distance of one way is chosen as the variable and the equation is set up based on part-whole relationship between times on the two ways and the total time (TT). It is also possible to choose T_1 and T_2 as variables and set up simultaneous equations based on both $D_1 = D_2$ and $T_1 + T_2 = TT$.

Model method. Models are used not only to better understand the problem by having something visual in front (Van de Walle, 1993), but also to suggest a plan for the solution to the problem (Kho, 1987). It is also frequently used with other strategies (Van de Walle, 1993). Three kinds of models can be drawn as below.

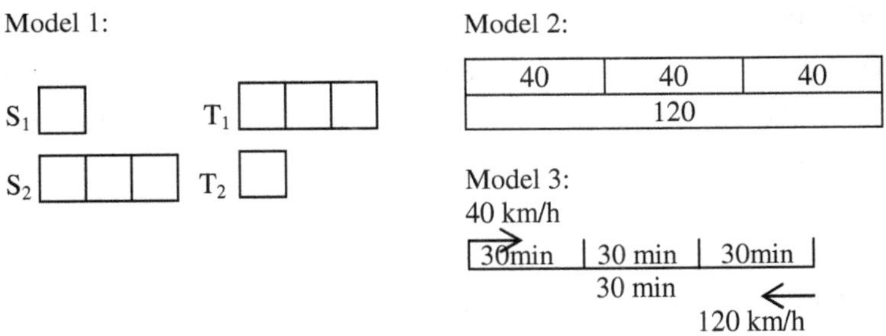

Figure 1. Model method.

Model 1 shows the ratio between S_1 and S_2 (1 : 3) and the ratio between T_1 and T_2 (3 : 1). Model 2 shows the inverse proportional relationship between speed and time because three 40s equals one 120. It also shows the ratio between T_1 and T_2. Model 3 directly shows that T_1 is three 30 minutes, and T_2 is one 30 minutes. The first two models involve unitary methods to find T_1 and/or T_2: "4 units = 2 h, 1 unit = $\frac{1}{2}$ h, 3 units = $1\frac{1}{2}$ h." After that, D_1, TD, and AS can be found. Model 3 is simpler compared to the first two models because there is no need to find T_1 and T_2.

Guess-and-check. Solutions 3 and 4 are two typical solutions using this strategy.

Solution 3:

T_1	D_1	T_2	D_2	D_1=D_2?
1	40	1	60	×
$1\frac{1}{2}$	60	$\frac{1}{2}$	60	√

40×1½ = 60 km, 60×2 = 120 km,

120÷2 = 60 km/h.

Solution 4:

$D_1 = D_2$	T_1	T_2	TT	TT=2h?
120 km	3	1	4	×
60 km	$1\frac{1}{2}$	$\frac{1}{2}$	2	√

60×2 = 120 km, 120÷2 = 60 km/h.

In Solution 3, T_1 is guessed and whether $D_1 = D_2$ is checked. Similarly, T_2 can be guessed followed by checking whether $D_1 = D_2$. In Solution 4, the distance of one way is guessed (120 km, half of 120 km, etc.) and whether TT = 2 hours is checked.

Logical reasoning. This is similar to the "Make suppositions" heuristic in the mathematics syllabus. Solution 5 illustrates this strategy. The distance of one way is taken as "1."

Solution 5:

If distance of one way be unit "1", then the total distance is 2.

Times on the two ways are $\frac{1}{40}$ and $\frac{1}{120}$ respectively.

Therefore, average Speed = $\dfrac{2}{\frac{1}{40}+\frac{1}{120}}$ = 60 km/h.

It was assumed that few students could know the formula $AS = \dfrac{2 \times S_1 \times S_2}{S_1 + S_2}$. Therefore, a piece of extra information (TT = 2 hours) is given. If a student could ignore it and set any number as the distance of one way or the total distance, the correct answer can be found.

Unitary method. Solution 6 is a typical solution using this method. After T_2 has been found, there are also three ways to find TD: (a) doubling D_1, (b) doubling D_2, and (c) figuring out D_1+D_2.

Solution 6:

Let time on the way back be 1 unit, then time on the way forth is $120 \div 40 = 3$ units, Total time = $3 + 1 = 4$ units,

4 units = 2 hours, 1 unit = $\frac{1}{2}$ h, $120 \times \frac{1}{2} = 60$ km.

$60 \times 2 = 120$ km. $120 \div 2 = 60$ km/h.

Proportion method. This involves using direct and/or inverse proportions. Solution 7 is a typical solution using this strategy. In this solution, $T_1 : T_2 = 3:1$ is deduced from the inverse proportionality between speed and time for the two parts of the journey. Ratio can then be used to find T_1 and/or T_2. There are also three ways to find TD after T_2 has been found.

Solution 7:

Because 40:120 = 1:3, the ratio of times on the two ways is 3:1.

$2 \times \frac{3}{1+3} = 1\frac{1}{2}$ h, $2 \times \frac{1}{1+3} = \frac{1}{2}$ h. $120 \times \frac{1}{2} = 60$ km.

$60 \times 2 = 120$ km. $120 \div 2 = 60$ km/h.

5 METHOD

In the study, 1002 Singapore students at three grade levels (345 P6, 315 S1, and 342 S2) were involved. Slightly more than 300 students at each grade level were chosen from intact classes. The students were from four primary schools and six secondary schools. The sample was quite representative of the student population at these levels (Jiang, 2005).

A test including the three problems was administered in April to July in 2001. All the students had already completed the topic of speed before the test was administered. The testing procedure began with a brief introduction of the purpose of the study by their mathematics teachers. The students were told to work hard on the test and to use any strategies that they preferred. No calculators were allowed.

Students' responses to the speed problems were analysed in two steps: (a) Score using a 0-1-2 scale. Two points were given to each correct answer or an incorrect answer where all the necessary steps were included but with only minor computational errors. One point was given to each answer that solved part of the problem. Zero point was given to answers that were completely wrong and no solutions. (b) Identify strategies used by students to solve each problem.

6 RESULTS

6.1 *Relative Performance*

From Problem 1 (Mean = 1.26, SD = 0.71) through Problem 2 (Mean = 0.45, SD = 0.81) to Problem 3 (Mean = 0.20, SD = 0.58), the students' performance became worse (see Figure 2). Problems 2 and 3 are at the same difficulty level, which is higher than Problem 1 (Jiang, 2005). Mayer (1981) classified them into the same category because they do not involve relation propositions. The result obtained in the study indicates that it is also necessary to take into consideration the solution methods that could be used to find the unknowns when effort is made to classify word problems.

A one-way ANOVA analysis indicate that there were significant performance differences ($p < 0.001$) among the students at the three grade levels on each problem (Problem 1: $F = 12.85$; Problem 2: $F = 13.57$; Problem 3 : $F = 10.58$. df = 2). Post hoc pair-wise comparisons indicate that the P6 students performed significantly better than the S1 students on all the three problems ($p < 0.001$). The P6 students performed significantly better than the S2 students on Problems 1 and 2 at the 0.001 significant level and on Problem 3 at the 0.02 significant level. However, there were no significant performance differences between the S1 and S2 students on Problems 1 and 2. The performance difference between the S1 and S2 students on Problem 3 is significant at the 0.02 level.

The performance comparison result is a bit surprising. Next section will provide some explanations from the strategy use.

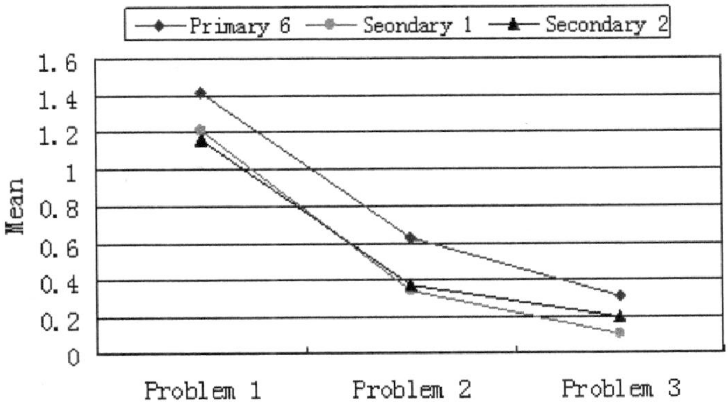

Figure 2. Mean scores for the problems by grade levels.

6.2 *Use of Strategies*

This section reports on the percentages of the students who used the various strategies and how successfully they were. For these purposes, two terms were introduced. One is Strategy Percent (SP), which is defined as the percent of a specific group of students using a specific category of strategy. If the Strategy Percent Difference (SPD) between different groups of students is more than 5%, it is taken as high. A high SPD is probably due to the teaching of problem solving in schools. The other is Success Rate (SR), which is defined as the proportion of the specific group who could use the strategy to get the correct answers to a problem. If the Success Rate Difference (SRD) between different groups of students is more than 10%, it is taken as high. If the success rate of a strategy is lower than 30%, the strategy is taken as inappropriate for solving a specific problem because a majority (70%) of the students using that strategy could not reach the correct answer. The statistical data for the three problems are presented in Table 4. If one of the SPs is lower than 5%, the comparison in SRs will not be discussed because it is not practically meaningful.

Problem 1. As expected, a majority of the students used arithmetic strategies to solve this problem. The SPs for the students at the three grade levels are quite close. However, the P6 students could use them more successfully than the S1 students (SRD = 10%). The SRDs between

Table 4

Students' Strategy Percents (SP), Success Rates (SR) in Solving the Three Problems

Strategies		Primary 6 (*n* = 345)	Secondary 1 (*n* = 315)	Secondary 2 (*n* = 342)	Total (*n* = 1002)
Problem 1					
Arithmetic strategies	SP	86.38[a]	84.13	85.09	85.23
	SR	43.29	32.83	37.46	38.06
Algebraic strategies[b]	SP	—[c]	—	1.46	0.50
	SR	—	—	80.00	80.00
Model method	SP	12.75	11.11	4.68	9.48
	SR	54.54	37.14	31.25	44.21
Guess-and-check	SP	—	—	0.29	0.10
	SR	—	—	0	0
No strategy		0.87	4.76	8.48	4.69
Problem 2					
Arithmetic strategies	SP	54.20	46.35	43.27	48.00
	SR	0	0.68	0	0.21
Algebraic strategies	SP	—	0.32	11.70	4.09
	SR	—	100.00	65.00	65.85
Model method	SP	7.83	3.81	2.05	4.59
	SR	40.74	41.67	14.29	36.96
Guess-and-check	SP	21.16	11.11	6.43	12.97
	SR	90.41	74.29	72.73	83.08
Logical reasoning	SP	1.45	1.90	0.58	1.30
	SR	100	50.00	100	76.92
Unitary method	SP	2.90	0.95	1.17	1.70
	SR	40.00	100.00	100.00	64.70
Proportion method	SP	5.22	3.81	1.46	3.49
	SR	83.33	75.00	40.00	71.43
No strategy		7.25	31.75	33.33	23.85
Problem 3					
Arithmetic strategies	SP	53.04	40.95	38.30	44.21
	SR	0	0	0	0
Algebraic strategies	SP	1.45	1.27	13.45	5.49
	SR	80.00	0	43.48	43.64
Model method	SP	8.12	2.54	1.46	4.09
	SR	0	12.5	0	2.44
Guess-and-check	SP	15.94	6.67	1.17	7.98
	SR	80.00	57.14	75.00	73.75
No strategy		21.45	48.57	45.61	38.22

[a]The % marks of all the numbers were omitted.

[b]The strategies highlighted are taken as appropriate because their SRs are more than 60%.

[c]No Primary 6 students used algebraic strategies.

P6 and S2 and between S1 and S2 students are not great. As for the algebraic strategies, only five S2 students (1.5%) used them, but no P6 and S1 students did so, though it is most effective for solving this problem (SR = 80%). About 10% of the students used the model method. A higher percentage of the P6 and S1 students used it compared to the S2 students (SPDs > 5%). For this method, the P6 students used it much more successfully than the S1 (SRD = 17%) and the S2 students (SRD = 23%). The SRD between S1 and S2 students was not great. The S2 students had the highest percentage of no strategy.

Problem 2. About half of the students used the arithmetic strategies to solve this problem. However, the success rates were very low (SR < 1%). The low success rates supported the assumption that this problem cannot be easily solved using arithmetic strategies. A higher percentage of the P6 students used this inappropriate strategies compared to the S1 and S2 students (SPD > 5%). As for the algebraic strategies, no P6 students used them, only one S1 student used them, and about 12% of the S2 students used them. About two-thirds of these S2 students used the algebraic strategies correctly. About 5% of the students used the model method, especially by the P6 students. The P6 and S1 students used this method much more successfully than the S2 students (SRDs > 25%). The P6 and S1 students could use this method almost equally successfully. When the model method was used, most students actually drew the simplified model which did not show the ratio(s) of speeds or times. This might account for the lower SRs of the model method.

Guess-and-check strategies are most effective for solving this problem. A higher percentage of the P6 students used this method compared to the S1 students (SPD = 10%), which is also marginally higher than that of the S2 students. The P6 students used this method much more successfully than the secondary students (SRDs > 15%). The S1 and S2 students could use this method almost equally successfully. The SPs for the proportion methods were low (SP < 6%). The P6 and S1 students used this method more successfully than the S2 students (SRDs > 35%), the SRD between the P6 and S1 students was not great. About 3% of the students used the appropriate unitary methods (SR = 65%) and logical reasoning method (SR = 77%). The SPs of the S1 and S2 students who had no strategy were much higher than that of the P6 students (SPD > 24%).

Problem 3. Similar to Problem 2, about 44% of the students used the inappropriate arithmetic strategies (SR = 0). A higher percentage of the S2 students used the appropriate algebraic strategies (SR > 40%) compared to the P6 and S1 students (SPDs > 12%). Most of the students who used algebraic strategies set up the equations based on average speed (i.e., TD/TT = AS), which are more complicated than the equations based on the part-whole relationship among times or the double relationship between distance of one way and the total distance. About 4% of the students used the inappropriate model method (SR < 3%). Most of them drew the simplified model, then used irrelevant procedures as those who used arithmetic strategies. About 8% of the students used the guess-and-check strategies. A higher percentage of the P6 students used this method compared to the S1 students (SPD = 9%), which is also higher than that of the S2 students (SPD = 5%). The P6 and S2 students could use this method more successfully than the S1 students (SRD > 18%). Among the strategies used to solve this problem, the guess-and-check methods were most effective. Higher percentages of the S1 and S2 students had no strategies compared to the P6 students (SPDs > 24%). Similar to Problem 2, there is also a piece of extra information (1 hour) in this problem. The students could ignore it and set any number as the distance of one way or the total time to find the answer. However, no students used logical reasoning methods. This also indicates that no students have detected the extra information.

Many students actually applied irrelevant procedures to solve the problems. It is commonly recognised that students may take the average of individual speeds as average speed without realising that time durations for different speeds are different (Gorodetsky et al., 1986; Thompson, 1994). This error was found in the solution of all the three problems, especially in Problem 2. Two hundred and fifty-eight students (25.7%) provided a solution like "120 + 40 = 160, 160 ÷ 2 = 80" to Problem 2. It seems that they hold another misconception of taking average speed to be the sum of the speeds on the two ways. This misconception is rarely mentioned in literature, but it was found to be prevalent in this study. For example, in solving Problem 3, eighty students (8.0%) provided a solution like "36 – 24 = 12." As no interview was conducted due to the time limit, this interpretation is purely speculative. Further investigation is needed.

7 DISCUSSION AND CONCLUSION

The strategy analyses revealed that the P6 students performed better than the secondary students because the P6 students could use the arithmetic strategies (Problem 1), model method (Problem 1), and guess-and-check method (Problem 2 and 3) more successfully than the secondary students, and lower percentages of the P6 students had no strategies compared to the secondary students. It is possible that the secondary students (especially S2) were not motivated to try their best. Some implications for the teaching and learning of problem solving and speed are discussed below.

First, the model method is recommended in mathematics syllabi. It is also the most frequently used method for solving worked examples in the textbooks used in primary and lower secondary schools (Ng, 2002; Zhu, 2003). However, it is not appropriate for solving the three problems in the study, especially for Problems 2 and 3, where the simplified model cannot reveal the relational relationships (ratio) between speeds or times on the two parts of a journey. Teaching of the model method needs to further explore how it can be used and when it may not be helpful. Only when we are aware of the limitations of the various problem-solving strategies can the teaching of problem solving become more effective.

Second, the guess-and-check method is the second most frequently used in solving Problems 2 and 3. This method had the highest success rates. Thus, it is a powerful method for tackling problems when students do not have much knowledge of formal algebra. As the grade level increased, lower percentages of the students used the guess-and-check methods. Higher percentages of the S2 students used algebraic strategies in Problems 2 and 3 than the P6 and S1 students. However, the percentages of the S2 students who used algebraic strategies were still low (SPs < 15%). Therefore, teachers of lower secondary levels need to explore how to use the guess-and-check method as the bridge between arithmetic and algebra (Jiang, 2005).

Third, Problem 2 can be solved using proportion methods (inverse proportion). However, less than 4% of the students were found to do so. This is consistent with the findings in the study of Hart (1981). Hart found that it took the U.K. secondary students a long time to develop proportional thinking. Proportion is included as a single chapter in S1 mathematics textbooks, and under the chapter of Arithmetic Problems and Standard Form in S2 mathematics textbooks. The findings suggest including more problems on proportion in the S2 textbooks.

Fourth, Problems 2 and 3 involve a piece of extra information about time. Very few students ignored this piece of information and used logical reasoning methods. This might be because the students were not frequently exposed to problems with extra information (Zhu, 2003). Hence, more problems with extra information should be used to improve students' abilities in solving real-world problems which often involve much irrelevant information.

Fifth, when algebraic strategies were used, the students often set up equations based on TD/TT = AS when the average speed (AS) was given, as in Problems 1 and 3. These fractional equations are more difficult to solve than equations constructed based on TD = TT × AS. The ability to choose appropriate equivalent statements to build equations needs to be developed.

Sixth, some students used the average of individual speeds as the average speed in solving the three problems. One possible reason may be because there are very few problems of this type in the textbooks (Jiang, 2005). Thus, the textbooks may include more round-trip problems.

The students probably confused the premises "when the times of the two parts are equal" with "when the distances of the two parts are equal" for the conditions under which the average speed of a journey consisting of two parts was the average of the speeds. Therefore, the teaching of the concept of average speed needs to expose the students to the different effects of these two premises on the average speed. At the P6 level, the following two examples with numbers can be presented:

(a) Car A traveled 480 km. It traveled half of the journey on the highway at a speed of 120 km/h and another half of the journey at a speed of 60 km/h. Find the total time. (Answer: $480 \div 2 = 240$ km, $240 \div 120 = 2$ h, $240 \div 60 = 4$ h, $2 + 4 = 6$ h).

(b) Car B traveled 6 hours. It traveled half of the time on the highway at a speed of 120 km/h and another half of the time at a speed of 60 km/h. Find the distance covered. (Answer: 6÷2 = 3, 120×3 = 360 km, 60×3 = 180 km, 360+180 = 540 km).

The two speeds are the same. However, the two cars traveled different distances in the same amount of time using the two speeds for half of the journey and for half of the time. Therefore, their average speeds are different, and the average speed for Car A is not the average of the two given speeds. Such a discussion could evoke students' cognitive conflicts to explore their understanding of average speed. Examples can also be constructed to show that it will take different time durations to cover the same journey using two different speeds for half of the journey and for half of the time. At the lower secondary levels, after the students have learnt the algebraic manipulations, the following two problems can be presented.

(c) Car A traveled half of the journey on the highway at a speed of S_1 and another half of the journey at a speed of S_2. Find the average speed. (Answer: $\frac{2\times S_1 \times S_2}{S_1 + S_2}$).

(d) Car B traveled half of the time on the highway at a speed of S_1 and another half of the time at a speed of S_2. Find the average speed. (Answer: $\frac{S_1 + S_2}{2}$).

The two average speeds in Problems (c) and (d) are equal only when the two speeds are equal.

Also common is another misconception of average speed as the sum of the individual speeds. The students probably oversimplify the concept of "average" as the "sum" of several numbers from the computational algorithm "add-them-all-up-and-divide" of average (Cai, 2000). Average speed is different from the general meaning of "average" in statistics. It is also different from the "sum" of individual speeds. The teaching of average speed needs to discuss the different meanings of "average" and the differences in algorithms used to find average of a set of numbers and average speed. Being exposed to the concept of average speed in such a broader sense, students will probably understand the concept of average speed better. Further experiments in this aspect need to be carried out.

To conclude, research and experimental studies in mathematical problem solving continue to be an important area in mathematics education. This study has also revealed that more research is needed for exploring students' development of speed-related concepts.

References

Bell, A., Fischbein, E., & Greer, B. (1984). Choice of operation in verbal arithmetic problems: The effects of number size, problem structure and context. *Educational Studies in Mathematics, 15*(2), 129-147.

Bowers, J. S., & Nickerson, S. N. (2000). Students' changing views of rates and graphs when working with a simulation microworld. *Focus on Learning Problems in Mathematics, 22*(3-4), 10-25.

Cai, J. (2000). Understanding and representing the arithmetic averaging algorithm: an analysis and comparison of U.S. and Chinese students' responses. *International Journal of Mathematical Education in Science and Technology, 31*(6), 839–855.

Cross, R. T., & Mehegan, J. (1988). Young children's conception of speed: Possible implications for pedestrian safety. *International Journal of Science Education, 10*(3), 253-265.

Cross, R. T., & Pitekethly, A. (1988). Speed, education and children as pedestrians: A cognitive change approach to a potentially dangerous naive concept. *International Journal of Science Education, 10*(5), 531-540.

Cross, R. T., & Pitekethly, A. (1989). A curriculum model to improve young children's concept of speed to reduce their pedestrian accident vulnerability. *School Science and Mathematics, 89*(4), 285-92.

Fischbein, E., Deri, M., Nello, M. S., & Marino, M. S. (1985). The role of implicit models in solving verbal problems in multiplication and division. *Journal for Research in Mathematics Education, 16*(1), 3-17.

Fong, H. K. (1999). Strategic model for solving ratio and proportion problems. *The Mathematics Educator, 4*(1), 34-51.

Fong, H. K., & Hsui, V. (1999). Strategy preferences and their association with hierarchical difficulties of fraction problems. *Science, Mathematics and Technical Education, 5*, 3-12.

Gallagher, A. M., Lisi, R. D., Holst, P. C., Lisi, A. V., McGillicuddy-De, Morely, M., & Cahalan, C. (2000). Gender differences in advanced mathematical problem solving. *Journal of Experimental Child Psychology, 75*(3), 165-190.

Gorodetsky, M., Hoz, R., & Vinner, S. (1986). Hierarchical solution models of speed problems. *Science Education, 70*(5), 565-82.

Greer, B. (1992). Multiplication and division as models of situations. In D. A. Grouws (Ed.), *Handbook of research on mathematics teaching and learning* (pp. 276-295). New York: Macmillan.

Hart, K. M. (1981). Ratio and proportion. In K. M. Hart (Ed.), *Children's understanding of mathematics: 11-16* (pp. 88-101). London: John Murray.

Jiang, C. (2005). *Strategies and errors in solving speed problems: A comparative study between Chinese and Singapore students.* Unpublished doctoral dissertation, National Institute of Education, Nanyang Technological University, Singapore.

Kho, T. H. (1987). Mathematical models for solving arithmetic problems. In *Mathematics Education in the 1990's. Proceedings of Fourth Southeast Asian Conference on Mathematical Education* (ICMI-SEAMS) (pp. 345-351). Singapore: Institute of Education.

Krulik, S., & Rudnick, J. A. (1988). *Problem solving: A handbook for elementary school teachers.* Boston: Allyn and Bacon.

Mayer, R. E. (1981). Frequency norms and structural analysis of algebra word problems into families, categories, and templates. *Instructional Science, 10*(2), 135-175.

Ministry of Education. (2000a). *Primary mathematics 6A* (3rd ed.). Singapore: Author.

Ministry of Education. (2000b). *Primary mathematics 6B* (3rd ed.). Singapore: Author.

Ministry of Education. (2001a). *Lower secondary mathematics syllabus, Special/Express, Normal academic, Normal technical.* Singapore: Author.

Ministry of Education. (2001b). *Primary mathematics syllabus.* Singapore: Author.

Ministry of Education. (2006a). *Mathematics syllabus: Primary.* Singapore: Author.

Ministry of Education. (2006b). *Secondary mathematics syllabuses.* Singapore: Author.

Ng, L. E. (2002). *Representation of problem solving in Singaporean primary mathematics textbooks with respect to types, Polya's model and heuristics.* Unpublished Master's thesis, National Institute of Education, Nanyang Technological University, Singapore.

Teh, K. L., & Looi, C. K. (2002a). *New syllabus D mathematics 1* (5th ed.). Singapore: Shing Lee Publishers Pte Ltd.

Teh, K. L., & Looi, C. K. (2002b). *New syllabus D mathematics 2* (5th ed.). Singapore: Shing Lee Publishers Pte Ltd.

Thompson, P. W. (1994). The development of the concept of speed and its relationship to concepts of rate. In G. Harel & J. Confrey (Eds.), *The development of multiplicative reasoning in the learning of mathematics* (pp. 179-234). Albany: State University of New York Press.

Van de Walle, J. A. (1993). *Elementary school mathematics: Teaching developmentally*. New York: Longman.

Yuen, C. L. (1995). *Strategies and errors of children in solving fraction problems.* Unpublished Master's thesis, National Institute of Education, Nanyang Technological University, Singapore.

Zhu, Y. (2003). *Representations of problem solving in China, Singapore and US mathematics textbooks: A comparative study.* Unpublished PhD dissertation, National Institute of Education, Nanyang Technological University, Singapore.

Chapter 11

Review of Research on Mathematical Problem Solving in Singapore

FOONG Pui Yee

Since the adoption of problem solving as the goal of mathematics education in the local curriculum from primary to secondary schools in the early 90's, the theme "problem solving in mathematics" has captured the research interest of many teachers and researchers. This review looks at various aspects of research in problem solving in mathematics under three broad strands: (a) problem solving approaches and tasks, (b) teachers' beliefs and practices, and (c) students' problem solving behaviours. Within each strand, related researches are described under various sub-themes. The final section draws together the key findings and highlights recommendations for future research.

Key words: research review, mathematical problem solving, teachers' beliefs and practices, metacognition, problem-solving behaviours, textbooks, assessment

1 INTRODUCTION

The essential goal for research in mathematics education is to produce critical and new knowledge about teaching and learning of mathematics relevant to the localised context of each country and the international community at large. Research can also document the current situations of how teachers are teaching, what curriculum materials schools are using, and how students are learning. What is the current status of mathematics education research in Singapore and are the findings of research known to teachers and the stake-holders here? My aim in preparing this review is to identify a major area of research in mathematics education in Singapore for review over the past decade. The first mathematics education research review ever done was in the *State-of-the-Art Review* by Chong, Khoo, Foong, Kaur, and Lim-Teo (1991) which surveyed the status of local research over a decade in the 1980s. Then there were only 42 studies of which 26 were in the categories of teaching and learning mathematics that focused on student understanding, learning strategies, and problem solving. The quantity of research in mathematics education in Singapore has been growing steadily such as in classroom investigations by school teachers that warranted a recent review by Foong (2007). This review shows a widening trend in the key research themes that are of significance to mathematics educators and researchers here. Two of these themes are problem solving in mathematics and cognitive aspects of mathematics learning, and there are substantial research done in these two areas in the past 20 years. It would seem that research interests in the 80's have seeded further research in these two areas. Basically these research interests have been driven much by the fundamentals in the Singapore mathematics curriculum that believes in "mathematics as an excellent vehicle for the development and improvement of students' intellectual competence in logical reasoning, spatial visualization, analysis and abstract thought and that students develop numeracy, reasoning, thinking and problem solving skills through the learning and application of mathematics" (MOE, 2007, p. 11). Hence, for the purpose of this chapter, I shall concentrate on reviewing research in various aspects relating to problem solving in mathematics.

2 PROBLEM SOLVING IN MATHEMATICS

The framework of the Singapore Curriculum since 1990 has embodied *Mathematical Problem Solving* in its core. Mathematical problem solving, as stated in the framework, includes using and applying mathematics in practical tasks, in real life problems and within mathematics. It advocates that problems should cover a wide range of situations from routine to non-routine mathematical challenges in unfamiliar context as well as open-ended investigations that require heuristics and thinking processes. According to Foong (2006), there have been many attempts in the literature to define the term "problem solving" in mathematics education. However, the key issue is still on how to go about finding a solution when faced with a problem that would involve the use of mathematical skills, concepts, and processes. Nevertheless, most researchers would agree that problem solving is not about doing repetitive exercises on basic mathematical concepts and skills that are commonly found in school mathematics workbooks. Review of the problem solving literature has identified some key areas that research in mathematical problem solving can be categorised such as student cognitive processes including metacognition; teacher role in promoting problem solving; problem-solving approaches; affect and assessment of problem solving. The present review attempts to cover these key areas under three broad strands: (a) problem-solving approaches and tasks, (b) teachers' beliefs and practices, and (c) students' problem-solving behaviours. Within each strand, related researches are described under various sub-headings.

2.1 *Problem-solving Approaches and Tasks*

Problem-solving approaches refer to different experiences that teachers choose to engage students' problem solving in the learning of mathematics. This section will examine the types and roles of tasks for different problem-solving approaches, cooperative learning, mathematics word problems, and the exploration of problem posing. Teaching through problems serves as a mean for students to construct mathematical concepts and to develop problem-solving skills. Problems lead students to use heuristics and thinking skills in deepening or applying the mathematical knowledge they possess towards solutions. Foong (2002) broadly classified most problems as "closed" or "open-ended" in structure. Closed problems would include content-specific routine multiple-step challenge problems as well as non-routine heuristic-based problems where there is usually a known answer. Open-ended problems are considered ill-structured problems for they often lack clear formulation as there maybe missing data or assumptions, and there is no fixed procedure that would guarantee a fixed answer. This type of problems is commonly associated with approaches such as mathematical investigations, project work, problem-based learning involving real-world situation and/or simply open-ended approach.

2.1.1 Problem-Solving Instructions

Wong and Lim-Teo (2002) examined the effectiveness of heuristics instruction on a class of 39 Primary 6 students' problem-solving processes. The study involved an intervention instructional programme that focused on teaching general strategies of problem solving and five specific heuristics: *draw a diagram, systematic listing, look for pattern, use tabulation* and *make a supposition.* Not unexpectedly, the findings showed that there was merit in teaching heuristics in mathematical problem solving as the performance of the students improved and that their choices of heuristics were associated with successful performance. It is interesting to note that among many mathematics educators, there are still debates as to the pros and cons of explicit instruction on specific heuristics targeted at certain types of non-routine problems. Some had criticised such instructional approaches as drilling the heuristics. In an approach that differed from direct instruction on heuristics, Ng (2006) experimented with an Ancient Chinese Mathematics Enrichment Programme on 177 Secondary 2 students. It was hoped that by introducing relevant elements of ancient Chinese mathematical classics with real-world problems, teachers would be able to foster creative problem-solving strategies naturally. In this pilot study, Ng found that the experimental group who had undertaken the course on Ancient Chinese Mathematics over a period of seven months did significantly better than the control group in the overall academic achievement in mathematics.

2.1.2 Open-ended Approaches

Some teachers had conducted and presented their M Ed research on open-ended approaches to promote students' problem solving and higher-order thinking in their mathematics classrooms. Chang (2005) explored how the open-ended approach advocated by Becker and Shimada (1997) could provide an environment for students to be engaged in higher-order thinking in mathematics. In this approach, an incomplete question was presented first. The lesson followed with using many correct answers to the given problem to provide experience in finding something new in the process. A class of 37 Secondary 2 boys participated in this study, and the topics were linear and quadratic graphs. These boys were able to demonstrate the skills of analysing, synthesising and evaluating in the lessons using the open-ended approach. On the whole, these students gave positive feedback towards the open-ended approach. They enjoyed learning through the approach and commented that they were challenged to think out of the box and engaged in higher-order thinking. In another study, Seoh (2002) conducted eight problem-solving sessions for 52 Secondary 5 Normal (Academic) stream students with a focus in helping these weaker and mathematics avoidance students to develop conceptual understanding and interpret knowledge critically through open-ended problems that were set in the real world context. Although students' ability to make inferences and critical thinking showed only slight improvement, the open-ended cooperative pair approach had enhanced the students' motivational orientations towards mathematics. They were more willing to engage in cooperative learning and more willing to help their weaker peers to improve their mathematics performance together.

At the primary level, Ng H.C. (2003) studied the potential benefits of using investigative tasks on 29 Primary 6 students from the Gifted Programme. One example of the tasks was for the students to unravel the general logic behind the construction of any $n \times n$ magic square. He found that such open-ended investigative experience promoted flexibility and inventiveness in student problem solving, and at the same time created opportunities for students to analyse data, explain their strategies, and share their mathematical perspectives. Gaps in the students' knowledge also surfaced during the sharing. In a conference paper, Chan (2005a) shared his experience of using contextualised open-ended mathematics problem tasks that had characteristics of simplified real-life situations. Working in small groups, the Primary 6 students made assumptions to the related problems and developed various options to help them make informed choices towards their goals and apply appropriate strategies of listing, categorising and graphing.

2.1.3 Cooperative Learning and Problem Solving

Many of the studies using open-ended approaches as reported above have used collaborative practices such as cooperative group or pair strategies to promote communication, reflection, metacognition, and reasoning between and among students while they engage in problem-solving experiences. Of special mention here are three studies that made cooperative learning a focus of their research. Tan (2002) designed a ten-week problem-based project for 70 Secondary 2 students (37 Express stream and 33 Normal (Academic) stream) to work on a business proposal in groups of four. For both groups of students, the group problem-based project work had a significant and positive effect on mathematics achievement. They improved in their teamwork skills, problem-solving abilities, communication skills, and social skills. Unlike the Express students, the Normal students believed that their self-directed learning skills had been enhanced after doing the project.

In the second study, Ho G.L. (2007) conducted in-depth observations on a group of four Secondary 3 students as they worked cooperatively with role assignments. She found that the students exhibited more metacognitive behaviours such as: ask for clarification, give suggestions, evaluate solutions, praise and encourage each other as they solved the problem in groups. In the last study, Chen (2001) paired 32 Primary 6 students in a cooperative learning environment to solve non-routine problems that required the different types of heuristics as recommended in the primary mathematics syllabus. Using a pre- and post-test method, the study found improvement in the use of a wider repertoire of heuristics by the students. In the initial stage, many of these students were not used to cooperating with each other in solving mathematical problems. Only later into the programme did they begin to open up and communicated with their partners in mathematical terms. This prompted the researcher to suggest more cooperative learning activities during normal classroom lessons to encourage communications among students who may otherwise just keep to themselves.

2.1.4 Problem Posing

In promoting problem solving in mathematics with the goal of developing students to be good problem solvers, it is also desirable that they become good problem posers. Thus, problem posing is another problem-solving experience that can promote engaged learning and enhance students' metacognitive process. Chua and Fan (2007) examined how a class of Secondary 3 Express students with different profiles of mathematical achievements responded to problem posing tasks in the learning of angles. An example of the tasks was: *List as many different questions that have an answer* $x = 60°$. High Ability students in the study were better problem posers, they could pose more complex problems in the goal specific situation and more solvable problems in the non-goal specific situation. However, the high number of unsolvable problems posed by all groups indicated that the students were not very adept in posing questions. This led the researchers to suggest that problem posing, like heuristics, should be taught explicitly as problem posing skills were not natural skills possessed by the students.

Problem posing was utilised as a tool to help lower primary students overcome language difficulties in solving word problems in a study by Yeap and Lee (2002). They identified 14 Primary 2 students who spoke mainly a language that was different from the instructional language (English) and conducted ten problem-posing lessons with them on place value concepts and arithmetic operations. Results were positive as these students made some improvement in basic skills, especially those that were language-related. They were able to use different semantic structures in posing their word problems. One of the benefits, although not reflected in the data analysis, was that the problem-posing activities had shifted the students' perception of mathematics as a "one-correct-answer" activity to a "many-possible-answers" activity.

Ho, Lee and Yeap (2001) also studied the relationship between students' mathematics achievement and their ability to pose problems. A cohort of 115 Primary 5 students was asked, for example, to pose a ratio problem sum to a pictorial stimulus. Half of the sample was unable to pose solvable problems prompting the researchers to pose a further question for research, namely, to identify the types of difficulties that students encountered in problem posing. Yeap's (2002) PhD thesis on the same research, including 330 Primary 3 students and 330 Primary 5 students, found that tasks with constraints and tasks that included a text were more difficult for students to pose solvable problems. This was particularly true for students in the lower grade level and students with low problem-solving ability. It was also found that tasks that were open and tasks that contained rich information tended to encourage students to pose semantically complex problems.

On the other hand, Quek (2002) inquired into the cognitive characteristics and contextual influences associated with the activity of mathematical problem posing among pre-service teachers. Through observing the participants as they engaged in problem posing, the researcher found that contextual and cognitive factors impinged on the construction of meaning and motive of the activity for particular individuals. These characterize the nature of mathematical problem posing in classroom practice as dynamic and continuously changing. Hence, the implication is for teachers to organise mathematical problem posing as an *activity system.*

2.2 *Mathematics Word Problems*

In Singapore, word problems have long constituted a major part of school mathematics both within the instructional programme as well as formal assessment. English is the main medium of instruction in Singapore schools, and students are required to solve mathematics word problems that are presented in this language. Research has indicated that both language and semantic structures play a part in determining students' performance in the solving of mathematics word problems.

Chan's (2005b) study aimed to determine if Primary 5 students' language proficiency and the rewording of mathematics word problems according to some semantic structures could affect their problem-solving process. With a repeated-measure counterbalanced approach, students sat for the Original Word Problem Test (OWT), and the Reworded Word Problem Test (RWT). Students' English proficiency did not affect performance in solving the word problems. However, certain rewording constructs like changing the chronological order of events and repositioning the givens in the structure of the word problems helped the low and average ability students. The students' failure to solve word problems was not due to their lack of arithmetic ability but to their inability to construct an appropriate problem representation as a result of the way the problem was structured. This finding recommends that when students have difficulty, teachers should consider rewording the problem semantic structure by rearranging the order of events according to real time or repositioning the given data in a sequential processing order to help weaker students form a clearer mental representation of the problem situation.

2.2.1 The Model Method

Particular to Singapore is a heuristic strategy called the "model method." It involves drawing bar diagrams to help students make sense of a word problem by organising the given information in a visual form that may lead them toward solutions. It is a powerful visual method for solving challenging multiple-step word problems, and it serves as a link to algebra. Ng and Lee (2005) asked 151 Primary 5 students to solve five arithmetic and algebraic word problems using the model method. Because of its visual nature, the models drawn by the students allowed the researchers to infer the nature of difficulties they had with word problems and to consider implications on the teaching of word problems in primary classrooms. It was found that students used the model method more successfully with arithmetic word problems than the algebraic word problems with two additive relationships which were non-homogenous. Students' errors in model drawing were clustered around relational phrases such as "more than," "less than," and "*n* times as many." When using the model drawing to solve problems involving fractions, the students showed a good command of how they visualised the problem.

In another study, Ng S.F. (2003) investigated whether learning the model method in primary mathematics had any effect on how students solve word problems after they had learned formal algebra in secondary mathematics. This study was motivated by a certain perception that some lower secondary students, although they had been taught algebra, continued to use the model method to solve problems. However, the 145 Secondary 2 Express students in her study preferred algebra to the model method, and they were able to identify the limitations of the model method in solving algebraic word problems. In one test item, students had difficulties identifying how unknowns were represented in the model method. This finding prompted the researcher to suggest that primary teachers, when teaching the model method, should be more explicit in emphasising that the rectangle bars are also used to represent the unknowns as presented in the word problems.

2.2.2 Suspension of Sense Making in Word Problems

A lack of sense making would develop if standard word problems at school are considered by students to be unrelated to reality. Foong and Koay (1997) designed eight pairs of items adapted from Verschaffel, De Corte and Lasure (1994). Each pair consisted of a standard word problem typically found in textbooks and a realistic word problem where the students need to consider the realities of the context of the problem statements. These were administered to 156 Secondary 1 (79 Normal and 77 Express stream) and 148 Secondary 2 students (73 Normal and 75 Express stream). The majority of the 300 students did not give realistic answers to the realistic word problems; they performed straightforward arithmetic operations without considering the realistic constraints of the problem situations. Consider this realistic word problem: *Ahmad's best time to run 100 m is 13 sec. How long will it take for him to run 800 m?* About 80% of the students did the straightforward operation and gave the expected answer. Only about 4% gave realistic answers for they were aware that it was impossible to run 800 m and 100 m at the same speed. Realistic answers given were: "*By the time he finishes then, he would be tired and will slow down*" and "*Ahmad's speed might not be constant as he has 800 m to run.*" The findings also indicate that attainment in mathematics did not appear to be related to the ability to activate real-world knowledge in solving arithmetic word problems for there was no difference in the performance between the Express and Normal stream students in this aspect.

This study was replicated by Chang (2004) who used a modified instrument to test 307 Primary 5 students from the EM1 and EM2 streams. Similar result was obtained, about 60% of the primary students solved the realistic problems without considering the realities of the context of the problem situation. Both studies indicate that where solving word problem sums are concerned, what is important to the students is to recognise familiar key words, to select the appropriate operation, and do some computations from the given data to produce an answer, no matter how unreasonable it might be. This kind of stereo-typed thinking hinders the development of problem solving and critical thinking skills which are given important emphasis in the Singapore school mathematics curriculum. Similar studies around the world (Verschaffel, Greer & De Corte, 2000) have also found this lack of sense making in students' arithmetic word problem solving.

Yeap and Abdul Ghani (2001) carried out a 6-month research on the instructional programme called *Think-Things-Through* to enhance students' sense making in mathematics word problems. The main purpose of this programme was to shift the students' attention from preoccupation with carrying out algorithms to choosing suitable techniques. The instruments used were parallel word problem tests where *standard* and *non-standard* problems were designed. Standard word problem can be solved by the direct application of procedures and algorithms, whereas non-standard problems require students to consider some contextual factors. About 400 Primary 3 students were involved. The project examined if they were more able to make sense when they solved word problems at the end of the programme and if they were able to transfer this ability to solve other problems not similar to those used in the programme. The students showed improvement in handling non-standard items, but they were not able to solve non-standard items that were not discussed in the programme. Thus, success in standard problems does not necessarily lead to engagement in sense making during problem solving.

2.3 *Looking Ahead*

The above review has examined various problem-solving approaches and the types of tasks for promoting them. Non-routine and routine problems such as mathematics word problems prevalent in the local mathematics syllabus, problem posing tasks as well as investigative open-ended tasks have been used in the studies described here. However, absent among the local researches is the promotion of *mathematical modeling* as a problem-solving experience. In the latest revised mathematics curriculum framework (Ministry of Education, 2007), mathematical modeling is newly included as an important process. Ang (2001, 2006) has recommended that the secondary mathematics curriculum should teach and integrate this process into relevant mathematics topics, and explicitly for the H3 Mathematics, a new syllabus at Junior College level. Hence, future research on problem solving should explore mathematical modeling as a formal process of formulating, representing, and solving real world problems, even starting at the primary level. Nevertheless, the review of existing studies points to the importance of examining the nature of the mathematical problem-solving tasks and attention to the classroom processes surrounding the tasks. Although many of the researches reviewed in this strand have reported positive results in students' performance while using the various problem-solving approaches, there are insufficient data on how the classroom processes are actually played out, especially when students are not engaged and the desirable outcomes of the problem-solving experiences escape these students. The mere presence of problem-solving tasks in the classroom might not automatically result in students' engagement in the experience. This implies that the teacher must create the suitable classroom environment for students to experience the process of problem solving. Teachers may have a misguided belief that if they teach students a set of procedures to solve mathematics problems, then problem solving is achieved. If teachers were to have this belief in a process-oriented problem-solving classroom, they will be reducing a high level problem to applying procedural skills. Hence, there is a need for further research into pedagogical beliefs and actions of teachers and the factors that bear upon them to broaden the investigations already initiated by the above studies.

3 TEACHERS' BELIEFS AND PRACTICES ABOUT PROBLEM SOLVING

The primary and lower secondary mathematics curricula for Singapore schools since the 1990's present a vision for the teaching and learning in mathematics that reflects changes from an emphasis on rote learning to meaningful understanding of concepts and from teaching by telling to engaged learning through mathematical problem-solving activities, group work, and communication in mathematics. It is important to recognise that the key figures responsible for changing the ways in which mathematics is taught and learned in the classroom are the teachers. How mathematics curriculum reform is implemented depends on teachers' conception of the mathematics they are teaching. With the adoption of problem solving as a goal of mathematics education, there has been in the later years of the problem solving bandwagon, a shift in focus of research from students' performance to the teachers: their beliefs and actual practices in the classroom. Research on teachers' beliefs and concerns has worked on the premise that to understand teaching from the teachers' perspectives, one has to understand the beliefs with which they define their work. The studies presented here explore the impact of teachers' beliefs and concerns on their classroom practices especially with regard to the implementation of approaches that incorporate problem solving and investigations. This section also examines the prevalence of problem-solving presentations in Singapore mathematics textbooks and its influence on classroom practices.

3.1 *Teachers' Beliefs, Concerns and Practices*

Foong's (1993) study focused on primary school teachers' pedagogical beliefs with respect to teaching and learning of pupil problem solving in arithmetic word problems. The process of representing a real-world or verbally posed problem is a fundamental problem-solving skill and it is a major objective of the primary mathematics curriculum. Fifty-five in-service primary and 66 pre-service primary teachers took part in a questionnaire survey. The instrument used was based on constructivism that students construct their own meanings of mathematical concepts and procedures when they are given opportunity to become actively engaged in learning. Three subscales were developed: (a) Scale 1: teacher's belief about students constructing their own knowledge in problem solving, (b) Scale 2: teacher's belief about how mathematical problem solving should be taught, and (c) Scale 3: teacher's belief that problem solving can facilitate students' understanding and computational skills. There was no significant difference in beliefs of the pre-service and in-service teachers in Scales 1 and 2. Both groups tended towards a traditional didactic perspective, they believed that students might be able to construct their own knowledge but in classroom teaching, teachers had to instruct students to receive knowledge. For Scale 3, they also agreed on the role of the teacher as a facilitator for mathematical problem solving. However, they disagreed on whether students were able to solve simple arithmetic word problems on their own before the mastery of number facts. These Singapore teachers still held traditional didactic view about being explicit in their instruction to get students to follow rules and algorithms even though they believed in their role as facilitators to help students solve problems. It would be a challenge for such teachers to implement a curriculum that requires new teaching approaches to promote student-centred, engaged learning in mathematical problem solving.

During the early 1990's transition period for Singapore teachers to adapt to the revised mathematics curriculum that had problem solving as its central focus, Foong, Yap and Koay (1996) conducted a survey on samples of 173 primary teachers and 116 secondary teachers. The purpose of the survey was to identify the stages of concerns and the types of constraints faced by teachers that were brought on by the changes in the curriculum. Among the concerns expressed by the teachers, many were directed at the curriculum's focus on problem solving and meaningful learning strategies. Teachers were concerned about their own ability to put across new concepts and skills for students without confusing them with too many methods. A third of the secondary teachers felt inadequately prepared to teach mathematical problem solving especially with non-routine problems that had many possible solutions based on the list of heuristics recommended in the syllabus. It would seem that the teachers' lack of confidence at that time could be due to over reliance on textbooks and the limited range of problems for higher-order thinking and use of mathematical problem-solving heuristics. The study aimed to provide directions for curriculum developers and teacher educators to design appropriate mathematics education programmes for teacher professional development to implement the reformed mathematics curricula into the 21st century.

In 2004, a study was conducted by Foong (2004) to examine sixty projects implemented by the mathematics departments of primary schools. The aim was to identify a pattern of exemplary practices which schools had attempted to supplement the curriculum with more engaging and relevant learning activities for students. Noteworthy were the thirteen schools that had implemented projects to make mathematical problem solving a feature in their enrichment programmes. Generally, there were two schemes, namely non-routine tasks and problems that required the "model method" as a strategy. To prepare teachers who were not too familiar in the use of the model drawing for the more challenging problems in the upper primary grades, workshop materials in the form of booklets for the various grades were developed and given to teachers. The motivation was to upgrade the teachers' skills so that more problem solving could then be incorporated into the classroom. For the non-routine problems requiring other heuristic strategies, teachers developed item banks of such questions and their solutions. In some schools, non-routine problems were given out during assembly periods as quizzes to be followed up by classroom teachers to help their pupils in sharing various strategies. In this way, teachers had to get involved in non-routine problem solving with their pupils. Other activities such as "Young Mathematician Awards," "A-Problem-A-Day," term quizzes, and on-line quizzes were opportunities for schools to expose non-routine challenging questions to teachers and pupils. This study showed that collaborations among teachers with a common goal can make curricula reform a reality.

3.2 *Teachers' Enactment of Mathematical Problem Solving in the Classroom*

Ho K.F.'s (2007) PhD research inquired into what and how teachers' classroom practices enacted mathematical problem solving. He drew upon the case studies of four Primary 5 teachers, using Activity Theory to frame his study. Video recording and observation of lessons were analysed with a coding scheme, and these provided a glimpse of what and how the teachers enacted the curriculum. Generally, the teachers began a topic with the teaching of concepts and skills, and then moved on to word problems. The four teachers did not use the whole range of problem types and heuristics as suggested in the curriculum. Instead, mainly routine problems were used and the "use a diagram or model" heuristic predominated. The study also found some variations in the ways teachers mediated word problems. One teacher, who spent the most time in the heuristics-instruction category, had a traditional approach of teacher-led talk predominated with the Initiate-Respond-Evaluate (IRE) structure. Another teacher had a similar approach but talked about problem solving more through the *action* of going over assigned work. The third had a balance between teacher-led instruction and student presentation and discussion of their solutions. The fourth had a traditional IRE approach but included the use of some non-routine problems and group work.

Ho and Hedberg (2005) followed up this research by designing an intervention programme to raise the awareness of these primary teachers' mathematical problem-solving ideas and processes as well as how to teach problem solving in their classrooms. The research was directed by these questions: Given a reflective intervention, emphasising mathematical problem solving instruction, did teachers change teaching strategies and did it result in increased students' problem solving success? It was found that the traditional pattern of teaching a topic, explaining concepts and giving exercises to practice related skills, still prevailed after the intervention. One teacher stopped using group work and reverted back to whole class teaching. His reason was that his class was slipping in their regular tests and the group work had not benefited them. This teacher's concern reflected the beliefs many teachers shared, that is, it is more important to prepare students to do well in tests than to implement problem-solving lesson.

On a positive note, the teachers in the study had at least started to question their own views about problem solving and how it could be emphasised in their mathematics teaching. All the four teachers used more non-standard problems in class after the intervention as they realised that they had been providing limited experience for their students. Two of the teachers made qualitative changes to their teaching. The overall results seem to suggest that the intervention strategy was effective, although the benefits were small. The findings of this and Ho K.F.'s (2007) study should inform teacher educators, resource and curriculum developers, and policy makers, about pedagogical possibilities that could transform classroom practices.

Kay (2003) explored whether the belief system of four primary mathematics teachers aligned with their instructional practices. Data collection comprised structured interviews and two lesson observations of each teacher. During the interviews, each teacher did a concept mapping activity to illustrate the factors influencing their beliefs and instructional methods. The beliefs and practices fell into three categories: "*Traditional*" where students learned passively through individual work set up in textbooks and worksheet; "*Non-Traditional*" where learning is active through explorations, problem solving, and cooperative group, and "*Even Mix*" of traditional and non-traditional. It was found that there was no obvious difference in the new and experienced teachers' beliefs about the nature of learning and teaching of mathematics. Generally, the four teachers' instructional practices were not aligned to their beliefs. An important finding to take note of was that all the teachers stressed that helping students to succeed in doing worksheet was still important in an examination-oriented curriculum. This point was shared by other studies such as Ho and Hedberg (2005) that surfaced the teacher's dilemma of integrating problem solving in lessons and preparing students for tests and national examinations.

Foong (2005) described three cases of how primary teachers implemented open-ended problem solving activities with varying degrees of success. The objective of the study was to elicit from these experiences the classroom-based factors that could support or inhibit higher-order thinking and creativity in the students' work. Data for this study were drawn from the case reports of the teachers. The teachers reported on how they set up and implemented the task. They had to observe students at work, analyse their written work, and take notes of students' behaviours and cognitive processes. Only one teacher was able to successfully implement the lesson: maintain students' engagement in higher-order thinking and sustain pressure on students to provide meaning, explanation and justification to demonstrate their understanding of the mathematics embedded in the task. For the other two teachers who used the same task, their students' engagement in problem solving declined to lower levels of processing. One was too procedural in her instruction as she believed that asking leading questions would help students to think through the steps and be systematic in their approach. The other teacher had limited understanding of the mathematical thinking embedded in the task and the kind of cognitive demands to be made of the students. Her students were not held accountable for the required high cognitive processes as she accepted all the unsystematic trials of her students.

These three classroom experiences have implications for the role of the teacher in implementing open-ended tasks. Teachers must make a paradigm shift towards a more process-based approach where getting a correct answer to a problem is not the main criterion for success.

4 SINGAPORE MATHEMATICS TEXTBOOKS

Teachers in Singapore had indicated that the mathematics textbooks were their major print resource for teaching by ways of illustrations of worked examples, assignment of exercises for class seatwork, and homework (Fan & Zhu, 2000; Kaur, Low & Seah, 2006). Most teachers had confidence in the recommended textbooks they were using because these had been approved by the Ministry of Education. The findings in the Kaur, Low, and Seah (2006) pointed to the essential role of the textbook from the perspectives of both the teachers and students, namely textbook was a mean to concretise the curriculum. The core of the curriculum has its focus on mathematical problem solving for the development of concepts, skills, processes, metacognition, and attitudes. So what and how are the school mathematics textbooks representing these ideas in their contents for teachers to implement the intended curriculum?

Two studies were found to examine the Singapore primary and secondary textbooks. Fan and Zhu (2000) examined how two widely used Singaporean school mathematics textbooks at lower secondary level represented problem solving, and Ng (2002) investigated the "Primary Mathematics Project" series for the primary level. They developed similar frameworks for classifying problem types and problem-solving procedures. Each study used a different coding scheme to code and analyse data from the textbooks to obtain a profile of the representation of problem solving in these textbooks.

For the secondary series, Fan and Zhu (2000) found that the textbooks presented a good foundation for students to develop their abilities in problem solving. Its strength was developing students' logical and higher-order thinking skills through use of multiple-step and challenging problems, exposing students to a variety of heuristics, and leading students to new concepts and algorithms through problem solving. The study also recommended that more non-routine problems of various types from open-ended projects to authentic real-life problems should be incorporated into future series.

The primary series examined by Ng (2002) had been used by all primary schools until its phasing out in 2005. In that series, 41.5% of the problems were application problems. Ninety-nine percent of the problems were routine problems, while 2.6% of the problems were open-ended problems and 72.6% of the problems were single-step problems. In the light of these findings, there should be a representation of the full range of the heuristics listed in the mathematics syllabus so students would have the benefit of learning how to apply all the different heuristics.

4.1 *International Comparison of Textbooks*

When comparing problem solving in the Singapore textbooks with that in China and US, Fan and Zhu (2007) noted that the secondary textbook series they examined in these three countries had displayed strengths in the modeling of various problem-solving procedures and using a wide range of specific heuristics. At a deeper analysis, they found some weaknesses in the Singapore series. There were inadequate demonstrations of the heuristics: "thinking of a related problem" and "looking back," which the syllabus placed great emphasis on in developing students' ability in metacognition. The results also showed that Singapore textbook authors might have a more formalistic view of problem solving for they tended to equate problem solving with a list of heuristics, and in the textbooks, specific problem-solving heuristics were introduced separately from other topics. This could have negative impacts on students' learning of problem solving for it might instill in them that problem solving is drilling of heuristics through practices targeted at certain types of non-routine problems. Ideally, problem solving should be integrated into regular mathematics teaching; this was evident in the Chinese textbooks. According to Fan and Zhu, Chinese standards introduced heuristics in implicit ways with explicit labeling of the general problem-solving stages to allow the problem solvers to distinguish and identify the different stages in the problem solving process. The researchers viewed this as an advantage in integrating the problem-solving experiences of students in regular mathematics lessons.

Findings from studies of mathematics textbooks can certainly provide much food for thoughts to curriculum developers and textbook authors to understand the possible ways to improve the problem-solving representations in school textbooks. Since textbooks are key component of the intended curriculum, their influence on teachers who use them as tools to promote and develop desirable mathematical problem-solving behaviours in students cannot be underestimated.

5 STUDENTS' PROBLEM-SOLVING BEHAVIOURS

Students' problem-solving behaviours generally refer to the cognitive and metacognitve processes as well as affective manifestations in various problem-solving experiences. As students explore non-routine and challenging questions, they would employ a range of heuristic strategies, including developing the process of abstraction, analysis, reasoning and generalising, and attitudes towards the problem-solving experiences. Metacognition comes into play when the problem solvers become aware of their own problem-solving process and take control of this process by monitoring and self-regulating their own thinking and learning. This section describes research that focuses on cognitive-metacognitive processes in problem solving, problem solving tasks as alternative assessment for general mathematical ability, and the relationship of problem-solving ability with attitude towards mathematics learning.

5.1 *Cognitive and Metacognitive Behaviours in Problem Solving*

According to Chong et al. (1991), the study by Foong (1990) was considered a pioneering research on problem solving in Singapore and the topic was then relevant and timely in view of the new aims of mathematics education which were set for the 1990s. Using information processing theory and protocol analysis methodology, the study developed a framework from thinking aloud data that can analyse the process of mathematical problem solving in terms of definable behaviours. This framework (Foong, 1992a) identified 28 behaviours which include cognitive, metacognitive, and affective behaviours manifested in individuals during problem solving. Such taxonomy of problem solving behaviors would be helpful in providing a common framework in research for the evaluation of problem solving performance through a scoring scheme. Foong had used the framework to identify differences in the processes between successful and unsuccessful problem solvers as they worked on non-routine process problems (Foong, 1994). The unsuccessful solvers were not able to apply their mathematical knowledge in the non-routine problems and they lacked knowledge of useful strategies in tackling unfamiliar tasks. Negative emotional expressions such as frustration and confusion were frequently expressed by these unsuccessful solvers. On the other hand, successful solvers planned their solutions in greater details and used more metacognitive processes which were task directed, showing greater awareness of their solution paths and where they should be going in the process.

A modified version of the Foong's taxonomy was used by Yeap (1997) to identify the types of metacognitive behaviours that secondary students manifested in his study. He identified five types of metacognition among the students: (a) State a plan of action, (b) Clarify task requirements, (c) Review progress, (d) Recognise errors, and (e) Detecte new development. Another problem-solving process framework was developed by Yeo (2006, 2007) to analyse the mathematical problem-solving behaviours of 621 Secondary 2 students. In this study, the students were asked to complete in writing the statement "When I am given a mathematics problem to solve, this is what I do….." The data were coded under the four phases of problem solving: Understand/Represent the Problem, Find a Way to Solve the Problem, Solve the Problem, and Check the Solution. The Secondary 2 students were found to rely on their own individual problem-solving frameworks.

To determine the types of heuristics that students are likely to use successfully to solve various types of problems, Wong and Tiong (2006) examined the written solution by a sample of Primary 5 and Secondary 1 students. They used five types of heuristics: systematic listing, guess and check, equations, logical argument, and diagrams. However, the heuristics used were dependent on the nature of the problem and prior experience of the students. Kaur (1995) used a double interview technique that included the *Newman Error Analysis Guideline* to study 139 students who had difficulties in solving word problems. The interview data revealed that it was not necessarily true that unsuccessful problem solvers lacked relevant mathematical skills. Students were not successful at arriving at the solution for various reasons, such as lack of comprehension of the question posed, lack of schema and strategic knowledge, and inability to translate the problem into mathematical form.

5.1.1 Disposition and Metacognition

Quek et al. (2002) interviewed eight selected high academic ability students from a cohort of 130 Integrated Programme secondary and Junior College (JC) students to identify their dispositions towards and skills in mathematical problem solving. The students had undergone a specially designed programme that aimed at sensitising students to the different dimensions (*Resources, Heuristics, Control,* and *Beliefs*) of mathematical problem solving proposed by Schoenfeld (1985), and went through the Polya's phases of solving problems. Video-taped interviews were analysed. Students who showed success in problem solving had all the four components, *Resources, Heuristics, Control,* and *Beliefs,* working hand in hand to support students' ability to solve a mathematics problem. Students poor in any one of the components were adversely affected in the problem solving process. Overall, there seemed to be a slight dependence on *Resources*, indicating to teachers that they cannot neglect it during the planning of curriculum and the teaching of mathematics.

In another study on JC students, Teo (2006) examined the effects of metacognition and beliefs as students worked on Sequence and Series problems. She recommended that students be exposed to metacognitive strategies in the classroom so as to train them to be self-directed learners. Results from her study indicated that inefficient control behaviour often caused the students to rush into solving the problems without much analysis. Students with good control and who were efficient in accessing their resources demonstrated their competence in approaching problems.

Wong (2007) examined metacognitive awareness among a group of Primary 5 and Secondary 1 students. A pre- and post-test questionnaire design on metacognition was used, asking students about what they did during problem solving and their levels of enjoyment and confidence while solving problems. The students' responses were fairly general, and lacking in deep awareness of personal metacognition. This prompted the researcher to recommend clinical interviews as method of data collection on metacognition. Through such method, rich data can be gathered to uncover the complexity and dynamic nature of metacognition to better inform educators for the appropriate training of students and teachers in the metacognitive process.

5.1.2 Technology as a Support in Problem Solving

Research has also focused on technology as an aid to provide opportunity for students to interact with a new range of problem-solving situations.

Ahmad Ibrahim (2002) conducted a study with 45 Primary 6 students using a computer-supported collaborative problem-solving program called *Knowledge Forum*. The students accessed a web-stream video to help them plan the budget for a class outing and come up with a transportation proposal. Students who used the computer support in a scaffold communication environment tended to perform better than those in a control group.

Hung (2001) also used a computer-mediated collaborative problems-solving environment to track over a period of time how two students engaged in mathematical problem solving. Using a distributed networked environment, each student constructed mathematical symbols through a shared whiteboard with a chat box facility provided by the system. Using this technology, the researcher was able to describe in-depth the problem-solving process between the two students as they conjectured, negotiated, reflected, and appropriated mathematical meanings in order to solve the given problem.

Teong (2002) demonstrated that a computer environment can benefit low achievers' mathematics word problem solving coupled with explicit metacognitive training. Eight students aged 11 to 12 years, from two primary schools were grouped into four pairs and assigned to two groups. One group had explicit metacognitive training before solving word problems with the software WordMath (treatment), while the other group solved word problems with WordMath without metacognitve training (control). A framework modified from Artzt and Armour-Thomas (1992) was used to analyse these lower achievers' think aloud protocols. The treatment group benefited from the metacognitive training as they were more aware of their cognitive processes during word problem solving and more successful in their performance as compared to the control group.

Leong and Lim-Teo (2003) reported the effects of Geometer's Sketchpad on spatial ability and achievement in transformation geometry among Secondary 2 students. The software was employed differently in three classes. In Class A, the approach adopted by the teacher was that of guided-inquiry where students explored concepts and made conjectures with extensive hands-on experience with the software. In Class C, the teacher's predominant role was that of an expositor and the software was used to demonstrate dynamically the properties of transformations. In Class B, the "in-between" class, the teacher adopted a guided inquiry in whole class discourse but with the teacher manipulating the objects on the projected screen as directed by the students. The results showed that spatial ability improved for all three classes with no significant difference between the classes, although Classes A and B performed significantly better than Class C in the transformation achievement test.

5.2 *Problem Solving as Assessment Tools*

Curriculum and assessment in mathematics education in Singapore are moving away from emphasis on mastery of content to one that will give students more opportunity to acquire thinking and problem-solving skills (Ministry of Education, 2004). There is a growing trend in schools to use problem-solving activities as alternative assessment tools to enable teachers to obtain in-depth and more qualitative information about students' cognitive and metacognitive processes in, as well as their attitudes toward, mathematics learning. Research has found that problem-solving ability can provide a reliable indicator of general mathematical ability. Many of the studies cited in this chapter are small scale and classroom based. The researchers explored students' responses to problem solving assessment items and in some cases related these to responses on traditional test items, and at times correlated them with other measures of students' attitude and motivation. The predominant research procedure of the studies reviewed here is based on task assessment. The nature of the tasks ranges from mathematics word problems to problem-based performance tasks. These test situations are often accompanied by interviews of a smaller sample to delve deeper into the students' thinking. For the more open-response performance tasks, rubrics are usually designed to score students' work. Pre- and post-tests data collection method is used in many of the studies to examine learner effects and effectiveness of certain pedagogy which the researchers have implemented.

5.2.1 Alternative Assessment

The following studies exemplify classroom research that attempt to use problem-solving tasks as alternative assessment. Rajaram (1997) designed a Problem-Solving Achievement test using non-routine tasks developed by the Shell Centre of Mathematics at Nottingham University in UK and an analytic scoring rubric to examine the differences in performance and choice of heuristics among 73 Secondary 3 Gifted students in his school. There was a range of problem-solving ability levels within even this gifted group of students. This has significant implication for teachers as to how mathematical problem solving should be taught to suit the differing types of students within the Gifted programme.

Teo (2003) explored an approach to assess constructivist learning in mathematics. The performance of the students was based on their ability to pose, formulate, and solve problems. One hundred and sixty-eight students from six intact classes participated and were grouped according to their elective study: Science, Technical, and Arts. Elective of study had a significant influence only on students' problem-posing ability which was in turn related to their mathematics ability. Surprisingly, problem posing ability had no significant relationship with problem formulating and problem-solving ability.

Chow (2004) found that there was an advantage in using open-ended problems as a form of alternative assessment. In her study involving Secondary 1 students, an experimental group was exposed to open-ended problem solving as an alternative assessment, while a control group was not given any exposure to open-ended problem solving. There was significant difference in the experimental group's overall performance with improvement particularly in cognitive areas like mathematical knowledge, strategic knowledge, and mathematical communication. From interviews with the students, it was found that the experimental group was quite receptive and positive towards open-ended problem solving.

Foo (2007) investigated the use of performance tasks on students' attitudes towards the learning of mathematics, problem-solving abilities, and its influence in the daily teaching and learning of mathematics. Forty students in one class served as the experimental group, while another class served as a control group. The overall findings suggest a positive influence of the interventions on the cognitive and affective domains of the students. Besides improving problem-solving skills, the authentic context of the performance tasks helped students to relate mathematics to real-life situations. On the whole, the teachers and students gave positive feedback about the integration of this new assessment strategy.

On the other hand, a team of NIE lecturers (Dong, Lee, Tay & Toh, 2002) were interested to find out if JC students were ready for the kind of non-routine problems that were exemplified in the revised syllabus for the *GCE Advanced Level Mathematics 9233* that took effect in 2002. They developed a 60-minute test consisting of six questions, with each question designed with a routine part and a non-routine part. Two intact classes from each of four Junior Colleges in Singapore took the test. There was considerable difference in the student achievement between the routine and the non-routine problems. In-depth analysis found that students were lacking in important skills needed for problem solving of the A level non-routine questions such as not being able to consider implied cases, to extend learnt techniques, and to check the conditions of the problem. Such results were indicative of the lack of readiness of JC students in general for the revised Mathematics A Level paper. JC teachers and school-based curriculum team would need to take note of this shortcoming and address the problem.

5.2.2 Mathematics Achievement and Problem-Solving Ability

Liu (2003) investigated the relationship between the academic mathematics achievement and problem-solving abilities of 124 Primary 5 children from a single school. Five non-routine problem items were used to assess problem-solving abilities, and the Semestral Assessment 1(SA1) results were used as a benchmark for the academic mathematics achievement. The study found only a weak correlation between achievement in academic mathematics and success in solving problems. In a different study, Tan (2005) gathered evidence to show that problem-based tasks crafted according to a process-oriented table of specification designed for her study could be used to assess primary students' higher-order thinking skills in mathematical problem solving. Three problem-based tasks were administered to 39 mixed-ability class of Primary 4 students. A scoring rubric with assessment profile was developed to evaluate students' responses to the problem-based tasks. Unlike Liu (2003), this study found that the mathematics achievement test scores were positively correlated with performance in the problem-based tasks. Students were also very positive to the use of problem-based tasks in mathematics lessons, prompting the researcher to recommend such problem-solving activity as formative assessment of students' higher-order thinking.

On a larger scale assessment of primary students' problem-solving ability, Ang, Boo and Toh (2005) assessed 540 students from 14 intact classes of five primary schools. Their objective was to find out Singapore students' strengths and weaknesses in approaching hands-on problems, also known as performance-based tasks, that were replicated from the TIMSS-95 Performance Assessment sub-study. Since Singapore did not take part in this TIMSS sub-study, the local study also aimed to use the results to make some comparison with the international sample. Overall the 540 Singapore students performed significantly better than the international sample. However, the Singapore Primary 4 students encountered varying degrees of difficulties such as in communicating their written answers to others; in making generalisations as well as recognising patterns beyond the observable.

5.2.3 Attitude and Problem-Solving Ability

Researchers have attempted to correlate problem-solving ability with constructs in the affective domain such as attitude, anxiety and even gender. For example, Foong and Loo (1998) investigated the different levels of reasoning in mathematical problem solving between two compatible groups of 147 male and 144 female Secondary 4 students from two non-coed schools. The Collis-Romberg Problem Solving Profiles (Collis, Romberg & Jurdak, 1987) was used to assess a student's level of problem-solving ability according to the four SOLO taxonomy levels of reasoning: *Unistructural, Multistructural, Relational,* and

Extended Abstract. The male students had a higher level of reasoning and performed better than the female students in all the content areas. The female students lacked reasoning skills to integrate indirect information, make a hypothesis, and generalise for more complex problems. The study has important implications for the teaching of higher order thinking in mathematics for the weaker students, especially females.

Mathivanan (1992) examined the relationship between the various constructs of attitude towards mathematics (using the Mathematics Attitude Scale developed by Sandman), mathematical achievement, and problem-solving achievement (a problem-solving test developed by the researcher). One hundred upper secondary male students participated in the study. Stepwise multiple regression analysis indicated that mathematical achievement and self-concept in mathematics were good predictors of problem-solving achievement. There was however, negligible relationship between anxiety in mathematics and problem-solving achievement. In a similar study, Yeo (2005) found marginal linear relationship between mathematics anxiety and performance on non-routine mathematical problem-solving test among 621 Secondary 2 students.

These findings on relationship between mathematics anxiety and problem solving performance based on self-report surveys could be contradicted by a study that used more qualitative data analysis of students' think aloud protocols while they solved non-routine problems as reported in Foong (1992b). She found that negative emotional responses from unsuccessful problem solvers tended to distract them in the process. Once they perceived the problem as difficult, anxiety and confusion arose and when their attempt to move on did not produce results, they became frustrated and showed lack of self-confidence. They either continued in a confused manner or gave up, thus avoiding further anxiety.

More qualitative studies could be conducted to look into students' state of feelings and anxiety while they are solving mathematical problems. Hence, teachers should not be interested solely in knowing if students can get the solution or not, but they should also promote positive emotions towards problem solving.

6 CONCLUSIONS

For the past one and a half decades, more than sixty local researches on various issues relating to mathematical problem solving have surfaced and are reviewed in this chapter. This corpus of research can be considered a major development not only in terms of significance to the Singapore curriculum but also in term of quantity, even though I cannot claim that this review is exhaustive. It is hoped that useful knowledge from this review can seed more research by teachers, teacher educators, and researchers to continue to improve teaching and learning in mathematics. What is left in this section is for me to draw together some of the key issues arising from the findings and highlight recommendations for future research relating to mathematical problem solving.

The attempts by researchers who are mostly practitioners in the classroom to use problem-solving approaches to teach mathematics from primary to JC level appear to have produced some positive outcomes. Teachers and researchers believe that it is necessary to reconstruct the culture of mathematics teaching through increasing their repertoire of instructional strategies for more student-centred learning with non-routine and open-ended as well as problem-based tasks. It suggests that teaching via problem solving is a feasible alternative to the traditional didactic style of whole class discourse that is much controlled by the teachers as knowledge givers. Hence, there have been several researches carried out for this purpose.

When the research is focused on what and how student learn during the various problem-solving experiences, again positive results are reported of those students who become more strategic, reflective, and communicative in and about their processes and interactions with their peers. Students are more able in using a variety of heuristics and with increased metacognition. These researches experimenting with mathematical problem solving as alternative approaches or assessment are mostly isolated small scale using intact classes and exploratory. But taken collectively, it should give encouragement and motivation for more research to be done in the classrooms within a school or within a cluster to convince more teachers of its multi-faceted values in changing teaching culture and maximising students' potential to be independent thinkers and problem solvers.

However, when researchers go into the classrooms to study the teachers by interviewing and observing their current practices for match with the intended curriculum goals, there emerges a more traditional picture of teachers teaching through exposition, followed by students practicing routine exercises or problem-solving procedures. Does this mean that there is no teacher practice of problem-solving approaches as advocated by the intended curriculum? At a more micro level of observing classrooms in the Learner's Perspective Study by Kaur et al. (2006), it points to the fact that when the teachers use textbook resources that are aligned with the intended curriculum, evidence of problem solving has been found in teachers giving multi-step challenging problems and concrete illustrations that demonstrate how abstract mathematical concepts are used to solve problems from different perspectives.

Many teachers in Singapore believe that they do teach "problem solving" in mathematics all the time, especially through mathematics word problems and the use of the "model method" as a heuristic for challenge sums. This interpretation of problem solving can be viewed from Schroeder and Lester's (1989) conception of the different roles of problem solving: teaching *for* problem solving, teaching *about* problem solving, and teaching *via* problem solving. According to Foong (2002), teachers in Singapore from primary to secondary level used problems for its role in teaching *for* problem solving, where the emphasis is on learning mathematics for the main purpose of applying it to solve problems after learning a particular topic. The success of Singapore students in the TIMSS studies could be due to this practice. Explicit heuristics instructions as teaching *about* problem solving are being practiced in the mathematics classroom where emphasis is on teaching the heuristics recommended in the syllabus and targeting them at certain non-routine problems.

Future research will need to provide teachers a clearer picture of these roles of problem solving, especially teaching mathematics *via* problem solving and how they can be incorporated into the regular mathematics classroom rather than as a one-off exercise. Researchers, who use problem-based investigative and open-ended tasks as discussed in this review, are adopting the concept of teaching *via* problem solving. They have shown the effectiveness of such approaches and the positive impact on student learning. But these researchers have also reported on constraints in implementing them into the current structure of the mathematics curriculum and assessment requirement. Teachers perceived teaching *via* problem solving as more time consuming to implement. It takes more time to evaluate students' solutions for problem-solving tasks than the usual worksheet questions that teachers would normally give. Further, there is this overarching need to finish the syllabus content within limited curriculum time to prepare students for high stake achievement assessment. Teachers constantly encounter this dilemma of integrating problem solving in lessons as they believe it is worthwhile and yet there is pressure to produce results for achievements in tests and national examinations.

Policy makers and school leaders need to focus on this teacher dilemma to give the necessary supports for teachers to truly implement the intended curriculum. And this is where more future research is needed to identify the various factors across domains of time, syllabus, teacher development, student engagement, assessment modes, and environment that can hinder or enhance integrating teaching via problem solving in the mathematics classrooms and how these can be tackled within the structure of the education system. In particular, the full range of teachers' beliefs, pedagogical content knowledge and enactment in the social-cultural aspects of the classrooms need to be explored through more qualitative participant observation kind of research. Last but not least, not more but improved high quality teacher professional development is required. There is a need to explore different forms of in-service teacher development such as the use of Lesson Study method in which teachers learn in a community of practice, to complement the usual menu of courses and workshop offerings. Presently with the setting up of the Centre for Research in Pedagogy and Practice (CRPP) four years ago in the National Institute of Education, teams of researchers and teachers with different perspectives have been working together to develop new research questions and methodologies. This can be the direction for future research into this important area in mathematics education.

References

Ahmad Ibrahim, E. (2002). *Computer-supported collaborative problem solving and anchored instruction in a mathematics classroom: An exploratory study.* Unpublished Master's dissertation, National Institute of Education, Nanyang Technological University, Singapore.

Ang, K.C. (2001). Teaching mathematical modeling in Singapore schools. *The Mathematics Educator, 6*(1), 62-75.

Ang, K.C. (2006). Mathematical modeling: Technology and H3 mathematics. *The Mathematics Educator, 9*(2), 33-47.

Ang, K.C., Boo, H.K., & Toh, K.A. (2005). Assessing mathematics learning outcomes through performance-based tasks. *The Mathematics Educator, 8*(2), 19-33.

Artzt, A.F., & Armour-Thomas, E. (1992). Development of a cognitive-metacognitive framework for protocol analysis of mathematical problem solving in small groups. *Cognition and Instruction, 9,* 137-175.

Becker, J.P., & Shimada, S. (1997). *The open-ended approach: A new proposal for teaching mathematics.* Reston, VA: National Council of Teachers of Mathematics.

Chan, C.M.E. (2005a, August). *Engaging students in open-ended mathematics problem tasks: A sharing on teachers' production and classroom experience (Primary)* Paper presented at the third ICMI-East Asia Regional Conference on Mathematics Education Conference, East China Normal University, Shanghai, China.

Chan, C.M.E. (2005b). Language proficiency and rewording of semantic structures in P5 pupils' mathematical word problem solving. *The Mathematics Educator, 9*(1), 84-99.

Chang, C.Y. (2005). *An open-ended approach to promote higher order thinking in mathematics among secondary two express students.* Unpublished M.Ed. dissertation, National Institute of Education, Nanyang Technological University, Singapore.

Chang, S.H. (2004). *Sense-making in solving arithmetic word problems among Singapore primary school students.* Unpublished M.Ed. dissertation, National Institute of Education, Nanyang Technological University, Singapore.

Chen, A.N. (2001). *Cooperative learning to enhance mathematical problem solving performance among primary six students.* Unpublished M.Ed. dissertation, National Institute of Education, Nanyang Technological University, Singapore.

Chong, T.H., Khoo P.S., Foong, P.Y., Kaur, B., & Lim-Teo, S.K. (1991). *A state-of-the-art review on mathematics education in Singapore.* Singapore: Institute of Education.

Chow, Y.P. (2004). *Impact of open-ended problem solving as an alternative assessment on Secondary One Mathematics students.* Unpublished M.Ed. dissertation, National Institute of Education, Nanyang Technological University, Singapore

Chua, P.H., & Fan, L.H. (2007, June). *Mathematical problem posing characteristics of Secondary 3 Express students in Singapore.* Paper presented at the fourth East Asia Regional Conference on Mathematics Education Conference Conference, Universiti Sains Malaysia.

Collis, K.F., Romberg, T.A., & Jurdak, M.E. (1986). A technique for assessing mathematical problem-solving ability. *Journal for Research in Mathematics Education, 17*(3), 206-221.

Dong, F., Lee, T.Y., Tay, E.G., & Toh, T.L. (2002). Performance of Singapore Junior College students on some nonroutine problems. In D. Edge & B.H. Yeap (Eds.), *Proceedings of Second East Asia Regional Conference on Mathematics Education & Ninth Southeast Asian Conference on Mathematics Education* (pp. 71-77). Singapore: Association of Mathematics Educators.

Fan, L., & Zhu, Y. (2000). Problem solving in Singapore secondary mathematics textbooks. *The Mathematics Educator, 5*(1/2), 117-141.

Fan, L., & Zhu, Y. (2007). Representation of problem-solving procedures: A comparative look at China, Singapore, and US mathematics textbooks. *Educational Studies in Mathematics, 66*, 61-75.

Foo, K.F. (2007). *Integrating performance tasks in the secondary mathematics classroom: An empirical study.* Unpublished M.Ed. dissertation, National Institute of Education, Nanyang Technological University, Singapore.

Foong, P.Y. (1990). *A metacognitive heuristic approach to mathematical problem solving.* Unpublished doctoral dissertation, Monash University, Australia.

Foong, P.Y. (1992a). Development of a framework for analysing mathematical problem-solving behaviors. *Singapore Journal of Education, 13*(1), 61-75.

Foong, P.Y. (1992b). Mathematical problem solving behavioral traits. *Proceedings of the 6th Annual Conference of the Educational Research Association, Singapore,* 175-184.

Foong, P.Y. (1993). Teachers' beliefs in a constructivist approach to teaching mathematics in Singapore primary school. *Proceedings of the Sixth South East Asia Conference on Mathematics Education (SEACME 6) and the Seventh National Conference on Mathematics* (pp. 433-442). Indonesia

Foong, P.Y. (1994). Differences in the processes of solving mathematical problems between successful and unsuccessful solvers. *Teaching and Learning, 14*(2), 61-72.

Foong, P.Y. (2002). Roles of problem to enhance pedagogical practices in the Singapore classrooms. *The Mathematics Educator, 6*(2), 15-31.

Foong, P.Y. (2004). Engaging mathematics curriculum: Some exemplary practices in Singapore primary schools. *Teaching and Learning, 25*(1), 115-126.

Foong, P. Y. (2005). Developing creativity in the Singapore mathematics classroom. *Thinking Classroom, 6*(4), 14-20.

Foong, P. Y. (2006). Problem solving in mathematics. In P.Y. Lee (Ed.), *Teaching primary school mathematics: A resource book* (pp. 54 - 81). Singapore: McGraw-Hill.

Foong, P.Y. (2007). Teacher as researcher: A review on mathematics education research of Singapore teachers. *The Mathematics Educator, 10*(1), 3-20.

Foong, P.Y., & Koay, P. L. (1997). School word problems and stereotyped thinking. *Teaching and Learning, 8*(1), 73-82.

Foong, P.Y., & Loo, C.F. (1998). Mathematical reasoning: Do males and females think alike? In M.L. Quah & W.K. Ho (Eds.), *Thinking processes-going beyond the surface curriculum* (pp. 198- 204). Indonesia: Simon & Schuster (Asia).

Foong, P.Y., Yap, S.F., & Koay, P.L. (1996). Teachers' concerns about the revised mathematics curriculum. *The Mathematics Educator, 1*(1), 99-110.

Ho, G.L. (2007). A cooperative learning programme to enhance mathematical problem solving performance among secondary three students. *The Mathematics Educator, 10*(1), 59-80.

Ho, K.F. (2007). *Enactment of Singapore's mathematical problem-solving curriculum in Primary 5 classrooms: Case studies of four teachers' practices.* Unpublished PhD thesis, National Institute of Education, Nanyang Technological University, Singapore.

Ho, K.F., & Hedberg, J.G. (2005). Teachers' pedagogies and their impact on students' mathematical problem solving. *Journal of Mathematical Behavior, 24*, 238-252.

Ho, S. Y., Lee, S., & Yeap, B. H. (2001). Children posing word problems during a paper-and-pencil test: Relationship between achievement and problem posing ability. In J. Ee, B. Kaur, N. H. Lee & B. H. Yeap (Eds.), *New 'Literacies': Educational responses to a knowledge-based society* (pp. 598-604). Singapore: ERA.

Hung, D. (2001). Conjectured ideas as mediating artifacts for the appropriation of mathematical meanings. *Journal of Mathematical Behavior, 20*, 247-262.

Kaur, B. (1995). A window to the problem solvers' difficulties. In A. Richard (Ed.), *Forging links and integrating resources* (pp. 228-234). Adelaide: Australian Association of Mathematics Teachers.

Kaur, B., Low, H.K., & Seah, L.H. (2006). Mathematics teaching in two Singapore classrooms: The role of the textbook and homework. In D.J. Clarke, C.Keitel & Y. Shimizu (Eds.), *Mathematics classrooms in twelve countries: The insider's perspective* (pp. 99-115). Rotterdam: Sense Publishers.

Kay, H.G. (2003). *An exploratory study of the belief systems and instructional practices of four primary school mathematics teachers.* Unpublished M.Ed. dissertation, National Institute of Education, Nanyang Technological University, Singapore.

Leong, Y.H., & Lim-Teo, S.K. (2003). Effects of Geometer's Sketchpad on spatial ability and achievement in transformation geometry among secondary two students in Singapore. *The Mathematics Educator, 7*(1), 32-48.

Liu, Y.M. (2003). *Primary school students' academic mathematics achievement and problem-solving abilities.* Unpublished M.Ed. dissertation, National Institute of Education, Nanyang Technological University, Singapore.

Mathivanan, K. (1992). *The relationship between problem-solving attitude and achievement in mathematics among male upper secondary school students.* Unpublished M.Ed. dissertation, National Institute of Education, Nanyang Technological University, Singapore.

Ministry of Education. (2004). *Assessment guide to primary mathematics.* Singapore: Author.

Ministry of Education. (2007). *Mathematics syllabus.* Singapore: Author.

Ng, H.C. (2003). *Benefits of using investigative tasks in the primary classroom.* Unpublished M.Ed. dissertation, National Institute of Education, Nanyang Technological University, Singapore.

Ng, L. E. (2002). *Representation of problem solving in Singaporean primary mathematics textbooks with respect to types, Polya's model and heuristics.* Unpublished M.Ed. dissertation, National Institute of Education, Nnayang Technological University, Singapore.

Ng, S. F. (2003). How Secondary 2 Express stream students used algebra and the model method to solve problems. *The Mathematics Educator, 7*(1), 1-17.

Ng, S.F., & Lee, K. (2005). How Primary 5 students use the model method to solve word problems. *The Mathematics Educator, 9*(1), 60-83.

Ng, W.L. (2006). Effects of an ancient Chinese mathematics enrichment programme on secondary school students' achievement in mathematics. *International Journal of Science and Mathematical Education, 4*, 485-511.

Polya, G. (1957). *How to solve it.* Princeton: Princeton University Press.

Quek, K.S. (2002). *Cognitive characteristics and contextual influences in mathematical problem posing.* Unpublished PhD dissertation, National Institute of Education, Nanyang Technological University, Singapore.

Quek, K.S., Tay, E.G., Choy, B.H., Toh, T. L., Dong, F.M., & Ho, F.H. (2002, June). *Mathematical problem solving for integrated programme students: Beliefs and performance in non-routine problems.* Paper presented at the fourth East Asia Regional Conference on Mathematics Education Conference, Universiti Sains Malaysia.

Rajaram, R. (1997). *Differences in performance and choice of heuristics in mathematical problem solving among Secondary Three male gifted pupils in Singapore.* Unpublished M.Ed. dissertation, National Institute of Education, Nnayang Technological University, Singapore

Schoenfeld, A. (1985). *Mathematical problem solving.* Orlando, FL: Academic Press.

Schroeder, T.L., & Lester, F.K. (1989). Developing understanding in mathematics via problem solving. In P.R. Trafton & A.P. Shulte (Eds.), *New directions for elementary school mathematics: 1989 yearbook* (pp. 31-42). Reston, VA: National Council of Teachers of Mathematics.

Seoh, B.H. (2002). *An open-ended approach to enhance critical thinking skill in mathematics among Secondary Five Normal (Academic) pupils.* Unpublished M.Ed. dissertation, National Institute of Education, Nanyang Technological University, Singapore.

Tan, B.B. (2005). *Problem-based tasks as a measure of higher-order thinking skills in primary school mathematics.* Unpublished M.Ed. dissertation, National Institute of Education, Nnayang Technological University, Singapore.

Tan, T. L. (2002). *Using project work as a motivating factor in lower secondary mathematics.* Unpublished M.Ed. dissertation, National Institute of Education, Nnayang Technological University, Singapore.

Teo, C.N. (2003). *A study on the WALK approach to assessing students' constructivist learning in mathematics in a secondary school.* Unpublished M.Ed. dissertation, National Institute of Education, Nnayang Technological University, Singapore.

Teo, O.M. (2006). *A small-scale study on the effects of metacognition and beliefs on students in A-level sequences and series problems.* Unpublished M.Ed. dissertation, National Institute of Education, Nanyang Technological University, Singapore.

Teong, S.K. (2002). The effect of metacognitive training on the mathematical word problem solving of lower achievers in a computer environment. In D. Edge & B.H. Yeap (Eds.), *Proceedings of Second East Asia Regional Conference on Mathematics Education & Ninth Southeast Asian Conference on Mathematics Education* (pp. 173-179). Singapore: Association of Mathematics Educators.

Verschaffel, L., De Corte, & Lasure, S. (1994). Realistic considerations in mathematical modeling of school arithmetic word problems. *Learning and Instruction, 4,* 272-294.

Verschaffel, L., Greer, B., & De Corte, E. (2000). *Making sense of word problems.* Lisse, The Netherlands: Swets & Zeitlinger B.V.

Wong, K. Y. (2007). Metacognitive awareness of problem solving among primary and secondary school students. *Proceedings of the Redesigning Pedagogy: Culture, Knowledge and Understanding Conference,* Singapore.

Wong, K.Y., & Tiong, J. (2006, May). *Diversity in heuristic use among Singapore students.* Paper presented at Educational Research Association of Singapore Conference: Diversity for Excellence: Engaged Pedagogies.

Wong, S.O., & Lim-Teo, S.K. (2002). Effects of heuristics instruction on pupils' mathematical problem-solving process. In D. Edge & B.H. Yeap (Eds.) *Proceedings of Second East Asia Regional Conference on Mathematics Education & Ninth Southeast Asian Conference on Mathematics Education* (pp. 180-186). Singapore: Association of Mathematics Educators.

Yeap, B.H. (1997). *Mathematical problem solving: A focus on metacognition.* Unpublished M.Ed. dissertation, National Institute of Education, Nanyang Technological University, Singapore.

Yeap, B.H. (2002). *Relationship between children's mathematical word problem posing and grade level, problem-solving ability and task type.* Unpublished PhD dissertation, National Institute of Education, Nanyang Technological University, Singapore.

Yeap, B. H., & Abdul Ghani, M. (2001, December). *Facilitating sense-making in primary mathematics through word problems.* Paper presented at the AARE-International Education Research Conference, Fremantle, Australia.

Yeap, B.H., & Lee, N.H. (2002). Writing word problems as a learning tool: An exploratory studying. In D. Edge & B.H. Yeap (Eds.), *Proceedings of Second East Asia Regional Conference on Mathematics Education & Ninth Southeast Asian Conference on Mathematics Education* (pp. 187-193). Singapore: Association of Mathematics Educators.

Yeo, K.K. (2005). Anxiety and performance on mathematical problem solving of Secondary Two students in Singapore. *The Mathematics Educator, 8*(2), 71-83.

Yeo, K.K. (2006). Mathematical problem-solving heuristics used by Secondary 2 students. *Korean Journal of Thinking & Problem Solving, 16*(2), 53-69.

Yeo, K.K. (2007). Mathematical problem-solving behaviors of eighth-grade students. *Korean Journal of Thinking & Problem Solving, 17*(2), 5-18.

Chapter 12

Use of ICT in Mathematics Education in Singapore: Review of Research

NG Wee Leng **LEONG Yew Hoong**

Significant technological advances in the last two decades have created new possibilities in the work of mathematics education. A review of the literature yields a picture of promise for technology in improving mathematical instruction while cautioning against simplistic over-claims of its potential to transform classroom experiences. In Singapore, research efforts on the use of technology in mathematics education are mainly along one or more of the following strands: (1) how teachers can use technological tools to replace or complement traditional media; (2) how students may benefit from learning with technology; and (3) how technology interacts with other elements of instruction. This chapter provides details of some of these works.

Key words: technology in mathematics education, graphing calculator, sketchpad, instructional environment

1 INTRODUCTION

Information and communications technology, or ICT, generally refers to digital devices that are used to store, process, and communicate information. It has pervaded almost every facet of our lives. At workplace and in educational institutes, the use of ICT is ubiquitous. Against this backdrop, the role of ICT in education has been a topic of special interest for the past decades. How ICT tools could take their place alongside textbooks and chalkboards as aids for the teacher and the effectiveness of these tools in the framework of an existing educational system are of particular interest to educators and policy makers alike.

In Singapore the education authorities have, over the last decade, taken concrete steps to encourage the use of computers to enhance teaching and learning. Much resources, in the region of S$2 billion, were channeled into infrastructure, computer hardware and software, and teacher training in the first phase of the information technology (IT) master plan (MP1) from 1997 to 2002 with the target that every student would have access to technology in learning (Ministry of Education, 1997). In the second phase of the master plan (MP2), which is ongoing at the time of writing, the aim is to further harness the power of ICT in bringing together key areas of education such as curriculum, assessment, instruction, and professional development to build school environments that are conducive for engaged and holistic learning (Ministry of Education, 2002).

In the mathematics classroom, the goals of the MP1 translate into a vision of "*integration* of ICT to enhance the mathematical experience" (University of Cambridge Local Examination / Ministry of Education, 2000, emphasis added). This statement of intent points to ICT policy objectives in at least the following ways:

- As "integrate" would imply, ICT should not merely be a bit-part player, but rather, it should feature prominently in mathematics classrooms.
- "Integrate" also implies that ICT should not be viewed or acted upon as an isolated part of instruction, but it should be weaved tightly into other components of teaching practice to form a well-coordinated whole.
- Students are expected to use technological tools directly in order to "enhance the mathematical experience."

The aim of using ICT to enhance mathematical learning continues with MP2: "Make effective use of a variety of mathematical tools (including information and communication technology tools) in the learning and application of mathematics" (Ministry of Education, 2006, p. 1). The focus, however, seems to have shifted from "integration" in MP1 to "effective use" in MP2. In other words, the challenge is not merely to weave ICT *tightly* into mathematics teaching, it is also to integrate it *effectively* for students' learning.

2 LOCAL RESEARCH ON USE OF ICT IN MATHEMATICS EDUCATION

Since the implementation of MP1, when schools become well equipped with ICT infrastructure and resources, there has been a steady increase in the number of studies on the use of ICT in the teaching and learning of mathematics in Singapore classrooms. Although research in this area is still developing alongside the fast pace of technological changes, there are already discernible trends emerging in the field. The list below shows the broad agendas which most local research directions are channeled towards.

- ICT-use as a "better" way for teaching mathematics
- ICT-use as a "better" way for learning mathematics
- ICT-use in relation to other factors in the instructional environment

Here ICT-use refers to the use of ICT tools and the term "better" refers to research studies that share common claims of ICT-use as being superior in some visible ways. As to what these purported improvements were, they will be discussed in greater detail under the respective categories below.

3 ICT-USE AS A "BETTER" WAY FOR TEACHING MATHEMATICS

The literature of research in this direction tends to focus on how teachers can harness certain features of ICT to improve the quality of mathematics teaching. The advantageous attributes of ICT in these settings are usually contrasted against conventional teaching tools and media to highlight how the same mathematical ideas would be better conveyed via the technological aids.

An example of such a technological tool in teaching is the random function in spreadsheets. In Tay (2004), a fictitious lottery game was designed using *Excel* where 4-digit numbers can be easily generated using the software's inherent random function. The intention of using the interactive interface is to convince students via multiple simulations of the lottery game that "near misses" — a permutation of the same 4 digits — is not as unlikely as commonly perceived by the students. The ease of random generation of numbers, which is so critical in the instructional setup of simulation, would be compromised if other conventional media such as pencil and worksheet were used in the lesson instead. In other words, the use of *Excel* is better for the purpose of dispelling "near miss" myths in teaching because traditional equipment is incapable of producing the quick and random generation of numbers afforded by the software.

Wu (2002) similarly highlighted this power of random-numbers generation in *Excel* as being advantageous for replicating experiment-like conditions. In her case, she used the feature for the simulation of throwing an unbiased die as demonstrated in Drier (2001). With the press of a button on the keyboard, the tabulated numbers and frequencies of 100 tosses of the die can be refreshed immediately. In addition to the speed in which the experiment can be simulated, another feature of the software that is not available in static media is also demonstrated by means of a bar chart that represents the frequencies in the table. Changes with the data in the table are simultaneously updated in the bar chart. To her, the dynamic inter-modal links between the number mode shown in the table and the diagram mode shown in the statistical graph is important in strengthening the connection between different representational modes in teaching.

On the other hand, in a case study by Wu and Wong (2007) which investigated the impact of a spreadsheet exploration on the understanding of statistical graphs, it was emphasised that "the design of computer-based activities for students should be based on sound principles in order to promote meaningful learning of mathematics concepts" (p.380). In this study, ten pairs of secondary school students of average ability used four *Excel* templates, constructed using certain features of statistics, technology and pedagogy, to explore four aspects of statistical graphs: zero in scale, effect of different scales, size pictogram, and cumulative line graph. An analysis of pair discussion, field notes, and online entries captured by macros revealed that students used an isolated approach and a relational approach in their *Excel* exploration. The study further concluded that the templates had helped the students extend their understanding of statistical graphs.

Apart from spreadsheets, graphing software is also featured widely in the literature. Ang (2006) demonstrated that although analytic solutions of certain differential equations, usually taught at Junior Colleges, can be obtained easily, graphing tools make it very easy to portray the solutions in graphical forms. The speed and accuracy in which graphing software produces the graphs is an advantage as the focus of instruction can shift from drawing the graphs to the more important work of interpreting the graphs as part of the modeling process. Moreover, there are some differential or difference equations that are not easily solved through analytical methods. By using graphing software, students can easily view the (approximate) solutions of these equations graphically, which allow them a peek into the overview of the numerical methods even though the analytic mathematics may be beyond them. To Ang, graphing tools do not merely supplement the symbolic representations by displaying the corresponding graphical representations quickly and precisely, they can also bring mathematical instruction to a more meaningful level where abstract representations alone cannot reach.

Yu, Lam, and Mok (2004) also focused on the speed and accuracy that graphing technology can afford. They described their use of hand-held graphing calculators in teaching the transformation of graphs and the sketching of polar curves. Due to the ease in displaying multiple graphical outputs, graphing calculators are able to help students focus on the gestalt changes and features of the graphs, such as the translation from graph of $y = f(x)$ to graph of $y = f(x+c)$ and the changes in the number of "petals" of polar graphs, rather than risk being sidetracked by the 'nitty-gritties' of sketching those same graphs on normal static media like the whiteboard or overhead transparencies.

Another type of computer program that features prominently in some local studies is dynamic geometry software (DGS). There are a few members within the DGS family but the *Sketchpad* is almost the only member in this class of software used widely in Singapore schools, partly due to the support provided by the Ministry of Education in the use of this software application. *Sketchpad* allows the user to imbue geometric properties such as perpendicularity and parallelism into constructed objects on the computer screen. Using click-and-drag, the on-screen display shows virtually infinite number of ways of varying the object while retaining the in-built invariant attributes. Ho (2002) used this drag-mode in a primary classroom to focus her students' attention on the symmetrical properties of the snowflakes that they tried to construct. Ong (2002) harnessed this feature of dragging to help her Secondary 3 students observe angle properties in circles. In both these studies, the authors argue that the dynamic feature of *Sketchpad* helps students see the underlying geometrical relationships in ways which are difficult to achieve with conventional static drawings.

Another commonly-used feature of *Sketchpad* is the animation function. This tool allows the programming of movement patterns which can be repeatedly performed by clicking a button. Toh (2004) described how he illustrated certain phenomena that are inherently motional using the animation tool of the *Sketchpad*, such as simple harmonic motions and the relative motions of two boats crossing the same river from the same point at the home bank but towards different points on the far bank. Similarly, Leong and Lim-Teo (2003) utilised the movement buttons in *Sketchpad* to show the translation, reflection, rotation, and dilation motions corresponding to the respective concepts in transformation geometry. *Sketchpad* is seen as a better tool in these instructional situations that call for simulations of dynamic movements because of the animation function embedded in the software.

The literature reviewed above is aimed at showing how mathematics teaching can be "better" with the aid of particular features of relevant software when they are suitably harnessed. The common emphasis in these studies is the aspects of ICT that a teacher can use to enhance the teaching of certain topics or ideas in mathematics.

4 ICT-USE AS A "BETTER" WAY FOR LEARNING MATHEMATICS

A number of experimental studies were conducted to study the effects of ICT use in terms of students' achievement scores. Lee and Pereira-Mendoza (2002) studied the impact of the *Logo* software in an intact Primary 4 class. The students used the program to learn angle concepts in normal class time over two weeks. In terms of the school's semestral examination results, the percentage passes of students in the treatment class were ranked against those of the other eight classes in the same grade level that did not have access to *Logo*. Similar rankings were also done using the class scores on particular items that directly tested angle concepts. It was found that students who had learning experience using *Logo* were in similar rankings as the overall examination performance in two out of three items. The authors interpreted the results to mean that computer-based learning can be included in mathematics lessons with "no loss in performance on the 'standard' angle objectives as identified in the curriculum" (p. 426).

Yeo (2006) also investigated the effects of computer use on students' learning. He focused on students' procedural knowledge and conceptual understanding of exponential and logarithmic curves. He taught two Secondary 4 intact classes using similar instructional approaches over a period of ten 40-minute lessons. One class (treatment group) had access to a software known as *Livemath* in five out of the ten lessons. *Livemath* is a type of computer algebra system (CAS) which performs algebraic operations as well as displays updates of changes in functional values graphically. Students interacted with the software by changing values of certain variables on the templates and observing the corresponding variations on the related graphs. The control class was not given the software but instead had printouts of the necessary instructional materials. Two paper-and-pencil tests that assessed procedural knowledge and conceptual knowledge respectively were used to measure treatment effects. On both tests, the treatment class did significantly better that the control class.

A similar study was done by Ong (2002). Instead of *Livemath*, she used the *Sketchpad*. She taught two intact classes over a period of twelve

35-minute lessons on angle properties of circles. While both classes shared access to similar instructional resources, only one of them had opportunities to work on the *Sketchpad* templates designed by the teacher. Students were given worksheets with the templates as a way to guide them towards the intended geometrical results and theorems. A test comprising items that were similar to those contained in the textbook was administered to both groups. Significant treatment effects were found which "seemed to indicate that [computer-based] mode of instruction appeared to enhance the learning of angle properties of circles in terms of achievement [scores]" (p. 54). There is also a noteworthy difference in Ong's study. While the subjects in the studies by Lee and Pereira-Mendoza (2002) and Yeo (2006) were students in high-ranking schools, the students in Ong's study were from a school in a below-average performance band.

In another study, Ng (2004) utilised the TI-92 CAS graphing calculator in a CAS Intervention Programme (CASIP) for Secondary 3 students under a quasi-experimental design. The programme was conducted for approximately 6 months in the second semester of 2002 and involved training teachers to use TI-92 calculator prior to carrying out the lessons. A total of 81 students were involved, 41 students from the experimental group and 40 students in the control group who used scientific calculators. Data were collected from students' scores in mid-year examination prior to CASIP, three class tests during the programme, and a post-CASIP final examination. The study did not confirm any advantages or disadvantages in the use of CAS calculators over scientific calculators. However, post-CASIP data showed that students in the TI-92 group had heightened interest in exploring mathematical concepts and were pleased to be able to utilise the calculator in the classroom and for homework.

Apart from outcomes in terms of mathematics test scores, several studies attempted to assess effects in the affective domain. However, these inquiries into affective outcomes tend not to be at the core of the research agendas, with priority given to cognitive outcomes. This imbalance is not surprising when one understands the heavy emphasis of Singapore's education goals on examination results. Nevertheless, there are some efforts at finding out students' interest level and emotional responses with respect to learning in computer environments. Ong (2002) also included questionnaires to elicit students' attitudinal responses to various aspects of their instructional experience. Her analysis showed ambivalent findings in that while students in the control group indicated greater interest in the topic on angle properties of circles, there was no significant difference about interest on mathematics in general.

Yeo (2003) also used questionnaires to assess students' attitude towards mathematics in general, the particular topic of mathematics taught, and the experience of computer-based learning. He found no significant difference between the control and treatment groups. There was, however, a moderate positive effect towards the use of computers in learning.

Another study by Ng (2005a) examined students' achievement and attitude involving the Texas Instruments Voyage 200 (V200) CAS calculator. There were 27 first year Junior College students in the experimental group. The programme included 15 lessons of 40 minutes each on Mathematics and Further Mathematics at A Level, with each student having access to a V200 unit. Lessons were infused with specific techniques using the V200. The control group comprised students from the previous year who underwent similar training with a TI-83 graphing calculator instead of the V200. Achievement scores were obtained from formal assessments prior to and after the programme. The control group performed slightly better than the experimental group, although after controlling for initial differences, the results were not statistically significant at the 5% level. The results show that the programme had a positive influence on students' perceptions of CAS as a whole through practical exposure to V200 use, particularly through students' journal entries. Students were therefore not adverse to the integration of the V200 into the curriculum and perceived some merits in the use of technology, although there were some reservations about usage due to concerns regarding over-reliance on the calculator.

In another study, Ho (2002) asked her Primary 5 students to write freely about what they felt about the use of *sketchpad*. The students expressed excitement at the possibilities provided by the *Sketchpad,* but they made very little reference to the actual geometrical ideas that they explored.

The obvious limitations to the studies reported above are mainly in the scale and the duration of the research. Because each study involved relatively small sample sizes of a class or two of students, it is difficult to generalise the results to the wider school population with respect to the effects of ICT use in mathematics classrooms. Moreover, these studies were conducted over a short period. These short treatment durations may partially explain why mixed results were reported in different studies and why changes in the affective domain were not detected. Nevertheless, while positive effects cannot be guaranteed, a common finding seems to be that thoughtful ICT use does not adversely affect students' learning, at least not in achievement scores and interest inmathematics. This holds promise for ICT use in mathematics teaching because it seems to contradict many teachers' mindset that the introduction of technology into the classroom takes away time from other productive instructional activities and hence sacrifices students' academic performance. With the confidence that ICT use does not necessarily result in falling standards, teachers can boldly experiment ways to tap into the rich potentials of technology to improve the quality of teaching without undue fear of compromising students' mathematics achievement and interest.

Caution is also mentioned by a number of authors (Ang, 2006; Wong, 2002; Wu, 2002) against an over-optimistic view of ICT use in improving mathematics education. It is a naïve assumption that technology in itself possesses some intrinsic qualities that can enhance students' learning by virtue of its mere inclusion in classroom teaching. Quality computer-based instruction certainly involves careful weaving of ICT tools together with other important components of successful teaching practice. It is thus important to understand the role of ICT within the wider context of other factors influencing teaching and learning, which is dealt with in the next section.

5 ICT-USE IN RELATION TO OTHER FACTORS IN THE INSTRUCTIONAL ENVIRONMENT

This third broad trend of local research on ICT use in mathematics instruction is different from the previous two areas in that it shifts the focus of inquiry away from the technological tools and their effects to how these tools interact with other elements during the instructional process.

Leong and Lim-Teo (2003) studied the relation between *Sketchpad* use and the instructional approach adopted in the classroom. To them, the act of bringing *Sketchpad*, or other ICT software, into the classroom alone will not ensure a superior instructional environment; rather, it is *how* computing technology can be used in conjunction with pedagogical approaches that might influence the quality of instruction. Three intact Secondary 2 classes – labeled A, B, and C – participated in the study. The students were taught the same topics in transformation geometry over a period of three weeks. However, *Sketchpad* was used differently in the three classes. In class A, the teacher used a guided inquiry pedagogy, where the students tested their own conjectures with the given *Sketchpad* templates. Class C modeled the learning environment of a regular classroom, where the teacher was an expositor; students did not have access to the software. *Sketchpad* acted as a demonstration tool in the hands of the teacher to supplement the direct teaching method. The mode of *Sketchpad* use in class B was somewhat "in between" that of class A and class C. The teacher controlled the software but invited conjectures from the students. He used the computer display to conduct whole-class discussions.

All the students were given paper-and-pencil tests and five students from different performance bands were randomly selected from each class for one-to-one verbal assessment. Although the test scores did not reveal any significant differences in conventional achievement between the classes that were indicative of effects of different modes of *Sketchpad* use, the verbal assessment did surface some qualitative differences between the responses of students from class A and those from the other two classes. The students in class A were better able to identify the transformations in object-image pairs and more competent in providing the full descriptions of the transformations. In addition, their answers revealed the use of more formal language, such as "perpendicular" and "parallel" rather than using on-screen finger pointing of "this line" and "this point", which was more prevalent among the students from the other classes. The outcomes of the individual assessment task indicate that students who used *Sketchpad* in a guided-inquiry and exploratory setting (class A) tended to develop stronger concept images of the underlying geometrical ideas. In the other two classes where the method of classroom instruction did not suitably harness the advantageous features of the technology, such as the ease in supporting students' own experimentation in *Sketchpad*, there was a comparative lack of depth in students' learning. These findings support the need to consider IT use, in relation to other instructional approaches.

Some writers have highlighted the problems when ICT is viewed against the backdrop of other complex instructional issues. Ang (2006) surmised that, although there are many ways IT can be utilised in classroom teaching, teachers are also required to look into other aspects of teaching, such as examination-relevance. As most of the common software is currently not permitted in local examinations, he questioned teachers' motivation in wanting to use these software tools in normal classroom teaching

Technical glitches associated with ICT use is also not a trivial problem. In Chua's (2006) study of students using video-conferencing to interact with other students across geographical boundaries and time zones, lapses in hardware or software can "cause considerable frustration to students and impinge on their learning when they are unable to keep up with a disrupted lesson" (p. 94). Even in classroom situations when technology plays a less crucial role than in video-conferencing, anecdotal evidences abound of cases where failed programs, damaged hardware, and "hanged" monitors pose significant hindrances to effective use of ICT. The stability and robustness of computer systems is therefore another important consideration when implementing technology-based lessons.

In situations where ICT promises transformations from traditional ways of teaching into reform-oriented modes of instruction, there remain challenges too for the teacher. *Sketchpad*, for example, is seen by a number of scholars (e.g., Laborde, 1995; Olive, 1998) to have the potential to shift teaching towards more inquiry-based because it is suitable for learning by exploration and experimentation. A conducive experimental environment is one where students can make accurate observations leading to conjectures, test conjectures, do modifications quickly, and retest conjectures. All these requirements are supported by

Sketchpad. Yet, Leong and Lim-Teo (2002), through interviewing students who worked with *Sketchpad* under such experimental-like environment, found that students' adaptations to new instructional situations brought about by ICT use were not straightforward experiences. Issues confronting the teacher included the time required for students to gain sufficient mastery of the software before they could use it as a tool for exploration. For students who are more used to direct instruction from the teacher, a switch to inquiry learning requires not only new mental tools but also mindset changes. The teacher too, has to adjust to the new role of guiding and monitoring the students' disparate experimental pathways and yet connecting them to the curricular agendas. In other words, when a teacher brings technological tools into the classroom, the changes to the instructional milieu may be far more complex than originally intended. It is not merely a replacement of one part of teaching by another. Because of the complex nature of the teaching activity, changes in one part, such as the inclusion of ICT, can change the instructional setup substantially. This change can pose significant challenges for the teacher and the students.

The close relation between ICT use and other complex instructional elements in teaching could explain why, despite strong moral and actual support through MP1, there is yet little evidence to suggest a widespread integration of information technology to enhance the mathematical experience in Singapore classrooms. Leong (2003) conducted a small-scale survey among 41 mathematics teachers from ten secondary schools. The teachers completed a questionnaire about their use of *Sketchpad* and their preference for the mode of use in their instruction. It was encouraging that 33 out of the 41 teachers indicated that they had used *Sketchpad* at some parts in their teaching. However, only three teachers indicated their preferred mode of *Sketchpad* use as one to encourage students' learning through participation in joint discovery. The other teachers preferred teacher-controlled demonstration. Thus, the full power of *Sketchpad* and its potential to transform classrooms into lab-like places for students' inquiries were generally not realised among the schools that participated in the survey. If the responses from these schools were at all reflective of the overall sentiments of other mathematics teachers, then the picture of ICT use may be less of "integration" but more of ICT as one standing outside the classroom and invited in only once in a while to fulfill its bit-part role as and when called upon by the teacher.

Another instructional element is the attitude of teachers towards ICT, as teachers' beliefs about educational change directly affect implementation of new initiatives (Fullan, 1991). To ascertain teachers' attitude towards CAS, Ng (2003) developed a 40-item CAS Attitude Scale (CASAS) based on Gressard and Loyd's Computer Attitude Scale (Gressard & Loyd, 1986) and trialed it with a sample of 50 pre-service secondary school mathematics teachers. The high Cronbach alpha coefficients obtained for the subscales and overall scale of the CASAS provide evidence that the instrument is reliable. The CASAS therefore can be used as a platform to aid in managing teacher development in this area. In another study, Ng (2006) developed the Crucial Factors in the Integration of ICT Survey (CFS) to examine six factors pertinent to ICT integration, an expansion based on four technology integration predictor variables reported by Ritchie and Wiburg (1994). CFS is a 36-item, 4-point Likert scale questionnaire that covers usefulness and worthiness of technology, school departmental support, availability and accessibility of technology, professional development opportunities, leadership, planning and implementation, and partnerships with external organisations. A field test of the CFS was carried out with 60 pre-service secondary school mathematics teachers who had some experience in school for at least a month on a contractual basis, with explicit instructions that they responded to whether each statement would influence their use of ICT in the classroom. Results showed that professional development had the greatest influence, followed by availability of technology, leadership and support. Usefulness and partnerships with external organisations were of the least priority. While this does not reflect the perceptions of in-service teachers, it gives some indication to the direction in which the overall environment for ICT integration needs to be developed.

In the light of the perceived importance of professional development, Ng (2005b) also conducted a survey on the perceptions of 90 in-service teachers after a course on the use of graphing calculators in the teaching of A Level mathematics. The in-service programme comprised three 3-hour weekly sessions that aimed to familiarise teachers with the TI-84 Plus (TI-84+) graphing calculator and its functionality, in addition to equipping them with the knowledge and skills required to solve A Level mathematics problems with the calculator. After the course, all the participants felt more confident in using the TI-84+ for teaching. They appreciated the discovery approach adopted in the course and the practical methods of utilising the calculator; their immediate concern was being able to impart relevant skills to their students in assessments. Thus, high quality professional development focusing on what the teachers need helps to increase teachers' confidence in utilising ICT in the classroom.

6 CONCLUSION

While there has been a growing body of local research on the use of ICT in the classroom, there are also many aspects of the use of ICT in mathematics education that have yet to be explored. Most of the completed studies have focused on the specific use of tools in a predefined context, with a fixed approach towards mathematics teaching and learning of a specific group of students and teachers. As such, variability in any context will likely produce a different result.

There is also great potential for research in ICT implementation in actual classroom. It is when ICT use is examined together with, not separately from, other classroom demands that a sufficiently in-depth understanding of the complexity of teaching can be realised. Teachers' struggles, problems, and even reluctance to use ICT can be accounted for when technology is viewed against the wider context of other classroom variables. Such an insight into the real constraints and challenges of ICT use can help inform the ongoing efforts for widespread integration of technology in mathematics education.

Although further study of the impact of different ICT tools and approaches on teaching and learning in both cognitive and affective domains may be investigated, it is virtually impossible to cover every facet of the field exhaustively. As educational professionals continue to pioneer new research and delve into variations of current research, teachers themselves need to take the initiative to discover what is appropriate for their students. The aim, therefore, is not to provide students with a new "technological toy" for the purpose of having fun in the classroom, but rather to create opportunities for active learning that enable the development of a wide variety of content knowledge, skills, processes, and attitudes that they may bring with them into the real world.

References

Ang, K. C. (2006). Mathematical modeling, technology and H3 Mathematics. *The Mathematics Educator, 9*(2), 33-47.

Chua, B. L. (2006). Video-conferencing for the mathematics classroom: Mathematics and music for secondary school students. *The Mathematics Educator, 9*(2), 80-96.

Drier, H. (2001). Teaching and learning mathematics with interactive spreadsheets. *School Science and Mathematics, 101*(4), 170-179.

Fullan, M. (1991). *The new meaning of educational change.* London: Cassell.

Gressard, C. P., & Loyd, B. H. (1986). Validation studies of a new computer attitude scale. *Association for Educational Data Systems Journal, 18*, 295-301.

Ho, S. Y. (2002). Using *Geometer's Sketchpad* with Primary five students. In D. Edge & B. H. Yeap (Eds.), *Mathematics education for a knowledge-based era* (pp. 390-393). Singapore: Association of Mathematics Educators.

Koh, T. S., Koh, I. Y. C., & Wu, W. T. (2004). Integration of information technology in Singapore school mathematics curriculum. In W. C. Yang, S. C. Chu, T. de Alwis & K. C. Ang (Eds.), *Technology in mathematics: Engaging learners, empowering teachers, enabling researchers* (pp. 17-26). Blacksburg, VA: ATCM Inc.

Laborde, C. (1995). Designing tasks for learning geometry in a computer-based environment. In L. Burton & B. Jaworski (Eds.), *Technology in mathematics teaching* (pp. 35-68). Sweden: Chartwell-Bratt.

Lee, C. M., & Pereira-Mendoza, L. (2002). Integrating the computer and thinking into primary Mathematics classroom. In D. Edge & B. H. Yeap (Eds.), *Mathematics education for a knowledge-based era* (pp. 421-426). Singapore: Association of Mathematics Educators.

Leong, Y. H. (2003). Use of *Geometer's Sketchpad* in secondary schools. *The Mathematics Educator*, *7*(2), 86-95.

Leong, Y. H., & Lim-Teo, S. K. (2002). Guided inquiry with the use of the *Geometer's Sketchpad*. In D. Edge & B. H. Yeap (Eds.), *Mathematics education for a knowledge-based era* (pp. 427-432). Singapore: Association of Mathematics Educators.

Leong, Y. H., & Lim-Teo, S. K. (2003). Effects of *Geometer's Sketchpad* on spatial ability and achievement in transformation geometry among Secondary two students in Singapore. *The Mathematics Educator*, *7*(1), 32-48.

Ministry of Education (1997). *Masterplan for information technology in education.* Retrieved July 20, 2006, from http://moe.gov.sg/edumall/mpite/overview/index.html

Ministry of Education (2002). *Overview of Masterplan II for information technology in education.* Retrieved July 20, 2006, from http://moe.gov.sg/edumall/mp2/mp2.htm

Ministry of Education (2006). *Secondary mathematics syllabus.* Singapore: Author.

Ng, W. L. (2003). Developing a Computer Algebra System (CAS) Questionnaire for teachers: A survey of pre-service teachers' attitudes toward CAS. *The Mathematics Educator*, *7*(1), 96-119.

Ng, W. L. (2004). Effects of Computer Algebra System on secondary students' achievement in mathematics: A pilot study in Singapore. *International Journal of Computer Algebra in Mathematics Education*, *10*(4), 233-248.

Ng, W. L. (2005a). Effects of a Computer Algebra System on Junior College students' attitude towards and achievement in mathematics. *The International Journal for Technology in Mathematics Education*, *12*(2), 59-72.

Ng, W. L. (2005b). Using a graphing calculator to explore pre-university level mathematics: Some examples given in an in-service course. In W. C. Yang, S. C. Chu, T. de Alwis & K. C. Ang (Eds.), *Technology in mathematics: Engaging learners, empowering teachers, enabling researchers* (pp.322-331). Blacksburg, VA: ATCM Inc.

Ng, W. L. (2006). Factors that influence the integration of information and communications technology into the classroom: Pre-service mathematics teachers' perceptions. *The Mathematics Educator*, *9*(2), 60-79.

Olive, J. (1998). Opportunities to explore and integrate mathematics with the *Geometer's Sketchpad*. In R. Lehrer & D. Chazan (Eds.), *Designing learning environments for developing understanding of geometry and space* (pp. 395-418). Mahwah, NJ: Lawrence Erlbaum.

Ong, M. F. (2002). *Effects of computer-assisted instruction on the learning of angle properties of circles among upper secondary students.* Unpublished Masters dissertation, National Institute of Education, Nanyang Technological University, Sinagpore.

Ritchie, D., & Wiburg, K. (1994). Educational variables influencing technology integration. *Journal of Technology and Teacher Education*, *2*(2), 143-153.

Tay, E. G. (2004). The psychology of "near-miss" in 4-digit lottery: A spreadsheet simulation. In W. C. Yang, S. C. Chu, T. de Alwis & K. C. Ang (Eds.), *Technology in mathematics: Engaging learners, empowering teachers, enabling researchers* (pp. 387-392). Blacksburg, VA: ATCM Inc.

Toh, T. L. (2004). Use of *Geometer's Sketchpad* to teach mechanics concepts in A level mathematics. In W. C. Yang, S. C. Chu, T. de Alwis & K. C. Ang (Eds.), *Technology in mathematics: Engaging learners, empowering teachers, enabling researchers* (pp. 429-436). Blacksburg, VA: ATCM Inc.

University of Cambridge Local Examinations Syndicate/Ministry of Education (2000). *General Certificate of Education, Ordinary level and Normal level: Mathematical subjects syllabuses. For examination in 2002.* Singapore: Author.

Wong, S. O. (2002). The effects of LOGO on curriculum changes: Content and pedagogy. In D. Edge & B. H. Yeap (Eds.), *Mathematics education for a knowledge-based era* (pp. 472-478). Singapore: Association of Mathematics Educators.

Wong, K. Y., Lim, Y. S., & Low. K. G. (1989). Research report RECSAM: Computers in education project, Singapore study. *Sinagpore Journal of Education, 10*(1), 66-70.

Wu, Y. (2002). The impact of *Excel* on the school mathematics curriculum. In D. Edge & B. H. Yeap (Eds.), *Mathematics education for a knowledge-based era* (pp. 469-485). Singapore: AME.

Wu, Y. K., & Wong, K. Y. (2007). Impact of a spreadsheet exploration on secondary school students' understanding of statistical graphs. *Journal of Computers in Mathematics and Science Teaching, 26*(4), 355-385.

Yeo, J. B. W. (2003). *The effect of exploratory computer-based instructions on Secondary Four students' learning of exponential and logarithmic curves.* Unpublished Masters dissertation, National Institute of Education, Nanyang Technological University, Sinagpore.

Yeo, J. B. W. (2006). Computer-based learning using *Livemath* for Secondary Four students. *The Mathematics Educator*, *9*(2), 48-59.

Yu, D. Y., Lam, Y., & Mok, W. H. (2004). The use of graphics calculator in teaching and learning transformations, polar curves and matrices. In W. C. Yang, S. C. Chu, T. de Alwis & K. C. Ang (Eds.), *Technology in mathematics: Engaging learners, empowering teachers, enabling researchers* (pp. 393-401). Blacksburg, VA: ATCM Inc.

Chapter 13

Mathematics Anxiety and Test Anxiety of Secondary Two Students in Singapore

YEO Kai Kow Joseph

This exploratory study identifies relationships between mathematics anxiety and test anxiety and analyses whether boys show greater mathematics anxiety and test anxiety than girls among 621 Secondary 2 students. Mathematics anxiety was measured using the Fennema-Sherman Mathematics Anxiety Scale (MAS) (Fennema & Sherman, 1978) which consisted of 12 items while Test Anxiety Inventory (TAI) was adopted for assessing test anxiety. The results showed a relationship between mathematics anxiety and test anxiety with a correlation coefficient of 0.39. The girls had significantly higher mathematics anxiety than boys, but there was no evidence that the girls had higher test anxiety than boys.

Key words: mathematics anxiety, test anxiety, relationship, gender difference

1 INTRODUCTION

Anxiety is one form of emotions. Some students experience anxiety or other negative emotions associated with mathematics (Buxton, 1984; Hembree, 1990; Ma, 1999) and with testing situations (Hembree, 1987) which are generally regarded as separate but related phenomena (Kazelskis et al., 2000). In the academic field, two most important types of anxiety traits have been identified: test anxiety and mathematics anxiety (Hembree, 1990). To study mathematics anxiety, it will be pertinent to look at test anxiety. In fact, a major negative consequence of mathematics anxiety is mathematics avoidance (Hembree, 1990; Ma, 1999) and limiting of career choices. Therefore, it is not surprising that mathematics-anxiety students take fewer mathematics elective courses and avoid taking mathematics as a major and careers that rely heavily on quantitative skills or mathematics (Ashcraft, 2002; Hembree, 1990), limiting their employment opportunities in an knowledge-based economy.

Many studies also indicate that primary school students are positive about mathematics but the likelihood of mathematics anxiety increases as students progress into the secondary schools (Renga & Dalla, 1993). In addition, findings from the TIMSS-R study (Ministry of Education, 2000a) presented some information on Singapore students, with respect to mathematics and their attitudes toward mathematics. This study showed that Singapore had a higher index of positive attitude towards mathematics compared to the other participating countries but had fewer students attaining a high self-concept in mathematics. This seems to show that while Singapore students had the ability and liking for mathematics, their confidence in the subject was not evident. Some studies on mathematics anxiety and test anxiety were carried out in Singapore. Foong (1987) conducted an exploratory study on mathematics anxiety and test anxiety of a sample of 206 Secondary 4 female students from a Singapore school. In addition, Tanzer and Sim (1991) and Lee (2003) also investigated primary school students' test anxiety. Even though mathematics anxiety has been the theme of more research than any other area in the affective domain (McLeod, 1992), little research is available on test anxiety and mathematics anxiety in Singapore schools. Furthermore, in a local study by Singapore's Ministry of Education (2000b) on a study of students' stress, it was found that mathematics was perceived to be the most stressful subject to Secondary 2 students. With the scarcity of local research in this area, the contribution of the present research is significant in providing local data that are pertinent in documenting the relationships among mathematics anxiety and test anxiety of Secondary 2 (13 to 14 years old) students in Singapore.

2 CONCEPTS OF TEST ANXIETY AND MATHEMATICS ANXIETY

2.1 *Test Anxiety*

In the field of test anxiety, Liebert and Morris (1967) were first to suggest that test anxiety consists of two elements (cited in Foong, 1984, p. 21):

- Worry: a cognitive element, consisting of self-deprecatory thoughts about one's performance,
- Emotionality: an affective element, including feelings of nervousness, tension, and unpleasant physiological reactions to testing situations.

Liebert and Morris indicated that these two elements of anxiety are empirically distinct. They also mentioned that although worry and emotionality elements are correlated, worry relates more strongly than emotionality to poor test performance, suggesting that worry is the primary performance indicator (Deffenbacher, 1980). Subsequent work by anxiety researchers (Hembree, 1988; Sarason, 1986; Wine, 1980) also reckoned that the worry or cognitive element of test anxiety interfered most with achievement performance. According to Wine (1980), students with high levels of test anxiety under evaluative situations divide their attention between task demands and personal concerns. However, those with low levels of test anxiety under evaluative situations seemed to devote a greater proportion of their attention to their task demands.

Spielberger et al. (1977) began in 1974 to design the Test Anxiety Inventory (TAI), which included subscales to measure worry and emotionality as elements of test anxiety. The final version was developed to measure individual differences in test anxiety as a situation-specific personality trait. Many researchers considered mathematics anxiety as a subject-specific manifestation of test anxiety (Bandalos, Yates & Thorndike-Christ, 1995; Brush, 1981; Hembree, 1990).

2.2 *Mathematics Anxiety*

Anxiety is a critical component and variable in the study of attitudes. From many research studies, it was observed that mathematics anxiety is a subset of mathematics attitudes (Schoenfeld, 1985). However, McLeod (1992) cautioned that the term attitude "does not seem adequate to describe some of the more intense feelings that students exhibit in mathematics classrooms" (p. 576), such as anxiety, confidence, frustration and satisfaction. In a less intense attitudinal response, mathematics anxiety could also be conceived as a feeling of dislike or fear when one was confronted with mathematics (Foong, 1984). There are an innumerable number of different affective states because of different situational evaluations. Several labels had been utilised to describe anxiety in mathematics, such as "Mathematics Anxiety," "Number Anxiety" and "Mathophobia." More important than the labels are the meanings behind the labels, what each of the term characterises. Fennema and Sherman (1978) referred to mathematics anxiety as feelings of anxiety, dread, nervousness and associated bodily symptoms related to doing mathematics. Tobias and Weissbrod (1980) stated that "the term was used to describe the panic, helplessness, paralysis and mental disorganisation that arises among some people when they are required to solve a mathematics problem" (p. 65). Cemen (1987) also defined mathematics anxiety as a state of anxiety which occurs in response to situations involving mathematics which are perceived as threatening to self-esteem. In her model of mathematics anxiety reaction, she described three antecedents that interacted to produce an anxious reaction with its physiological manifestations such as perspiring and increased heart beat. These antecedents are (1) environmental antecedents: negative mathematics experiences and lack of parental encouragement, (2) dispositional antecedents: negative attitudes and lack of confidence, and (3) situational antecedents: classroom factors and instructional format. Mathematics anxiety is often referred to as "the general lack of comfort that someone might experience when required to perform mathematically" (Wood, 1988, p. 11). Mathematics anxiety can take multidimensional forms including for example, dislike (an attitudinal element), worry (a cognitive element), and fear (an emotional element) (Bessant, 1995; Hart, 1989; Wigfield & Meece, 1988). Later, Hembree (1990) broadly defined mathematics anxiety as being in a state of emotion underpinned by traits of fear and dread. Recently, Ashcraft and Ridley (2005) provided a rather straightforward definition of mathematics anxiety "as a negative reaction to math and to mathematical situations" (p. 315).

The aforementioned reviews suggest that the concept of mathematics anxiety is often difficult to separate from test anxiety. It seems that test anxiety provides the main source of theoretical support for research on mathematics anxiety.

3 LITERATURE REVIEW

Mathematics anxiety has notably been well researched compared to other areas within the affective domain. In an earlier study, Carpenter (1980) reported that 21% of the nine-year-olds claimed that doing mathematics made them nervous. A cluster-analytic study was conducted using 157 college students by Jackson and Leffingwell (1999). Three clusters of grade levels in which students first experienced anxiety in their mathematics classes were evident: grade 3 and 4 in elementary level, grade 9, 10 and 11 in high school level, and freshman year in college level. Other studies also indicated that grade 4 is the first time the students experience mathematics anxiety (Tankersley, 1993). In a review of 151 studies done mainly in the USA, Hembree (1990) found that almost all studies indicated a negative correlation between mathematics anxiety and performance. At the college level, females reported more mathematics anxiety than males. Higher mathematics anxiety levels were related to mathematics performance and avoidance. Similarly, Tay (2001) found mathematics anxiety among 62 grade 9 "smart" students in Malaysia. About 70.7% of the students also exhibited great anxiety for test items. Boys were more anxious that girls when dealing with mathematical problem solving. Hembree concluded that a variety of treatments and techniques are effective in reducing mathematics anxiety. Improved mathematics performance consistently accompanies valid treatment. Some valid treatments included suitable teaching approaches and anxiety reduction programs. In another study, Loke and Vikaneswari (1999) investigated the level of mathematics anxiety among the grade 4 students. They found that there was no significant difference in the level of mathematics anxiety between boys and girls. Evans (2000) indicated that most of the research in the 1970s onwards is based on "a conception of mathematics anxiety as 'debilitating', that is, as having a negative relationship with performance" (p. 53). In other words, high anxiety hinders performance especially for the less able students. Foong (1987) indicated that some of the highly anxious, poor mathematics achievers, in their moments of frustration or resignation, felt that doing mathematics was a waste of time.

Numerous studies have also been done on mathematics anxiety in relation to test anxiety. Research studies (Ikegulu, 1998) have found that mathematics anxiety is related to test because it surfaces most dramatically when the students are under test conditions or perceive that they are under test conditions. Several research studies had used Spielberger Test Anxiety Inventory (TAI) to measure test anxiety. Foong (1984) used the TAI on a sample of 206 Secondary 4 girls from a Singapore school. She reported that the lowest level of test anxiety did not necessarily correlate with highest achievement, although it was observed that very high test anxiety correlated with low achievement. The highest level of mathematics achievement was found to be in the middle ranges of the test anxiety scores. Milgram and Toubiana (1999) conducted a test on 167 males and 187 females aged 13, 14 and 16 years using the TAI and stated that female students were more anxious about examinations than the male students. In the study by Dawson (2001), 61 post-Baccalaureate and graduate students were administered the Test Anxiety Inventory (TAI). High-test anxiety students in the presence of a high evaluative threat condition obtained a significantly lower mean score in the achievement tests than those high-test anxiety students under the low evaluative threat condition. In a most recent study conducted locally, Lee (2003) administered the TAI three weeks before 162 Primary 6 students sat for their Primary School Leaving Examination (PSLE). There was a significant relationship between the TAI and the PSLE results. Students with higher test anxiety obtained significantly lower grades in the PSLE than those with lower test anxiety. Tanzer and Sim

(1991) investigated 3713 Singapore students (Primary 4 and 6) and 584 eleven-years-olds and twelve-year-olds Austrian students' test anxiety using the German Test Anxiety Inventory (TAI-G). The TAI-G is a modified and elaborated four-component German version of Spielberger's Test Anxiety Inventory (TAI). Tanzer and Sim found that older students in Austria and Singapore were more test-anxious than younger students. While there appears to be no difference between genders in test anxiety in Singapore, Austrian girls showed more test anxiety. All the four subscales (worry, confidence, emotionality and interference) of the TAI-G correlated negatively with good academic performance in Austria while only three (confidence, emotionality and interference) had a negative correlation with good academic performance in Singapore. The worry subscale correlated positively with good academic performance in Singapore. Tanzer and Sim explained that these differences could be due to Singapore students' fear of failure that made good students worry more about their academic performance and thus work hard to produce better results. One other possibility that was also mentioned was Singaporean "kiasu syndrome" (fear of losing out) which had been deeply ingrained in the Singapore society.

4 METHODOLOGY

4.1 *Research Design*

As the present study required identifying relationships between mathematics anxiety and test anxiety, a quantitative approach was thought to be the most appropriate for this purpose (Robson, 1993). A quantitative approach would provide some indication of the relationships between mathematics anxiety and test anxiety. From the literature review presented earlier, the primary aim of the research was to identify any relationships between mathematics anxiety and test anxiety of Secondary 2 students in Singapore and to analyse if boys show greater mathematics anxiety and test anxiety than girls.

4.2 *The Sample*

According to Cohen and Manion (1994), "the correct sample size depends upon the purpose of the study and the nature of the population under scrutiny" (p. 89). The quantitative study is inferential in nature, so the sample should be large enough so that empirical generalisations can be made from the sample chosen to the population from which it comes (Robson, 1993). The school population in this study consisted of all the 160 secondary schools in Singapore. The study sample consisted of 10 schools, a stratified random sample from the population. One school was randomly selected from the top 16 schools, one from the next best 16, and the remaining from the rest of the population, according to the school ranking by the Singapore Ministry of Education which was based on their students' average performance in the GCE O level Examination. Subsequently, intact class groups were used to include students from all the "selected" schools. Although the data were collected from 644 Secondary 2 students, only a total of 621 Secondary 2 students with no missing data were included in the study, resulting in a response rate of 96.4%. This was to ensure that a consistent sample size was used in the various analyses throughout. As all the students were from the Express stream in Secondary 2, they would have undergone roughly the same mathematics curriculum.

4.3 *The Instruments*

Two paper-and-pencil instruments were adopted and used in the study; see Tables 1 and 2.

(1) Fennema-Sherman Mathematics Anxiety Scale (MAS), comprising 12 items.
(2) Test Anxiety Inventory (TAI), comprising 20 items.

4.3.1 Fennema-Sherman Mathematics Anxiety Scale (MAS)

The MAS, one of nine domain specific scales which make up the Fennema-Sherman Mathematics Attitude Scales, is a 12-item, 5-point Likert-format instrument. The scale was validated on US secondary school students. A pilot study was carried out to provide some indication of the reliabilities of the MAS for a sample of Singapore Secondary 2 students. Results of the responses of 70 Secondary 2 students to the MAS were used to compute the reliabilities of the instruments. The reliability using Cronbach's alpha for the pilot sample was 0.75. The Fennema-Sherman Mathematics Anxiety Scale was specifically designed for secondary school students and had a high reliability in the pilot sample. Furthermore, it had been used in Singapore by three researchers (Foong, 1984; Lenden-Hitchcock, 1994; Tan, 1990). It was thus decided that the Fennema-Sherman Mathematics Anxiety Scale (MAS) would be the instrument adopted for assessing mathematics anxiety in the study.

For the purpose of the study, the scores in the Fennema-Sherman Mathematics Anxiety Scale (MAS) were modified to range from 1 (strongly agree) to 5 (strongly disagree) instead of the other way. Half of the items were positively worded, while the other half were negatively worded. Scoring of negatively worded items was reversed so that a higher score would indicate higher mathematics anxiety. A student's score can thus range from 12 to 60. A high total score in the scale would reflect a high level of reported mathematics anxiety whereas a low total score would mean a low level of reported mathematics anxiety.

4.3.2 Test Anxiety Inventory (TAI)

The Test Anxiety Inventory is a self-report psychometric instrument developed to measure individual differences in test anxiety as a situation-specific personality trait (Spielberger, 1972; Spielberger, Gozalez, Taylor, Ross & Anton, 1977). The TAI consists of 20 statements pertaining to feelings and reactions while taking tests. Responses were obtained on a 4-point Likert scale with response categories as 1 (almost never); 2 (sometimes); 3 (often) and 4 (almost always). Scores may range from 20 to 80, and higher scores indicate higher levels of anxiety. The TAI indicates subscales of 8 items each for the Worry and Emotionality components of test anxiety. Items 3, 4, 5, 6, 7, 14, 17 and 20 belong to the Worry Subscale while items 2, 8, 9, 10, 11, 15, 16 and 18 are in the Emotionality Subscale and the other 4 items describe the general feelings of the students which contribute to anxiety. With subscales for measuring worry and emotionality, the TAI was designed to assess individual differences in anxiety inclination in test situations.

A pilot study was conducted using the TAI. The results showed the distribution of scores. The language used in the TAI was less culturally biased. After careful considerations, the TAI was used in this study as the items are easily comprehended by the Secondary 2 students and the questions asked are more focused with reference to the test anxiety of students. Furthermore, it had been used in Singapore by two researchers, Foong (1984) on Secondary 4 students and Lee (2003) on Primary 6 students.

4.4 *Procedure*

The Fennema-Sherman Mathematics Anxiety Scale (MAS) and the Test Anxiety Inventory (TAI) were administered together during the school hours in 2001. In the administration of the MAS and TAI, the teachers referred to the MAS as the Mathematics Attitudes Scale and the TAI as the Test Attitude Inventory, as printed on both forms and the word "anxiety" was avoided. The mathematics teachers were not present in class during the administration. The students were given enough time to respond to all the items. Though no time limit was set, the students completed the two questionnaires within 30 minutes. As the instructions and statements were clear, no oral questions were asked during the course of the administration of the questionnaires.

5 RESULTS AND DISCUSSION

The overall mathematics anxiety results measured by Fennema-Sherman Mathematics Anxiety Scale (MAS) are summarised in Table 1. The items have been arranged in descending order by the item means. High means indicate high levels of mathematics anxiety. On the 5-point scale, the highest mean showed that these students generally reported a moderately sinking feeling when they think of trying difficult mathematics problems.

Table 1
Items arranged in Descending Order by Means (Fennema-Sherman Mathematics Anxiety Scale)

Item	Mean	Standard Deviation
9) I get a sinking feeling when I think of trying difficult maths problems.	3.15	1.24
3) I do not usually worry about being able to solve maths problems.	3.10	1.16
5) I am usually at ease during maths tests.	3.06	1.18
4) I seldom panic during a maths test.	3.00	1.21
1) Maths does not scare me at all.	2.95	1.26
2) It would not bother me at all to take more maths courses.	2.91	1.20
11) A maths test would scare me.	2.76	1.23
10) My mind goes blank and I am unable to think clearly when working mathematics.	2.64	1.22
7) Mathematics usually makes me feel uncomfortable and nervous.	2.55	1.15
12) Mathematics makes me feel uneasy and confused.	2.50	1.15
8) Mathematics makes me feel uncomfortable, restless, irritable and impatient.	2.45	1.17
6) I am usually at ease in maths lessons.	2.24	1.01

Table 2
Items arranged in Descending Order by Means

Item	Mean	Standard Deviation
12) I wish examinations did not bother me so much.	2.95	1.05
10) I start feeling very uneasy just before getting a test paper back.	2.80	1.02
16) I worry a great deal before taking an important examination.	2.43	0.98
9) Even when I'm well prepared for a test, I feel very nervous about it.	2.32	0.95
11) During tests I feel very tense.	2.29	0.86
8) I feel very jittery when taking an important test.	2.21	0.85
7) Thoughts of doing poorly interfere with my concentration on tests.	2.20	0.94
17) During tests I find myself thinking about the consequences of failing.	2.20	1.06
1) I feel confident and relaxed while taking tests.	2.19	0.74
18) I feel my heart beating very fast during important tests.	2.19	0.99
15) I feel very panicky when I take an important test.	2.18	0.93
3) Thinking about my grade in a course interferes with my work on tests.	2.17	0.87

20) During examinations I get so nervous that I forget facts I really know.	2.15	0.90
19) After an exam is over I try to stop worrying about it but I just can't.	2.07	0.99
2) While taking examinations I have an uneasy, upset feeling.	2.03	0.77
4) I freeze up on important exams.	1.93	0.90
5) During exams I find myself thinking about whether I'll ever get through school.	1.90	0.92
14) I seem to defeat myself while working on important test.	1.88	0.74
6) The harder I work at taking a test, the more confused I get.	1.83	0.81
13) During important tests I am so tense that my stomach gets upset.	1.57	0.81

The overall test anxiety results are summarised in Table 2 as measured by the Test Anxiety Inventory (TAI). The items have been arranged in descending order by the item means. High means indicate high levels of test anxiety. On the 4-point scale, the first two highest means indicated that quite often these students were bothered by the examinations and they also felt very uneasy just before getting a test paper back.

The means and standard deviations were computed for the Fennema-Sherman Mathematics Anxiety Scale and Test Anxiety Inventory (TAI). Table 3 gives a summary of the means and standard deviations of scores for the four variables: Mathematics Anxiety, Test Anxiety, Worry, and Emotionality for a sample of 621 Secondary 2 students.

Table 3
Mean Scores and Standard Deviations (SD) for the Mathematics Anxiety, Test Anxiety, Worry and Emotionality Scores.

Variables	Mean	Standard Deviation
Mathematics Anxiety (max = 60)	33.27	10.45
Test Anxiety (max = 80)	43.50	10.30
Worry (max = 32)	16.27	4.64
Emotionality (max = 32)	18.45	5.33

Scores on the mathematics anxiety scale show a mean of 33.27 and a standard deviation of 10.45, while the test anxiety scores gave a mean of 43.50 and a standard deviation of 10.30. The students scored a lower mean (16.27) on the Worry subscale than the Emotionality (18.45) subscale.

The relationship between mathematics anxiety and test anxiety was found to be moderately positive and related ($r = 0.39$, $p < 0.005$). Thus, there was a correlation between the two variables, with high levels of mathematics anxiety associated with high levels of test anxiety.

The results showed a relationship between mathematics anxiety and test anxiety with a correlation coefficient of 0.39. Analysis of mean scores also showed that there was a general tendency for higher levels of test anxiety to be associated with higher scores on mathematics anxiety. This showed that high test-anxious students exhibited high levels of mathematics anxiety more frequently than low test-anxious students.

According to Spielberger (1972), high trait-anxious personnel tend to be anxious in various circumstances and would frequently experience higher test anxiety for specific circumstances. They may perceive mathematics as difficult especially in test situations and feel inadequate to handle the task, while accompanied by the anticipation of failure as threatening to their self-esteem. However, the results also indicated that the statistical correlation between the mathematics anxiety and test anxiety was low at $r = 0.39$ ($p < 0.005$). This was consistent with the result of Hembree's (1990) study that indicated it is unlikely that mathematics anxiety is purely restricted to testing. Rather, the construct for mathematics anxiety comprises a general fear of contact with mathematics, including attending mathematics classes, homework, and tests. Similarly, Williams (1994) used the Test Anxiety Inventory (Spielberger, 1977) and the Math Anxiety Questionnaire (Wigfield & Meece, 1988) and found that the affective scales of both measures share only about 24% of their variance and the cognitive scales share about 13% of their variance, indicating a lack of convergence. In other words, the two anxiety measures' definitions of affective (emotionality) and cognitive (worry) anxiety may be somewhat different (Williams, 1994).

Table 4 shows a detailed comparison of the mean scores and standard deviations for the mathematics anxiety and test anxiety. The boys had lower mean scores in the mathematics anxiety and test anxiety than the girls. This difference was statistically significant ($t = -6.65$, $p = 0.001$). However, there was no significant difference in the test anxiety scores between the girls and the boys ($t = -1.41$, $p = 0.16$).

Table 4
Means and Standard Deviations (SD) of the Mathematics Anxiety and Test Anxiety

Gender	Mathematics Anxiety		Test Anxiety	
	Mean	Standard Deviation	Mean	Standard Deviation
Boys (N= 349)	30.89	10.08	42.99	10.30
Girls (N= 272)	36.33	10.12	44.16	10.29
Total Sample (N= 621)	33.27	10.45	43.50	10.30

However, the evidence on gender effects in mathematics anxiety is not so simple or straightforward (Ashcraft, 2005). This study concurred with Hembree (1990) who reported that "across all grades, female students report higher mathematics anxiety levels than males" (p. 45).

6 CONCLUSIONS

Although the ten schools in the present study were selected to include different types of schools in Singapore and the classes were selected to include students with differing achievement profiles, the findings in the present study cannot be generalised to the student population in Singapore schools. Also, only students from the Express stream in Secondary 2 were included. Although a large representative sample provides a comprehensive picture for mathematics anxiety and test anxiety among Express stream Secondary 2 students, the findings cannot be generalised to all Secondary 2 students in Singapore as other streams (e.g., gifted, Normal (academic), and Normal (Technical) streams) were not included in this study. Further studies are required to determine if the trend observed in the present study occurs with students in other grade levels. Another potential limitation of the present study was the design. The decision to use two paper-and-pencil instruments allowed written data to be collected from a large sample. There is therefore possible bias in self-reporting which is difficult to control.

With regard to the interaction between mathematics anxiety and test anxiety, this correlational study provides no basis to conclude which is the cause or effect. The causes of mathematics anxiety and test anxiety are not clear cut. However, Newstead (1998) indicated that some of the causes of mathematics anxiety could relate to teacher anxiety, societal, educational or environmental factors, innate characteristics of mathematics, failure and the influence of early-school experiences of mathematics. In fact, the beginnings of mathematics anxiety can often be traced to negative classroom experiences which appear to be particularly strong and well-documented (Stodolsky, 1985; Tobias, 1978). Thus, it is considered crucial to examine classroom practice and establish whether the roots of mathematics anxiety may be in instructional methods and in the quality of mathematics teaching. Another difficulty lies in being able to identify and empathise with the students as some mathematics teachers in the secondary school may have not experienced mathematics anxiety.

An implication from this study is the fact that teachers play an important role in helping students to overcome their mathematics anxiety. It is recommended that the Fennema-Sherman Mathematics Anxiety Scale (MAS) be conducted at the start of each academic year. Whatever the results of the survey measured by the Fennema-Sherman Mathematics Anxiety Scale (MAS), they are aimed at providing mathematics teachers with a deeper understanding of students' perception of mathematics learning in the classroom. Whether it is mathematics anxiety or otherwise, early detection and appropriate actions would assist students to overcome their mathematics anxiety. In particular, it is important for classroom teachers to have an awareness of the kind of remedial work, individual help and counselling that can assist students to reduce their mathematics anxiety in learning mathematics.

References

Ashcraft, M.H. (2002). Math anxiety: Personal, educational, and cognitive consequences. *Current Directions in Psychological Science, 11*(5), 181–185.

Ashcraft, M. H., & Ridley, K. S. (2005). Math anxiety and its cognitive consequences: A tutorial review. In J. I. D. Campbell (Ed.), *Handbook of mathematics cognition* (pp. 315-327). Hove: Psychology Press.

Bandalos, D. L., Yates, K., & Thorndike-Chirst, T. (1995). Effects of math self-concept, perceived self-efficacy and attributions for failure and success on test anxiety. *Journal of Educational Psychology, 87*, 611-623.

Bessant, K. C. (1995). Factors associated with types of mathematics anxiety in college students. *Journal for Research in Mathematics Education, 26,* 327-345.

Brush, L. R. (1981). Some thoughts for teachers on mathematics anxiety. *Arithmetic Teacher, 29*(4), 37-39.

Buxton, L. (1984). *Do you panic about maths?* London: Heinemann.

Carpenter, T. P. (1980). Students' affective responses to mathematics: Results and implications from national assessment. *Mathematics Teacher, 73*(7), 531-539.

Cemen, P. B. (1987). *The nature of mathematics anxiety.* (Report No. SE 048 689). Oklahoma State University. (ERIC Document Reproduction Service No. ED 287729).

Cohen, L., & Manion, L. (1994). *Research methods in education* (4th ed.). London: Routledge.

Dawson, R.H. (2001). Effects of test anxiety and evaluative threat on students' achievement and motivation. *The Journal of Educational Research, 94*(3), pp. 223-253.

Deffenbacher, J. L. (1980). Worry and emotionality in test anxiety. In I. G. Sarason (Ed.), *Test anxiety: Theory, research and applications* (pp. 111-124). Hillsdale, NJ: Erlbaum.

Evans, J. (2000). *Adults' mathematics thinking and emotions: A study of numerate practices.* London: Routledge-Falmer.

Fennema, E., & Sherman, J.A. (1978). Sex-related differences in mathematics achievement and related factors: A further study. *Journal for Research in Mathematics Education, 9*, 189-203.

Foong, P. Y. (1984). *Anxiety and mathematics performance in female secondary school students in Singapore.* Unpublished Master's dissertation, Monash University, Australia.

Foong, P. Y. (1987). Anxiety and mathematics performance in female secondary students in Singapore. *Singapore Journal of Education, 8*(2), 22-31.

Hart, L. E. (1989). Describing the affective domains: Saying what we mean. In D. B. McLeod & V. M. Adam (Eds.), *Affect and mathematical problem solving: A new perspective* (pp. 37-45). New York: Springer-Verlag.

Hembree, R. (1987). Effects of noncontent variables on mathematics tests performance. *Journal for Research in Mathematics Education, 18*(3), 197-214.

Hembree, R. (1988). Correlates, causes, and treatment of test anxiety. *Review of Educational Research, 58,* 47-77.

Hembree, R. (1990). The nature, effects and relief of mathematics anxiety. *Journal for Research in Mathematics Education, 21*(1), 33-46.

Ikegulu, T. N. (1998). *Mathematics anxiety-apprehension survey*. Retrieved October 20, 2003, from http://mathforum.org/epigone/math-teach/skimpplenkhand/r06mug8st251@legacy.

Jackson, C.D., & Leffingwell, R.J. (1999). The role of instructors in creating math anxiety in students from kindergarten through college. *Mathematics Teacher, 92*(7), 583-586.

Kazelskis, R., Reeves, C., Kersh, M. E., Bailey, G., Cole, K., Larmon, M., Hall, L., & Holiday, D. C. (2000). Mathematics anxiety and test anxiety: Separate constructs? *Journal of Experimental Education, 68*(2), 137-146.

Lee, K.M.C. (2003). *Test anxiety and academic performance: The effect of a class based stress management programme.* Unpublished Master's thesis, Nanyang Technological University, Singapore.

Lenden-Hitchcock, Y. P. (1994). *Gender differences in mathematics achievement among gifted secondary students in Singapore.* Unpublished Master's dissertation, Nanyang Technological University, Singapore.

Liebert, R.M., & Morris, L.W. (1967). Cognitive and emotional components of test anxiety: A distinction and some initial data. *Psychological Reports, 20*, 975-978.

Loke, M.L.T., & Vikaneswari, R. (1999). *Mathematics anxiety among the Form Four stream students in Malaysia.* Proceedings of Malaysian Educational Research Association (MERA) and the Educational Research Association (ERA) joint Conference 1999 (pp. 1258-1264).

Ma, X. (1999). A meta-analysis of the relationship between anxiety toward mathematics and achievement in mathematics. *Journal for Research in Mathematics Education, 30*(5), 520-540.

McLeod, D.B. (1992). Research on affect in mathematics education: A reconceptualisation. In D.A. Grouws (Ed.), *Handbook for research in mathematics teaching and learning* (pp. 575-596). New York: Macmillan.

Milgram, N., & Toubiana, Y. (1999). Academic anxiety, academic procrastination and parental involvement in students and their parents. *British Journal of Educational Psychology, 69*(3), 345-361.

Ministry of Education. (2000a). *Third International Mathematics and Science Study 1999 (TIMSS 1999): National report for Singapore.* Singapore: Author.

Ministry of Education. (2000b). *A study of pupils' stress (1999).* Singapore: Author.

Newstead, K. (1998). Aspects of children's mathematics anxiety. *Educational Studies in Mathematics, 36*(1), 53-71.

Renga, S., & Dalla, L. (1993). Affect: A critical component of mathematical learning in early childhood. In R. J. Jensen (Ed.), *Research ideas for the classroom: Early childhood mathematics* (pp. 22-42). New York: Macmillan.

Robson, C. (1993). *Real world research: A resource for social scientists and practitioner-researchers.* Oxford: Blackwell.

Sarason, I. G. (1986). Test anxiety, worry and cognitive inference. In R. Schwarzer. (Ed.), *Self-related cognitions in anxiety and motivation* (pp. 19-33). Hillsdale, NJ: Erlbaum.

Schoenfeld, A. H. (1985). *Mathematical problem solving*. Orlando, FL: Academic Press.

Spielberger, C. D. (1972). Conceptual and methodological issues in anxiety research. In C. D. Spielberger (Ed.), *Anxiety: Current trends in theory and research* (Vol. 2). New York: Academic.

Spielberger, C. D., Gozalez, H. P., Taylor, C. J., Ross, G. R., & Anton, W. D. (1977). *Manual for the test anxiety inventory.* Palo Alto, Ca: Consulting Psychologists Press.

Stodolsky, S.S. (1985). Telling math: Origins of math aversion and anxiety, *Educational Psychologist, 3*, 125-133.

Tan, O. S. (1990). *Mathematics anxiety, locus of control and mathematics achievement of secondary school students.* Unpublished Master's dissertation, National University of Singapore, Singapore.

Tankersley, K. (1993). Teaching math their way, *Educational Leadership, 50*, 12-13.

Tanzer, N. K., & Sim, C. Q. E. (1991). *Test anxiety in primary school students: An empirical study in Singapore* (Research Report No. 1991/6). Department of Psychology, University of Graz, Schubertstr.

Tay, B. L. (2001). Do smart kids have math anxiety? In K. Y. Wong & H. H. Tairab, *Energising science, mathematics and technical education for all. Proceedings of the Sixth Annual Conference of the Department of Science and Mathematics Education* (pp. 307-316). Brunei: Universiti Brunei Darussalam.

Tobias, S. (1978). *Overcoming math anxiety.* New York: Norton.

Tobias, S., & Weissbrod, C. (1980). Anxiety and mathematics: An update. *Harvard Educational Review, 50*(1), 63-70.

Wigfield, A., & Meece, J. L. (1988). Math anxiety in elementary and secondary school students. *Journal of Educational Psychology, 80*, 210-216.

Williams, J. E. (1994). Anxiety measurement: Construct validity and test performance. *Measurement and Evaluation in Counseling and Development, 27,* 302-307.

Wine, J. D. (1980). Cognitive-attentional theory of test anxiety. In I. G. Sarason (Ed.), *Test anxiety: Theory, research and applications* (pp. 349-385). Hillsdale, NJ: Erlbaum.

Wood, E. F. (1988). Math anxiety and elementary teachers: What does the research tell us? *For the Learning of Mathematics, 8*, 8-13.

Chapter 14

Positive Social Climate and Cooperative Learning in Mathematics Classrooms

LUI Hah Wah Elena TOH Tin Lam CHUNG Siu Ping

This chapter presents two research projects in empowering the low achievers in mathematics to learn the subject. Project 1 was a cross-discipline quasi-experimental study on Positive Social Climate for Enhancing Pupils' Mathematics Self-concept, carried out in four Singapore secondary schools in years 2003 - 2004. The objective of this first project was to find the attributes in the social climate which can account for the increase in self-concept of Secondary 2 pupils in mathematics remedial classrooms. The outcomes of Project 1 inspired a group of mathematics teachers in one of these four schools to mount their action research (Project 2), on adopting cooperative learning in teaching mathematics at the lower secondary level in the years 2005 – 2006. Project 2 aimed to identify ways to implement cooperative learning strategies to help low achievers in mathematics improve their performance in this subject. A framework to guide future implementation of cooperative learning in Singapore mathematics classrooms is proposed.

Key words: social climate, mathematics self-concept, cooperative learning, low achieving pupils

1 INTRODUCTION

Teaching mathematics to low achievers has always been a challenge faced by school teachers. Educational programmes worldwide have evolved to one that is more theory-based than emphasis on skills (Glass,

2003). This has put the lower achieving pupils at a disadvantage; these pupils are generally visual and kinaesthetic learners (Amir & Subramanian, 2007; Rayneri & Gerber, 2003).

Anecdotal evidence from schools has suggested that the lower achieving pupils generally show little interest in academic subjects, are not focused, and tend to be restless in classes. Most likely the teaching approaches might not be suitable for them. Thus, teachers need to be more sensitive to the individual differences of the pupils in the classroom. According to Myron and Keith (2007), teachers may be more successful if they take note of the different learning styles of their pupils and try different teaching methods.

This chapter presents two research projects in helping the low achieving mathematics pupils learn the subject. The first project "Positive Social Climate for Enhancing Pupils' Mathematics Self-concept" was a cross-discipline quasi-experimental study funded by a research grant of the National Institute of Education. The objective of this Project 1 was to find the attributes in the social climate which were accountable for the increase in self-concept of Secondary 2 pupils in mathematics remedial classrooms. The project was carried out in four Singapore secondary schools. The outcome of this project inspired a group of mathematics teachers to mount their action research on adopting cooperative learning in teaching mathematics. This was Project 2.

Project 2 aimed to identify ways to implement cooperative learning strategies to improve the learning of mathematics amongst the low achievers in the Express Course in a neighborhood school. Based on the research experience of these two projects, the authors propose a framework to guide future implementation of cooperative learning in Singapore mathematics classroom.

2 LITERATURE REVIEW

2.1 *Individualised Instruction*

Mathematics is an extremely powerful tool to represent the world. Its power comes from its abstraction. Because of its abstract nature, mathematics seems to develop without any particular context, but it can be applied to a variety of real world situations. Mathematicians are enchanted with the abstract beauty of mathematics. However, this abstraction is the source of much of the difficulty for learning and teaching mathematics (Weissglass, 1990).

In a large mathematics classroom, it is not possible for a teacher to use one set of instructions without either wasting the time of the faster pupils, who need not review many basic concepts in the lower order concepts in the hierarchy, or losing pupils who are weaker in foundation, who may need much time to review basic mathematical concepts.

Thus, a natural progression in mathematics education is the emergence of individualised instructions. Individualised instructions offer instructions that are appropriate for each pupil's learning needs.

2.2 *Social Interaction and Cooperative Learning of Mathematics*

Burns (1990) proposed that social interaction is one of the key points in learning mathematics. The more opportunities pupils have for social interaction with their peers, parents, and teachers, the more viewpoints they will encounter. Confronting pupils with others' thoughts provides them with a perspective on their own ideas, stimulating them to think through their own viewpoint.

This is in line with Vygotsky's Theory (1982) of social constructivism. According to Vygotsky, pupils learn well when they are involved in group activities. The increase in such interactions will open an avenue for a better use and the widening of the zone of proximal development for each pupil. Cooperative learning is based on this theory.

Davidson (1990) noted that it is difficult to give a precise definition of *cooperative learning* because of the large variety of learning settings that can be seen as facilitating cooperative learning. Cooperative learning involves pupils working in groups, with members helping one another on a task to achieve a common group goal (Seow, 1988). Leikin and Zaslavsky (1997) proposed four necessary conditions for a cooperative-learning setting: small cooperative groups, learning tasks that require pupils to mutually and positively depend on one other, a learning environment for equal opportunity to participate, and accountability for progress of the group.

Cooperative learning methods generally work well for all types of pupils, regardless of learning style, age, gender, and ability levels (Goodwin, 1999; Slavin, 1996). Leikin and Zaslavsky (1997) showed that through the use of cooperative learning, there is an increase in pupils' activeness in mathematics learning, and a shift toward pupils' on-task verbal interactions. Various opportunities are available for pupils of different levels to receive help, and many pupils have positive attitude towards the cooperative learning approach. They also noted better mathematics achievement under cooperative learning compared a control group. They supported the use of heterogeneous grouping and interaction among different groups.

King (1993) pointed out that in order to improve the efficiency of small groups, the pupils must be prepared to work and learn within their groups. Thus, the teachers should train pupils to get used to helping one another to learn. Furthermore, the group tasks should progress from straightforward to more complex and demanding ones.

While a majority of the studies have shown favourable results for cooperative learning, some studies have produced contradictory evidence. For example, King (1993) observed that in a heterogeneous cooperative learning group, although the low achievers were active in the learning process, it was still the high achievers who assumed dominant roles in carrying out the group tasks, including the frequency and quality of contributions to group efforts. He also found that the low achievers were not learning much from cooperative learning setting.

The discrepancy in findings has led researchers to seek conditions under which cooperative learning can be effective. This chapter offers some insights from Singapore.

3 PROJECT 1: POSITIVE SOCIAL CLIMATE FOR ENHANCING PUPIL'S MATHEMATICS SELF-CONCEPT

There are many international studies on pupils' academic achievement, but few are research on the relationship between the pupils' mathematics self-concept and social climate. This project was the first of its kind in Singapore. It was carried out in the years 2003 – 2004. This project had two phases, the first phase aimed to validate the instruments used in the intervention which was the second phase.

3.1 *Phase 1 – Instrumentation*

In Phase 1, more than 700 Secondary 2 pupils from four neighbourhood schools took part in the survey conducted in 2003. Pupils from all the three streams (Express, Normal Academic and Normal Technical) participated in this study. In Singapore education system, pupils are streamed into these courses based on their Primary School Leaving Examination (PSLE) results.

The instruments were the Self-Description Questionnaire (SDQ)II on Self-concept (Marsh, 1990) and "What Is Happening In This Class?" (WIHIC) (Fraser & Yan, 2000) on social climate. They were validated together with the Motivational Orientation scales and Intellectual Achievement Responsibility (IAR) questionnaire. Significant correlations were only found among the SDQII and WIHIC scales and subscales. Therefore, other two scales were not used in the pre-test and post-test of the intervention in Phase 2.

Marsh and Shavelson (1985) postulated the multi-faceted hierarchical structure of self-concept to include academic self-concept, non-academic self-concept, and general self-concept. The Self-Description Questionnaire SDQII has 102 items in three areas of academic self-concept (*Mathematics, Verbal, General School*), seven areas of non-academic self-concept (*Physical Abilities, Physical Appearance, Opposite-sex Relations, Same-sex Relations, Parent Relations, Honesty-Trustworthiness,* and *Emotional Stability*), and General Self-concept, which is derived from Rosenberg's self-esteem scale (Rosenberg, 1965). The sum of these 11 scales yields a Total Self-concept score. This scale reflects an adolescent's self-ratings in various areas of self-concept. Each item requires one of the six responses: False, Mostly False, More False than True, More True than False, Mostly True, or True. In this study, only four of the SDQII scales were used: Mathematics (MATHSC, 10 items), Parent Relation (PARSC, 8 items), General School (SCSC, 10 items), and General (GENSC, 10 items).

The "What Is Happening In This Class?" (WIHIC) questionnaire has seven scales (with 70 items): *Pupil Cohesiveness* (SSCOH), *Teacher Support* (TSUP), *Involvement* (INVOLV), *Investigation* (INVEST), *Task Orientation* (TASKOR), *Cooperation* (COOP), and *Equity* (EQUITY). Each item is responded on a 5-point scale: Almost Never, Seldom, Sometimes, Often, and Very Often. The WIHIC questionnaire was validated in a local study (Fraser & Yan, 2000). In that study, a sample of 2310 pupils from 75 Secondary 4 classes was used.

For the SDQII, the Cronbach's alphas of the four scales (*Mathematics, Parent Relations, General School*, and *General*) ranged from 0.79 to 0.87, and 0.90 for the total of the four scales. The Cronbach's alphas of the sevenWIHIC scales ranged from 0.83 to 0.89, and 0.96 for the total scale. These findings suggest that these scales are reliable for local secondary school pupils.

The inter-scale correlation coefficients of the four SDQII scales and seven WIHIC scales are all significant ($p < .05$, 2-tailed). See Table 1.

The four sub-scales of SDQII have correlation coefficients ranging from 0.235 to 0.746. The General Self (GENSC) and General School (SCSC) scales have very strong relationship ($r = 0.746$). The Mathematics Self-concept (MATHSC) scale's correlation with the seven WIHIC scales range from 0.078 (Cooperation) to 0.239 (Investigation). Mathematics Self-Concept has the strongest relationship with Investigation ($r = 0.239$) followed by Task Orientation ($r = 0.214$).

Table 1

Pearson Correlations of Self-Concept & Social Climate Scales

	MATHSC	GENSC	PARSC	SCSC
GENSC	.324			
PARSC	.235	.396		
SCSC	.244	.746	.309	
SSCOH	.099	.325	.248	.367
TSUP	.174	.313	.179	.321
INVOLV	.085	.272	.127	.317
INVEST	.239	.314	.155	.291
TASKOR	.214	.420	.271	.422
COOP	.078	.304	.306	.325
EQUITY	.163	.338	.242	.311

The significant and positive relationship between the SDQII (self-concept) and WIHIC (social climate) suggests that mathematics self-concept enhancement should focus on the interaction between teacher and pupils, and among pupils, in other words, social cohesion and teacher support. The strong relationship between mathematics self-concept with Investigation and Task Orientation indicates attention on these two areas of classroom interaction.

There were gender differences in both SDQII and WIHIC. For SDQII, the male mathematics self-concept mean score was significantly higher than that of the female (see Table 2).

Table 2
Mathematics Self-concept Significant Difference of Mean Scores by Gender

	Mean Scores		
	M (N=322)	F (N=360)	F-value
MATHSC	3.75	3.34	27.31 p < .001

In WIHIC, there were also significant gender differences in three sub-scales: Investigation in favour of males; in both Cooperation and Pupil Cohesiveness, females' mean scores were higher than the males' mean scores (see Table 3).

Table 3
WIHIC (3 scales) Significant Difference of Mean Scores by Gender

	Mean Scores		
	M (N=321)	F (N=356)	F-value
SSCOH	3.59	3.74	9.03 p = .003
INVEST	2.89	2.81	2.82 p = .009
COOP	3.50	3.81	3.56 p = .001

3.2 *Phase 2 – Intervention*

At the beginning of Phrase 2, in May 2004, a one-day workshop at NIE was specially designed and conducted for twelve mathematics teachers from two schools which were this study's experimental groups. The instructional design of this workshop was based on the data analysis of Phase 1 which indicated that amongst the subscales of the classroom climate instrument, the significant predictors of Mathematics Self-concept were investigation, task orientation and cooperation; the predictors of School Self-concept were task orientation, student cohesiveness and teacher support; and General Self-concept: task orientation, investigation, student cohesiveness, equity and cooperation. The contents of this workshop covered:

- Briefing on the research projects in "Positive Social Climate" and "Self-Concept"
- Cooperative Learning: Concepts & Application
- Helping low achieving pupils in mathematics
- Group discussion and sharing of helpful strategies in Positive Social Climate for Enhancing Students' Mathematics Self-concept

Subsequently, the teachers from one of these schools planned and carried out activities to enhance students' capabilities and confidence in learning mathematics in all its Secondary 2 classes, except the best class. The intervention, which took place in the 2004 July semester, included the following features:

- As a team, the teachers identified appropriate sub-topics and took ownership in developing lesson plans and activities.
- Leveraged on the school's professional sharing scheme to share, refine and reflect upon lesson plans.
- Mixed-ability, same-gender groupings were formed. The teachers created the initial groupings, pupils were allowed to request for change.

Guided by the lesson planning template provided by trainers from NIE, the teachers developed lesson resources and pedagogy to meet instructional objectives in five domains: Content, Affective, Social Skills, Cognitive and Meta-cognitive. These were in line with the focus of the research study.

There were significant increases from pre-test to post-test results for three variables: general self-concept (Gen SC), mathematics self-concept (Math SC), and equity (Equity). The summary statistics is shown in Table 4.

Table 4
Summary Statistics for the General Self-concept, Mathematics Self-concept and Equity

Variable	Pre-test	SD	Post-test	SD	t-value	df	*p*-value
Gen	3.90	0.94	3.96	0.72	2.19	235	.05
Math SC	3.26	1.21	3.42	1.10	2.65	236	.01
Equity	3.36	0.87	3.51	0.82	2.86	231	.01

The main achievements of Project 1 were:

(1) The two instruments from Australia "Self-Description Questionnaire SDQII" and "What's Happening in This Class?" were validated.
(2) Twelve mathematics teachers from two experimental schools (groups) benefited from the training provided in this project.
(3) Mathematics teachers in one school reported that they had developed team spirit and new teaching strategies, resources, and lesson plans.

The outcomes of Project 1 inspired six mathematics teachers in one school to mount an action research in their school, which will be described as Project 2 below.

4 PROJECT 2: COOPERATIVE LEARNING FOR LOWER SECONDARY MATHEMATICS PUPILS

Project 2 was a school-based action research on cooperative learning for lower secondary pupils who were weak in Mathematics. It started in July 2005 and ended in June 2006. This project had two phases as explained below.

4.1 *Participants*

During Phase 1, the study involved 39 Secondary 2 pupils from one low achieving Express class and 40 pupils from a Normal (Academic) class doing the Secondary 2 Express mathematics syllabus. These pupils were less academically inclined, and most of them tended to be restless and disruptive in mathematics classrooms.

In Phase 2, the study involved 113 Secondary 2 pupils from three mixed ability Express classes. Unlike Phase 1, the pupils in Phase 2 were more heterogeneous in terms of academic performance. Based on their performance in the continual assessments and teachers' feedback, these pupils showed varying degrees of mathematics learning ability.

Throughout the two phases of the study, three mathematics teachers were involved in the study. They were the mathematics teachers teaching the classes and the Head of Mathematics Department. The study was arranged such that two teachers played the role of observers while one teacher taught the class.

4.2 *Procedures*

Three cooperative learning activities were included in Phase 1 and four in Phase 2. These activities were conducted for different mathematics topics. The findings of this study were largely based on direct observation of pupils participating in the cooperative learning activities and reflection on the activity-based lessons.

The teacher observers were asked to take note of the following during the activities:

(a) What are my observations on how the pupils worked together during their cooperative learning activities?
(b) Did the pupils achieve what I hoped they would achieve from the activities?
(c) To what degree, if at all, did I observe an improvement in the implementation of the activities?
(d) What specific things did I do that worked well?
(e) What specific things did I do that did not work well?

Tests were conducted before and after the cooperative learning activities for each topic to detect the effects of these activities. The participating pupils were asked to record their feelings about cooperative learning in their journals. In the pupils' journal writing, structured questions were designed by the teachers to facilitate the pupils to record their feedback about the cooperative learning activities.

4.3 *Phase 1: Findings*

4.3.1 Activity 1: Investigating the Properties of Linear Graphs Using Graph Plotter

This activity began with explicit instructions and lesson objectives for the pupils written on the worksheet. The activity was conducted in the computer laboratory. It was not a regular place where the pupils had their mathematics lessons. It took some time to get the pupils to sit according to the seating plan. Each pupil was assigned to a computer. The pupils were engaged in using the computer application to complete the activity. Even though the pupils were encouraged to work in pairs and cooperatively, no explicit instructions on how to cooperate were given. There was very little discussion between partners on their investigation as required in the activity. After the session, the pupils were asked to complete their journal writing about their feelings about the activity. The journal writing showed that most of the pupils could not state one thing that they had learnt from their partner or they had taught their partner. They gave low rating for confidence in completing the activity sheet correctly. Thus, teachers who intend to use cooperative learning have to cultivate the spirit of cooperation among the pupils, teach the pupils the cooperative skills, prompt them to use these skills during activities, and include cooperative skills usage by group members in the activity evaluation checklist.

4.3.2 Activity 2: Interpretation of Distance-Time Graphs

This activity was arranged two weeks after Activity 1 was carried out. For the period between Activity 1 and Activity 2, normal classroom teaching (without cooperative learning in place) was carried out.

Similar to Activity 1, this cooperative learning activity had explicit instructions and clear purpose of the activity written on the activity sheet. In addition, there was also explicit briefing prior to the start of the activity, including instructions on how to cooperate and how the pupils' work would be evaluated. According to the two teacher observers, cooperation and discussions were observed this time. Most pupils were able to state in the journal at least one thing that they had learnt from their partner or had taught their partner. They gave higher rating for confidence in completing the activity sheet correctly if they had worked with a partner.

For pairs where both pupils were mathematically weak, little learning took place. A pupil had reflected in the journal that "both of us don't know how to solve the problem." Pupils who lack discipline should not be paired for activity as they could easily lose focus. For example, one pupil from a pair of weak pupils commented in the journal that "I cannot concentrate during pair work. We end up talking nonsense." Another observation was that the duration of the activity must not be too long because pupils would start to chat and lose attention when they continued to sit in pairs for post-activity discussions.

The challenge for teachers is that the higher the percentage of weak and less disciplined pupils in the class, the harder it is to achieve ideal pairing of pupils. The activity has to include engaging tasks, relevant contents, and short duration to keep pupils on task. From this study, the activity should not take too long to complete. Classroom experience by senior teachers seems to suggest that a duration of 15 minutes would be suitable for cooperative learning. The duration could be slightly longer if the class was more disciplined.

4.3.3 Activity 3: Application of Trigonometric Ratios

A group activity involving teams of four pupils was organised for the class after the pupils were taught trigonometrical ratios. The activity required the use of a clinometer and application of trigonometrical ratios to determine the height of various physical structures in the school. Kagan (1985) suggested that group interaction should be structured to make individuals accountable for the outcome. To achieve this, structured activity which had specific tasks was given to pupils performing certain role in the group. Scores were allocated for the completion of each task. Prior to the start of the activity, the pupils were briefed about the objectives of the activity, the need for cooperation, the tasks to be completed, the data to collect, and allocation of scores. Each group of four then assigned the roles among themselves and recorded the roles on their activity sheets.

When the pupils were in the open area for the activity, it became more difficult to manage them than when they were in the classroom. It was a very challenging task to ensure that the pupils were on task. The pupils had to be reminded repeatedly of their roles, the tasks they were to accomplish and to stay with their group members. In every group, there were one or two pupils who did not carry out the tasks properly. In some instances, the more responsible members took over the tasks. When the pupils returned to the classroom, it took them unduly long to settle down to discuss the findings. In an interview after the activity, some pupils said that they were clear about the steps and things to do. However, some did not take the activity seriously, even though they knew the work would be graded. They did the minimal and left the rest to other members to complete.

It appears to be more effective to limit the activity to *pair work* that is of reasonably short duration (of about 15 minutes) so that there is a higher chance of success in implementing cooperative learning.

4.4 *Preparation for Phase 2*

Phase 1 of this project provided the team with more refined understanding of the cooperative learning situation to better plan the action intervention for Phase 2.

The cooperative learning activities for Phase 2 were mostly investigative in nature to guide the pupils in the discovery of mathematical concepts. The following arrangements were put in place at the start of the school year in 2006 and reinforced along the way:

(a) Class rules were spelt out clearly at the beginning of the year. They were firmly and fairly enforced.

(b) The pupils were seated such that those who were better in mathematics will partner the weaker ones. This is to facilitate peer tutoring and minimise movement of pupils when they form pairs for cooperative learning.
(c) The pupils were encouraged to help one another in learning through peer tutoring. This helped to cultivate a spirit of cooperation in class.
(d) The pupils were prompted to use cooperative skills when offering help or asking another classmate for help.

4.5 *Phase 2: Findings*

4.5.1 Activity 1: Direct Proportion and Inverse Proportion

The pupils had prior knowledge on solving arithmetical problems involving proportion. To reinforce the learning of proportion, the cooperative learning activity required pupils to work in pairs to investigate direct relationship and inverse relationship between two variables.

The pupils could get into pairs very quickly for the activity. They were on task in completing the activity. Most of the pupils were active in discussion, while some who were less familiar with one another tended to work on the activity individually. To find out the number of pupils who had learnt from the activity, pre- and post-tests on inverse proportion were administered. Across the three classes, there was a significant percentage increase in the number of pupils who could solve the word problem correctly.

4.5.2 Activities 2 and 3: Congruent and Similar Figures

These two activities required pupils to identify the corresponding angles and corresponding sides in the given figures. The pupils worked with the same partner for both activities. They were more familiar with one another as compared to Activity 1. They were chatty but still on task in working together to complete the activities. The noise generated during the activities was at a controllable level. From the problems that they were able to solve after the activities, it was evident that they understood the idea of corresponding angles and corresponding sides.

4.5.3 Activity 4: Sum of Interior Angles of a Polygon

The briefing for this activity required teacher demonstration of the construction of lines to form as many triangles as possible for each of the given polygons. In one of the classes, the teacher used the Fun with Construction software and a Tablet PC. With the Tablet PC, the teacher was able to use different colours to highlight pertinent points. As a result, the pupils took shorter time to understand what they were required to do and more of them did the activity correctly as compared to the other two classes which did not use the Tablet PC for the activity briefing. This confirms that explicit explanation and clear instructions are important for successful implementation of cooperative learning. Tools that can create a visual impact (such as the Tablet PC) would complement cooperative learning by enhancing the learning of visual learners.

At the end of each cooperative learning activity in Phase 2, the key learning points were discussed with the whole class, and the pupils were taught the application of the mathematical concepts in problem solving. The pupils were given seatwork in class to practise solving some of the problems.

4.6 *Discussion of Findings*

Phase 1 of the study did not result in significant improvement in mathematics learning. However, there were findings about the essential elements for successful implementation of cooperative learning, such as setting a culture for cooperative learning, encouraging pupils to use cooperative skills, pairing the pupils according to their different abilities in mathematics, and providing explicit activity instructions. These concurred with what are discussed in the literature (Artzt, 1999; Johnson & Johnson, 1992; Kagan, 1994).

These elements were consciously put in place in Phase 2. Our key proposal was that when pupils are less disciplined and academically weak,

(a) activities should be more structured;
(b) interaction should preferably be confined to small group such as pairs;

(c) greater emphasis should be placed on individual accountability;
(d) duration of individual activities should be short; and
(e) activities should preferably be in classroom setting or otherwise, the teacher-to-pupil ratio should be increased to 2 or more for outdoor activities.

The implementation of cooperative learning in Phase 2 was more successful than in Phase 1. Several key factors had contributed to better outcomes in Phase 2. They include the following:

(a) Phase 2 started at the beginning of a new academic year. It was timely to start a class culture as well as to encourage pupils to use cooperative skills and to peer teach to support cooperative learning.
(b) There were more pupils with average academic ability in the class for Phase 2. It was easier to manage their learning and behaviours as compared to pupils in Phase 1, where most of them were academically weak and had difficulties staying focused during lessons.
(c) The Secondary 2 Express classes in Phase 2 were mixed-ability. There was a better mix of academically more able pupils to pair with the less able ones for cooperative learning.
(d) Peer tutoring worked better for mixed ability classes in Phase 2. For mixed-ability classes, there were more pupils who were keen to learn from one another.

After the series of cooperative learning activities in Phase 2, a survey was conducted. Due to schedule constraints, the survey was only conducted for two of the three participating classes. The pupils were asked in a survey whether they were able to help one another during cooperative learning. About 80% of the pupils gave positive responses. Their comments showed a high degree of positive interdependence and demonstration of peer tutoring. Some examples are:

We shared our thoughts and helped each other ... discovered other ways to solve the problem ...

My partner helped me to spot careless mistakes and common errors.

We "argued" over who was right and who was wrong .... We learned from one another in the process.

It's easier to communicate ... more confident in telling the problems to my partner....

A few pupils expressed negative responses, and the common comments were that they found it difficult to concentrate on the activity when working in pairs and they tended to talk about other things. There were fewer such comments in Phase 2 compared to Phase 1. Despite that, most pupils who participated in Phase 1 indicated in their journal that they wanted cooperative learning activities because they felt that working with peers could help them learn better.

5 A COOPERATIVE LEARNING FRAMEWORK

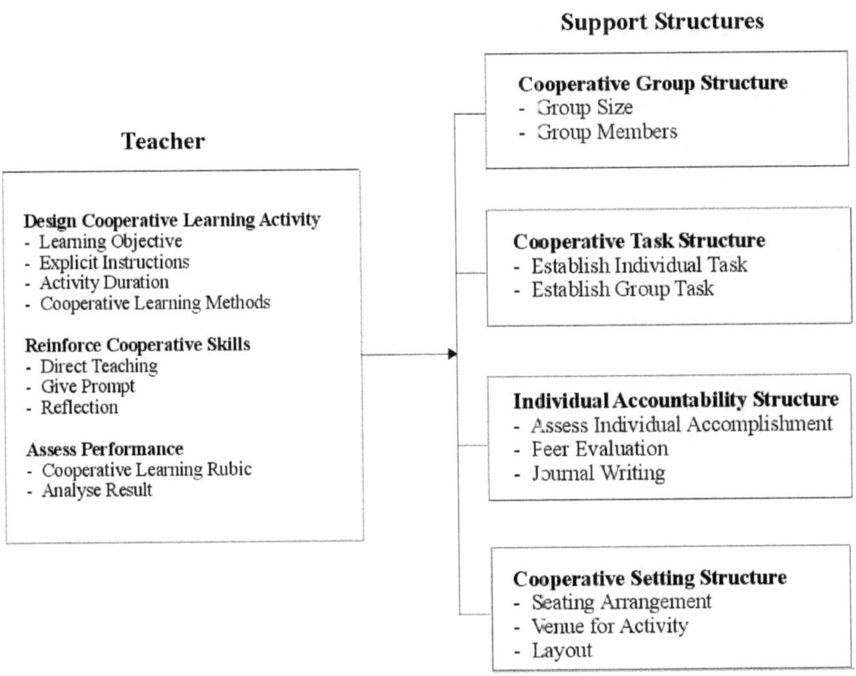

Figure 1. A Cooperative learning framework.

For the purpose of guiding future implementation of cooperative learning, we propose the cooperative learning framework shown in Figure 1. The framework consists of components that structure interaction based on the essential elements of cooperative learning gathered from Project 2 of this study.

6 CONCLUSION

In summary, the implementation of cooperative learning strategies requires careful attention to essential elements of cooperative learning. The proposed framework provides a guide on what the key elements are in structuring interaction for effective cooperative learning for Singapore mathematics classrooms. Equally important, as Artzt (1999) has stressed, the teachers need to believe that classroom activities can best be carried out through cooperative learning. Otherwise, the teachers might end up implementing the strategies of cooperative learning while preserving little of the intent of the designers of the cooperative techniques.

The mathematics teachers' personal quality and their relationship, and communication with pupils could make a difference to pupils' learning. This was supported by the findings of Phase 1 and also a research by Kaur (2003). Future studies in action research may look into the gender differences and draw the participating teachers' attention to these variances.

Other collaborators:
Project 1: Lim Kam Ming, and Liu Woon Chia.
Project 2: Loo Liat Siang, Ann Chiam, Paul Lee, and Geoffrey Cheang.

References

Amir, N., & Subramaniam, R. (2007). Making a fun Cartesian diver: A simple project to engage kinaesthetic learners. *Physics Education, 42*(5), 478 – 480.

Artzt, A.F. (1999). Cooperative learning in mathematics teacher education. *Mathematics Teacher, 92*(1), 11 – 17.

Burns, M. (1990). The math solution: Using groups of four. In N. Davidson (Ed.), *Cooperative learning in mathematics: A handbook for teachers* (pp. 1–11). California: Addison-Wesley.

Davidson, N. (1990). Small-group cooperative learning in mathematics. In T.J. Cooney & C.R. Hirsh (Eds.), *Teaching and learning mathematics in the 1990s* (pp. 52-61). Reston, VA: National Council of Teachers of Mathematics.

Fraser, B.J., & Yan, H.C. (2000, April). *Classroom environment, self-esteem, achievement and attitudes in geography and mathematics in Singapore.* Paper presented at the American Educational Research Association Annual Meetings, New Orleans.

Glass, S. (2003). The uses and applications of learning technologies in the modern classroom: Finding a common ground between kinaesthetic and theoretical delivery. *Educational Research Report. Information Analyses (070).*

Goodwin, M. W. (1999). Cooperative learning and social skills to teach and how to teach them. *Intervention in School and Clinic, 35* (1), 29-33.

Johnson, D. W., & Johnson, R. T. (1992). *Positive interdependency: The heart of cooperative learning.* Edina, MN: Interaction.

Kagan, S. (1985). *Cooperative learning: Resources for teachers.* Riverside, CA: Kagan Publishing.

Kaur, B. (2003, November). *Excellence in the mathematics classroom: A look at the teacher.* Paper presented at the Educational Research Association of Singapore Conference, Singapore.

King, L.H. (1993). High and low achievers' perceptions and cooperative learning in two small groups. *The Elementary School Journal, 93*(4), 399 – 416.

Leikin, R., & Zaslavsky, O. (1997). Facilitating pupil interactions in mathematics in a cooperative learning setting. *Journal of Research in Mathematics Education, 28*(3), 331 – 354.

Leikin, R., & Zaslavsky, O. (1999). Cooperative learning in mathematics. *Connecting Research to Teaching, 92*(3), 240 – 246.

Marsh, H.W. (1990). *Self-description questionnaire (SDQ)II: A theoretical and empirical basis for the measurement of multiple dimensions of adolescent self concept: An interim test manual and a research monograph.* San Antonio, TX: The Psychological Corporation.

Marsh, H.W., & Shavelson, R.J. (1985). Self-concept: Its multifaceted, hierarchical structure. *Educational Psychologist, 20*(1), 107 – 125.

Myron, H.D., & Keith, H. (2007). Advice about the use of learning styles: A major myth in education. *Journal of College Reading and Learning, 37*(2), 101 – 109.

Rayneri, L.J., & Gerber, B. (2003). Gifted achievers and gifted underachievers: The impact of learning style preferences in the classroom. *Journal of Secondary Gifted Education, 14*(4), 197 – 204.

Rosenberg, M. (1965). Society and the adolescent self-image. Princeton, NJ: Princeton University Press.

Seow, A. (1988). Maximizing pupil involvement through cooperative learning. *TELL Magazine*, *4*(2). Singapore: Curriculum Planning Division, Ministry of Education.

Slavin, R. E. (1996). Cooperative learning in middle and secondary schools. *Clearing House*, *69*(4), 200-204.

Vygotsky, L.S. (1982). Thought and language. In L.S. Vygotsky, *The collected works of L.S. Vygotsky* (pp. 5 – 36). Moscow: Pedagogika.

Weissglass, J. (1990). *Cooperative learning using a small-group laboratory approach.* In N. Davidson (Ed.), *Cooperative learning in mathematics: A handbook for teachers* (pp. 105 – 132). California: Addison-Wesley.

Chapter 15

Mathematics Curriculum for the Gifted in Singapore

KHONG Beng Choo

A mathematics programme for gifted students should reflect the rigour of the discipline and be responsive to the characteristics of the gifted students as well as the needs of the society. It should be differentiated from the mainstream curriculum. In Singapore, curriculum differentiation in the Gifted Education Programme is manifested in the content, process of teaching and learning, the kinds and quality of products expected from the students, and the learning environment. This chapter describes how various approaches are adopted by the programme to differentiate the mathematics curriculum for students in the programme.

Key words: gifted education, curriculum differentiation, Integrated Programme (IP), out-of-class enrichment

1 BACKGROUND

In Singapore, the Ministry of Education (MOE) identifies the top 1% of the age cohort of intellectually gifted students at Primary 4 and provides them with gifted education programming. The rationale of providing gifted education in Singapore is twofold. It is recognised that there should be special provisions to meet the needs of students who are intellectually gifted. Without special help, some gifted students might be frustrated, bored, and underachieve. Their outstanding talents may be lost to society. Gifted students must be stretched to realise their potential, and equally importantly, be nurtured so that they will take their place in society later on as morally upright and responsible individuals and citizens. Indeed, giftedness is a rare and invaluable human resource and should be carefully nurtured for the good of society and nation (Gifted Education Branch, 2004b).

The Gifted Education Programme (GEP) was first implemented in 1984 by MOE. It provides intellectually gifted students with appropriate opportunities to develop to their full potential. At Primary 3, students are invited to sit the Screening Tests in English Language and Mathematics after which about 4000 students are shortlisted to sit the Selection Tests in English Language, Mathematics, and General Ability. Admission to GEP is based solely on students' performance in the battery of selection tests. The 2-stage screening/selection identification process taps a wide range of intellectual abilities in verbal, numerical, and visual-spatial domains, and assesses students' readiness for the enriched GEP curriculum. Today, the GEP has classes from Primary 4 to 6 in nine primary schools with a total enrolment of about 1500: 500 each for Primary 4, Primary 5, and Primary 6.

The GEP is housed in different schools, and it is centrally monitored by the Gifted Education (GE) Branch in MOE. The GEP adopts self-contained classes for the gifted within a normal school so that gifted students can maintain regular contact with the other students. In such a model, while the gifted students are provided with an intellectually challenging curriculum, they are, at the same time, provided with full opportunities to socialise with their age peers.

At the secondary level, GEP students are served by the School-Based Gifted Education (SBGE) programmes in six Integrated Programme (IP) schools, which offer 6-year courses leading to either the A Level or the International Baccalaureate, or in the National University of Singapore (NUS) High School of Mathematics and Science, which offers a 6-year programme leading to the NUS High School Diploma. The IP was first implemented in 2004 in selected secondary schools and it provides a seamless upper secondary and junior college enriched education. Students in the IP do not sit the GCE O Level examination, a qualifying test for entry to junior college. The time "saved" by not having to prepare for the GCE O Level examination is used to develop students' intellectual curiosity, enrich their experiences, and provide a broad-based education that is more in tune with desired real-world competencies.

2 MATHEMATICS CURRICULUM IN THE GEP: GOALS

The GEP is a programme built on the mainstream curriculum. The mathematics curriculum in GEP expands on the mainstream mathematics framework in breadth and depth. The teaching approach takes into consideration the abstract structure of mathematics and processes involved in mathematical thinking, as well as the different needs of students of varying abilities in mathematics. While meeting these varying needs, the mathematics curriculum seeks to provide a firm foundation of concepts and thinking skills to enable students to effectively apply them in other disciplines and areas of specialisation. It also aims to provide appropriate guidance and training to promote and develop mathematical talents so that some may eventually become creative producers of mathematics.

The mathematics curriculum in the GEP aims to

- develop higher level critical and creative thinking skills in mathematics to enable the construction of relationships between and among mathematical topics and other disciplines as well as to emphasise concepts and principles rather than rules and procedures;
- empower students to become competent users and creators of mathematics and to provide opportunities to discover, recreate and find original and different solutions;
- develop students' abilities to communicate and reason mathematically using accurate use of mathematical terminology as well as to emphasise understanding of the logic behind new skills and concepts taught, and the awareness of metacognitive processes in solving closed and open-ended problems;
- develop students' disposition for mathematics, emphasising elegant solutions and multiple solutions to problems as well as the generation of a variety of products, including oral presentation, investigation and written work;

- nurture students' appreciation of the beauty and value of mathematics as well as to relate the content of mathematics to students' everyday experiences;
- provide opportunities for students to exercise leadership in pursuing special projects and investigations as well as provide opportunities for group and collaborative work.

Gifted Education Branch, Singapore, 2004a

3 CURRICULUM DIFFERENTIATION: APPROACHES

Instructional practices in classrooms for the gifted should emphasise the importance of concept development, thinking and reasoning, problem solving, and flexible accommodations for working with highly able learners (VanTassel-Baska, 2003). Effective teachers of the gifted emphasise higher order thinking skills that are integral to all content areas and in everyday life experiences (Paul, 1992). Learning activities for gifted students should provide them with appropriate level of challenge and opportunities to interact cognitively with the subject matter through problem-solving, experimentation, and inquiry where students are engaged in ideas generation, making conjectures and proving them rather than subscribing to the authority of a teacher's knowledge. Through such interaction, the student constructs his own knowledge base, which involves much deeper understanding and fluency in using the knowledge. Wheatley (1983) mentioned that much time is spent on how to use specific algorithms in solving problems. He argued that these basic skills can be taught to gifted students more effectively in the context of exploring concepts, solving problems, discovering new relationships, or conducting experiments.

The mathematics curriculum in the GEP is differentiated from the mainstream curriculum in the content, process of teaching and learning, the kinds and quality of products expected from the students, and the learning environment (Gifted Education Branch, 2004b). Differentiation not only allows teachers to build the learning experiences of gifted students beyond the core curriculum to provide them with breadth, depth and rigour of learning, it also respects the entry point of students who are not ready for the advanced content to learn at their own pace. Curriculum differentiation in the mathematics curriculum in the GEP is carried out via the multiple approaches as explained below.

3.1 *Problem Solving*

Gifted students are curious and creative. To them, problems should be the opportunity to act like "detectives." Good problems make students feel unsettled, surprised and even disturbed while attempting to solve them. Like detectives, they will need to explore all possibilities as each of these possibilities will provide them with some clue to solve the problem. Problem solving requires students to use reasoning to make connections between concepts, skills, and heuristics. In problem solving, gifted students generate their own ideas and ways of doing mathematics. These allow them to appreciate the beauty, power, and structure of mathematic and to enjoy the process of constructing meaning in mathematics.

While problem solving is the organising framework for the mathematics curriculum in Singapore, problem posing is a mathematical habit of mind that helps to sharpen students' mathematical inquiry skills. Eisner (2001) remarked that "the ability to raise telling questions is not automatic" (p. 370) and opportunities for the kind of thinking that yields good questions should be promoted. Students in GEP are provided with opportunities to formulate and frame problems in addition to working on well-defined tasks. Also, teachers can help students cultivate problem posing skills and habits of mind by thoughtful organisation of their classroom instruction, encouraging them to form their conjectures and challenge their conjectures.

3.2 *Level of Depth of Exploration of Concepts*

The GEP mathematics curriculum also focuses on proofs. Proving makes students take responsibility for their understanding and be more aware of not passing casual statements. Mathematical theorems are not fully understood by students if they are just cited or applied from books. To fully understand the theorems, students need to come up with a proof or at least be able to read and understand one. The discovery of a theorem usually proceeds in stages. The student has a hunch, makes a conjecture/guess, finds a counterexample, refines the guess, and tries to prove the conjecture. If unsuccessful, he/she modifies the conjecture, and repeats the process. Perseverance is required in this endeavor and the teachers' role is important here. In the process, students should experience success in mathematics and teachers should keep encouraging them to reinforce discipline and self-criticism, especially when students' numerous attempts to prove a theorem are not successful.

3.3 *Level of Complexity and Connectivity*

The GEP mathematics curriculum provides students with inter- and intra-disciplinary learning experiences. Real-world connections to the concepts and principles of mathematics are explored as part of or an extension to lessons in the mathematics classroom. Students are provided with opportunities to apply mathematics in other content areas and real-world problems. Traffic Statistics and Supermarket visits are two examples for Primary 4 students. In Traffic Statistics, students apply concepts and skills learned in the classroom to collect, analyse, interpret, and present data of traffic condition in the vicinity of their school. In the Supermarket Visit activity, students plan and decide on what items to purchase for a class party with a given budget. These activities are conducted during curriculum time.

3.4 *Level of Advanced Knowledge and Breadth of Content*

Gifted students learn at a faster pace compared to their mainstream peers. Teachers in the programme move more quickly through the core curricular materials and the extra time gained is used to cover topics in greater breadth and depth. Extensions take the form of applications of concepts and skills to everyday experiences, more challenging exercises with emphasis on complex problems and problem solving, introduction of new topics, further investigations and explorations of mathematical concepts and ideas, study of the history of mathematics and lives of great mathematicians, and integration with other subjects. For example, in the topic on Tessellation, students are introduced to the work of M.C. Escher and understand how a tessellating shape can be created by modifying a polygon using transformation. In the topic on Whole Numbers, lesson activities include different ancient numeration systems such as Mayan, Babylonian, Egyptian, Greek, and Roman numeration systems. Students compare these number systems with the Hindu-Arabic numeral system.

3.5 *Level of Independence in Learning*

GEP students are encouraged to read about the topics taught in class, and teachers follow up on the readings with activities and class discussions. Students are also encouraged to take up mathematics projects for their Individualised Research Study (IRS), a cornerstone of the GEP curriculum. IRS offers students the opportunities to expand their intellectual horizon and to develop their talents through the pursuit of areas of study of interest to them. They are also provided with opportunities to carry out other forms of research and experiments independently or with peers within and outside curriculum time. Primary 6 students in the programme are required to complete a research cum investigation and journal writing assignment individually. Each student chooses a topic from a list of topics (e.g., Card Tricks, Shortest Path/Least Cost Problem, and Board Game Design) suggested by their teachers and they research on the chosen topic to understand the mathematical ideas and to use a variety of problem solving heuristic and thinking skills to solve the problems. In addition to research on the topics, one of the requirements of the assignment is that students have to constantly reflect on their thinking and working process and to log them in their journal.

3.6 *Infusing Values in Mathematics*

There is a systematic infusion of values within the mathematics curriculum in the GEP. This is carried out via three approaches: real-world applications of mathematics with discussion of moral and social issues, life stories of mathematicians, and classroom activities. To illustrate the first approach, teachers use case studies with data drawn from sources such as national reports and daily news to teach topics on arithmetic and statistics and discuss the moral and social issues/implications from those cases. For the second approach, teachers use life stories of mathematicians to discuss the values of excellence and commitment. In addition, teachers also use classroom activities as an avenue to capitalise on the learning opportunities during their lessons to infuse values. They use questioning techniques to encourage clarity of thinking, group work to provide opportunities for students to work collaboratively, and student presentations to provide opportunities for students to articulate their thoughts and ideas or defend their arguments.

3.7 *Provision for Individualisation*

Gifted students are not homogeneous. They differ in their abilities, readiness to learn, interests, and styles of learning. Teachers in the programme design activities to allow students to discover ideas individually and provide opportunities for individual and group learning which accommodate individual or subgroup differences by allowing choice in materials and tasks.

3.8 *Pacing and Grouping*

Diagnostic pre-tests are used to assess students' entry knowledge. From the results, teachers can compact the lesson. Students are also grouped based on teachers' observation to assess students' ability in mathematics as well as their interests and attitudes towards mathematics learning. See exhibit in Figure 1.

3.9 *Variety of Teaching Strategies*

A variety of teaching strategies are used to facilitate and encourage higher order thinking such as reflective thinking, critical and creative thinking, and generalisation and abstraction. These are mathematical thinking essential for the general consumers of mathematics and critical for the training of potential creative producers of mathematics. With appropriate teaching strategies, mathematics is presented as a dynamic mental activity, with emphasis on concepts, principles, and the logic behind the new skills taught. Teaching strategies include Paul's Reasoning Model, inductive reasoning, deductive reasoning, and critical and creative thinking strategies. Students are often asked to think of alternatives and/or better solutions to problems and they are asked to assess and analyse their own solutions/proofs and synthesise methods of solution and the reasoning process. They are also provided with opportunities to compare and contrast ideas/solutions/methods, observe patterns to make generalisation from concrete data, and develop and elaborate their ideas.

Mathematics Classroom Observation Checklist

Instruction:

On a scale of 1 to 3, rate the frequency with which you observe the following characteristics in the student:
1 = Rarely
2 = Sometimes
3 = Always

	He/She…	Rating
1	can identify patterns and apply pattern recognition to reason mathematically.	
2	is fast and accurate in mental calculations.	
3	demonstrates ability and keenness in generalising an observed pattern.	
4	demonstrates ability to grasp mathematical concepts quickly.	
5	demonstrates interest and enthusiasm in figuring out new things, enjoys challenges and learning new things.	
6	has the end in mind and formulates a systematic approach before tackling a problem.	
7	examines alternative strategies to problem solving.	
8	approaches information provided with a critical mind by questioning its accuracy, reliability, or underlying assumptions.	

Figure 1. Teacher Observation Checklist
(Gifted Education Branch Mathematics Resource File)

3.10 *Level of Mathematical Communication in the Classroom*

Students in the GEP are engaged actively in asking questions and offering solutions and in explaining what they think or know about any problem. Students share mathematical ideas formally or informally with their peers and teachers. During formal presentations, they have to explain and defend their ideas.

3.11 *Out-of-Class Provisions*

Provisions for out-of-class enrichment are planned to inspire and motivate gifted students by introducing them to the beauty and variety of mathematics, to impart to them knowledge and skills for the early pursuit of mathematics, and to provide them with a supportive environment for interaction among peers who love mathematics. The Mathematics Masterclass and Advanced Mathematics Enrichment Class are organised to challenge students who have demonstrated exceptional ability in mathematics by introducing them to topics beyond the GEP mathematics curriculum. The Masterclass is a 3-day programme followed by a mentorship during which students work on a topic of their interest with a mentor. Topics in the Masterclass include Number Theory, Graph Theory, and Combinatorics and the focus is on developing problem-solving skills through the use of interesting and challenging problems. Participation in the Masterclass is dependent on the students' performance in a selection test conducted by the GE Branch.

Advanced Mathematics Enrichment Class is organised by each school offering GEP to cater to students who are mathematically precocious at Primary 5 or 6 level. The aim is to challenge them beyond the core curriculum. They participate in 2-hr classes weekly for a year conducted outside curriculum time. The advanced concepts taught in these classes complement the core contents in curriculum time. In the Enrichment Class, students learn different and efficient problem solving strategies including heuristics. Students who have demonstrated exceptional abilities in the Masterclass or Enrichment Class are selected to participate in mathematics competitions, e.g., Raffles Institution World Mathematics Contest and the International Mathematics and Science Olympiad.

Provisions for other out-of-class enrichment also include Mathematics Enrichment Day, Traffic Statistics and Supermarket Visit, and Mathematics Trail. The Mathematics Enrichment Day is a 1-day activity organised by the GE Branch and its aim is to introduce Primary 6 students to various problem solving strategies and to apply mathematical concepts to real-life problems. Students work in groups to solve mathematical puzzles at the station games which focus on mathematical communication and reasoning.

The Mathematics Trail is an activity for Primary 5 gifted students. The trail is usually conducted within the school compound, the school neighbourhood, or at parks or places of interest such as Sentosa and the Botanic Gardens. The Mathematics Trail provides a good opportunity for students to experience the relevance of mathematics in real life. Activities in the trail emphasise skills on estimation and mental calculation. It aims to debunk the myth that mathematics is all about formulae, precise answers, and doing "sums" with paper and pen.

4 MATHEMATICS CURRICULUM IN SCHOOL-BASED GIFTED EDUCATION (SBGE)

The School-based Gifted Education programmes of different IP schools have their origin in the GEP, as those IP schools have previously offered the Secondary GEP. IP schools work very closely with the GE Branch in tailoring their programmes to provide intellectual challenges and experiential learning to gifted students in their SBGE programmes. GE Branch also provides training for teachers of the gifted and consultancy services to these IP schools on the programmes and curriculum for the gifted. It also evaluates the SBGE programmes.

The curriculum design guidelines for SBGE mathematics curriculum are:

- Emphasis on conceptual understanding and substantive content;
- Emphasis on investigative processes in mathematics;
- Emphasis on mathematical habits of mind;
- Emphasis on critical and creative thinking;
- Emphasis on problem solving in mathematics;

- Opportunities for students to develop various advanced products;
- Diversity of resources & use of technology as a learning tool; and
- Provisions for out-of-class enrichment.

Gifted Education Branch, 2004c

5 MATHEMATICS CURRICULUM OF THE NUS HIGH SCHOOL

The National University of Singapore High School of Mathematics and Science (NUS High School, http://www.highsch.nus.edu.sg) provides an enriched and accelerated curriculum to engage bright students in mathematics and the sciences. The mathematics syllabus is compacted so that basic facts, concepts, and skills are taught in a shorter time and topics that are normally covered at higher levels are brought down to a lower level. It also adopts a modular system. Students can select modules that cover the same subject area to delve deeper into it and this allows students full rein to develop their interests and abilities. Alternatively, students can select modules across various subject areas so as to have a broad-based exposure. The curriculum also includes research and students are given the time and space to work on subject areas that they are interested in. In addition, each student in the programme is assigned a teacher-mentor who provides guidance and counselling to the student and also shares his or her aspiration, interest, and problems.

6 CONCLUSION

Gifted students have characteristics and learning needs that are different from mainstream learners. Their needs cut across affective, cognitive, and social domains. The GEP aims to meet these needs by a confluent approach that allows for advanced learning, with enriched and extended experiences. The mathematics curriculum in the GEP takes into consideration gifted students' advanced and varying abilities in mathematics and it is differentiated from the mainstream curriculum both in quality and quantity. The curriculum emphasises problem solving, both mathematical and in real-life world, exploration of concepts, advanced content, and more breadth in content. There are also various out-of-class provisions to cater to students with different interests and abilities. Through these approaches, the GEP mathematics curriculum seeks to provide a firm foundation of mathematics concepts and thinking skills to enable students to effectively apply these in other disciplines and areas of specialisation. It also seeks to develop mathematical talent in order that some students may eventually become creative producers of mathematics.

References

Eisner, E. (2001). What does it mean to say a school is doing well? *Phi Delta Kappan*, 82(5), 367-372.

Gifted Education Branch. (1991). *Philosophy of teaching mathematics to the Gifted.* Ministry of Education, Singapore.

Gifted Education Branch. (2002). *Guidelines for the Gifted mathematics programme.* Ministry of Education, Singapore.

Gifted Education Branch. (2004a). *Goals of the GEP mathematics curriculum.* Ministry of Education, Singapore.

Gifted Education Branch. (2004b). *A handbook for teacher in the Gifted Education Programme.* Ministry of Education, Singapore.

Gifted Education Branch. (2004c). *Mathematics curriculum design guidelines for school-based Gifted Education Programme.* Ministry of Education, Singapore.

Ministry of Education, Singapore. (2005). *Gifted Education Programme.* Retrieved March 10, 2007 from http://www.moe.gov.sg/gifted/

National Council of Teachers of Mathematics (NCTM). (2000). *Principles and standards for school mathematics.* Reston, VA: Author.

Paul, R. (1992). Critical thinking: What, why, and how. In C. A. Barnes (Ed.), *Critical thinking: Educational imperative* (pp. 3-24). San Francisco: Jossey-Bass.

VanTassel-Baska, J. V. (2003). *Curriculum planning and instructional design for gifted learners.* Denver, CO: Love Publishing.

Wheatley, G. (1983). A mathematics curriculum for the gifted and talented. *Gifted Child Quarterly, 27*(2), 77-80.

Chapter 16

Early Intervention for Pupils At-risk of Mathematics Difficulties

Fiona CHEAM **CHUA Wan Loo Jocelyn**

Learning Support for Mathematics (LSM) provides for the early identification of and intervention for pupils who are at risk of encountering mathematics difficulties. Pupils are systematically screened in the first year of school using the School Readiness Test (Mathematics). Individualised support is provided in small groups by specially selected teachers. A holistic approach to intervention is advocated in LSM, addressing the development of cognitive and metacognitive skills and pupils' motivation to learn, and capitalising on resources in the environment and the larger community. The effectiveness of LSM will be assessed through a monitoring and evaluation framework that has been designed specifically for LSM. To ensure that LSM teachers are well-equipped for their role, they are provided with training, mentoring, and on-going professional development opportunities.

Key words: Learning Support for Mathematics, School Readiness Test, evaluation framework, teacher training, metacognition, motivation, home environment

1 INTRODUCTION

The aim of the Singaporean education system is "to help our students to discover their own talents, to make the best of these talents and realise their full potential, and to develop a passion for learning that lasts through life" (Ministry of Education, 2006, p. 2). In the initial years of schooling, the focus of education is on developing language, literacy, and numeracy skills in pupils; these skills form the foundations for future learning.

For pupils who start school with limited numeracy skills, it is important that they are identified and provided with the necessary support (Gersten, Jordan & Flojo, 2005) as timely intervention has been shown to have positive effects (Fuchs, Compton, Fuchs, Paulsen, Bryang & Hamlett, 2005). Hence, a process of systematic identification and support of pupils who are at-risk of encountering mathematics difficulties has been put in place through the Learning Support for Mathematics (LSM) programme.

Learning Support for Mathematics (LSM) caters to pupils who start primary school with limited numeracy skills, with the aim of facilitating their access to the curriculum as early as possible. The key features of LSM are:

a. Systematic screening and identification of pupils early in the first year of school;
b. Provision of additional resources for schools to provide early intervention for identified pupils;
c. Teaching by specially trained LSM teachers.

Figure 1 below represents an overview of LSM, which includes processes for the systematic identification of pupils who require additional support, the provision of customised intervention, the

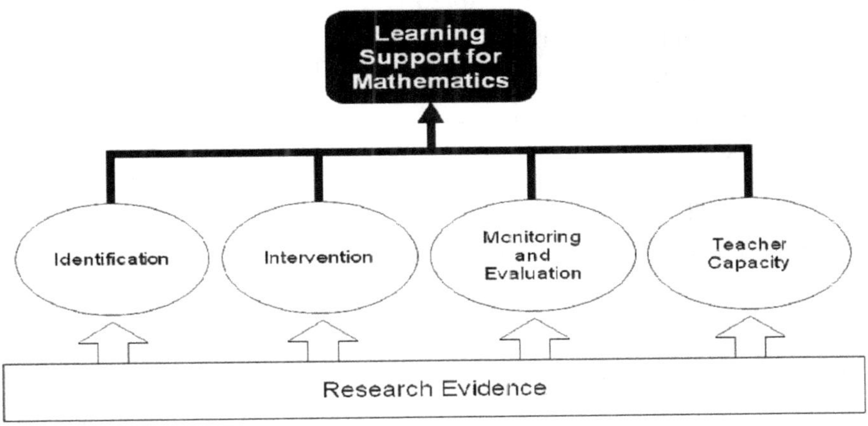

Figure 1. Components of Learning Support for Mathematics.

monitoring and evaluation of the support, and the building of teacher capacity to implement the support. The development of these components of LSM was guided by reference to evidence-based practices, as well as consultation with practitioners.

2 IDENTIFICATION

To facilitate the identification of pupils who require support, a School Readiness Test (SRT) (Math) was developed. A study was first conducted in 2000 with local children aged 4 – 8 years old to determine the developmental continuum for the acquisition of the basic numeracy skills. Reference was also made to the Singapore Mathematics Syllabus (Ministry of Education, 2001), British Ability Scales (Colin, Smith & McCulloch, 1996) and Quest (Robertson, Robertson, Fisher, Henderson & Gibson, 1997). Since 2000, the SRT (Math) has undergone two revisions to accommodate feedback from the content specialists and users, and also for the purpose of improving the psychometric properties of the instrument.

SRT (Math) is a criterion-referenced, group-administered, paper-and-pencil test. It assesses basic mathematical concepts and simple problem-solving skills:

a. Concept of numbers up to 20
b. Pattern recognition
c. Concepts of "more" and "less"
d. Addition and subtraction of numbers with and without picture cues

The SRT (Math) is administered to all Primary 1 pupils within the first month of school to identify promptly pupils who lack the skills to follow the curriculum, before they start to experience difficulties. The test is administered by a trained teacher, and is typically completed within half an hour. The SRT (Math) assigns pupils to 5 criterion levels, and pupils in the lowest 2 levels, Level 0 and Level 1, are considered to require additional support in Mathematics.

3 THE 4-PRONGED INTERVENTION APPROACH

Intervention in LSM is conducted by trained LSM teachers. Lessons are conducted in small groups of no more than 10 pupils, so that the teachers are able to provide more individualised support and attention. The schools receive an additional manpower allocation, which is over and above the standard provisions, to enable them to provide this customised support.

Intervention in LSM is guided by the 4-Pronged Intervention Approach (4-PIA). In recognition that learning is determined by factors that are both internal and external to a pupil, the 4-PIA targets support in four domains: *Cognition*, *Metacognition*, *Motivation*, and *Environment* (Figure 2). The first three domains focus on factors internal to the pupil (Mayer, 1998), while the fourth domain focuses on factors external to the individual (Bronfenbrenner, 1989; Ysseldyke & Christenson, 2002). Despite the distinction made between the internal and external domains, the domains are not mutually exclusive. Instead, they interact with one another to impact on a pupil's academic performance. LSM, in adopting the 4-PIA, seeks to simultaneously address factors in the four domains so that a pupil's needs are understood and attended to in a more holistic manner.

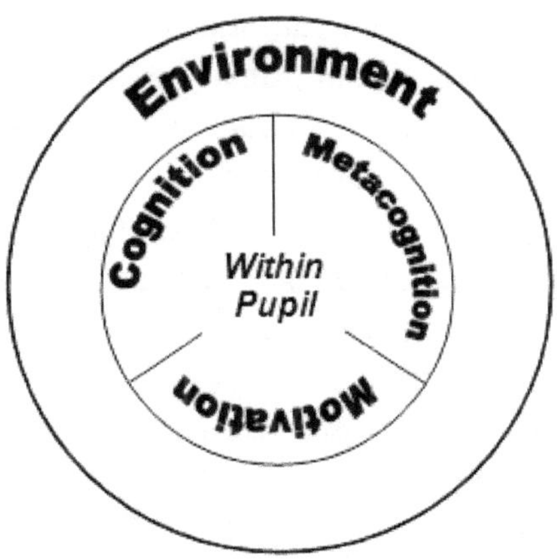

Figure 2. The 4-Pronged Intervention Approach targeting the Child's Cognition, Metacognition, Motivation, and the Environment.

3.1 *Prong One: Cognition*

Cognition refers to topic-specific knowledge, skills, and sub-skills necessary for the accurate performance of mathematical operations. This includes knowledge of language specific to mathematics. Teaching in mainstream classrooms has focused on the development of mathematics learning through three stages: Concrete, Pictorial and Abstract[1] (C-P-A). The 4-PIA adopts the same focus but with additional stress on

a. explicit bridging between C-P-A stages;
b. ensuring sufficient over-learning (Ford et al., 1998);
c. differentiation of instruction; and
d. active pupil engagement through experiential learning.

3.2 *Prong Two: Metacognition*

Research has shown that possessing basic mathematical skills is necessary but not sufficient for successfully solving higher level problems. In addition to having domain specific skills, pupils need to be able to manage their skills; this involves the application of metacognition (Kuhn, 2000; Mayer, 1998).

Metacognition is the ability to monitor one's thinking processes in problem solving. It has been shown to affect school performance and is a good predictor of mathematical abilities (Lucangeli & Cornoldi, 1997). Pupils as young as those in the primary grades who know why, when, where, and how to use different strategies have been found to perform better in mathematics compared to students who do not have such knowledge (e.g., Desoete Roeyers & De Clercq, 2003; Pappas, Ginsburg & Jiang, 2003). Weak language has an adverse impact on the metacognitive ability of a student to describe his/her thinking. This, in

---

[1] The *Concrete Stage*: Pupils require a collection of meaningful experiences and physical involvement relating to a concept to make sense of it (e.g., stacking blocks, counting coloured cubes, grouping shapes, etc.). The *Pictorial Stage*: Pupils begin abstracting and understanding mathematics in another representation, i.e., move from use of concrete models to pictures. The *Abstract stage*: Pupils apply mathematical concepts in the form of symbolic representations and to situations which may or may not be similar to their experiences.

turn, affects mathematical performance (Pappas, Ginsburg & Jiang, 2003). Therefore, both the *procedure* and *language* of self-reflection should be taught to pupils.

The 4-PIA encourages the development of metacognition through the explicit modelling of problem solving process and guided reflection using verbal and/or written prompts and questions (see Figure 3).

> Xiao Ming had four apples. He gave two to his sister and ate one. How many apples does he have left?
>
> Teacher: What is your answer?
> Pupil: One.
> Teacher: How did you get the answer 'one'?
> Pupil: Four minus two minus one.
> Teacher: Why did you minus?
> Pupil: Because...no more.... (gestured)
> Teacher: What words made you think there is 'no more'?
> (pointing to the question)
> Pupil: 'gave...sister'...'ate'.....

Figure 3. Example of teacher-questioning to support metacognition.

3.3 *Prong Three: Motivation*

Pupils who are intrinsically motivated have greater interest, better conceptual learning, greater cognitive flexibility, better memory and greater academic achievement (e.g., Kaplan, 1990; Lepper & Cordova, 1992). However, focus group discussions with LSM teachers have highlighted low motivation as a potential concern for LSM pupils. Poor motivation has implications on the pupils' task perseverance, help-seeking behaviours, and academic performance.

To address pupil motivation, the use of rewards has been a common practice among teachers. Whilst recognising the role of extrinsic rewards in motivating pupils, the 4-PIA places additional emphasis on encouraging intrinsic motivation through:

a. Promoting positive affect among pupils while learning mathematics by providing enjoyable activities that promote feelings of success.
b. Promoting a sense of control over academic outcomes by guiding pupils to focus on factors within their control when interpreting their experiences of success and failure (Ormrod, 1998).
c. Promoting a sense of self-efficacy by providing appropriately challenging tasks that are commensurate with the pupil's perceived level of competence (Schweinle, Meyer & Turner, 2006).
d. Designing interesting and meaningful tasks so that pupils are more motivated to engage in learning (Assor, Kaplan & Roth, 2002).

3.4 *Prong Four: Environment*

Based on the systems-ecological framework proposed by Bronfenbrenner (1989), a pupil's environment should to be taken into consideration when addressing learning needs. For the 4-PIA, a pupil's environment was conceptualised as comprising the home, the school, and the wider community. Consequently, intervention efforts are aimed at understanding and enhancing school-based instructional support, home support, and support from the community.

Within schools, instructional support could come in the form of infrastructural resources, IT resources, teachers, parent-volunteers and pupils within the school, as well as from community resources, for example, Voluntary Welfare Organisations and Self-Help Groups. Schools could implement a buddy support or peer tutoring programme by pairing pupils in LSM with pupils from the upper primary who are proficient in mathematics problem solving. Having a structured cross-age peer tutoring programme using mathematical games has been found to have a positive impact on attitude towards mathematics, mathematical skills and knowledge, and social and emotional factors for both tutors and tutees (Topping, Campbell, Douglas & Smith, 2003).

At home, the value of parental support in promoting positive learning outcomes has been widely documented (Desforges & Abouchaar, 2003). Schools are encouraged to conduct sharing sessions with parents or provide instructional resources to equip members of pupils' family with the skills, knowledge and materials to provide support to the pupils in their mathematics learning (see section on Engaging Parents). Some schools also collaborate with Self-help Groups to carry out such parent-centred programmes.

In addition to enhancing support within the school and home contexts, schools are also encouraged to develop a good understanding of the support programmes available in the community, such as those mounted by Self-help Groups. These programmes need not be limited to those specific to mathematics learning. A good knowledge of such community-based programmes allows schools to make appropriate recommendations that extend support for pupils' learning beyond the school and home.

4 MONITORING AND EVALUATION

In order to ascertain the effectiveness of support that is provided to pupils receiving LSM, an evaluation framework was developed and introduced to schools when LSM was implemented in all schools in 2007. This framework serves to articulate the expectations of schools in terms of pupil progress, as well as to facilitate self-evaluation by schools. In addition, it allows for exceptional schools to be identified: schools that are particularly successful and can share best practices, and schools that may require further support to enhance their effectiveness.

To inform the development of the LSM Evaluation Framework, a review of selected frameworks from educational systems within the UK, USA, Canada, and Australia was conducted to identify key principles and elements. In formulating the evaluation framework, the following criteria were adopted:

a. Measures of pupils progress should be objective and quantifiable.
b. Outcome measures should be standardised to allow for comparisons across schools and over time.
c. The collection of data should be feasible for teachers.
d. The data should be easily interpreted and used by school personnel for formative and summative purposes.

The resulting LSM Evaluation Framework is comprised of four key components: a Performance Indicator, a Performance Goal, a Performance Feedback Process and Performance Evaluation (Figure 4).

Performance Indicator To provide clarity on the nature of the desired outcomes for an initiative	**Performance Goal** To help set directions and formal strategic implementation plans; serves as benchmarks for implementation effectiveness
Norm-referenced Singapore Mathematics Achievement and Reasoning Test-Alternate (SMART-Alt) Score	Pupil Level: Pupil's Math age is commensurate with chronological age School Level: Proportion of LSM pupils who attain age-appropriate math skills
Performance Evaluation To inform decision-making and planning to enhance effectiveness	**Performance Feedback** To monitor the effectiveness of intervention
Schools not meeting benchmarks for the Performance Goal (at school level) will review their processes with support from MOEHQ	Provision of feedback to schools on national performance and the performance of similar schools at: Pupil Level: Pupil's Math age is commensurate with chronological age School Level: Proportion of LSM pupils who attain age-appropriate math skills

Figure 4. The LSM evaluation framework.

To measure pupil performance, schools typically conduct Semestral Assessment (SA) at the end of the school year. However, these assessments are school-based and not standardised across schools, making interpretations of the national data difficult. Hence, a standardised, norm-referenced instrument, the Singapore Mathematics Achievement and Reasoning Test-Alternate (SMART-Alt) is chosen to render a *Performance Indicator* in the form of a "math age." The *Performance Goal* would then be that pupils attain a math age that is commensurate with the chronological age, that is, by the end of the year, their mathematics skills are at a level that is typical for a child of their age. The aim of *Performance Feedback* is to provide objective and timely feedback to schools for both formative and summative purposes. Using information on the national performance and performance of comparable schools, the schools have an opportunity to assess their own performance in relation to others. Finally, *Performance Evaluation* facilitates the systematic identification and support of schools where intervention has been less effective.

In line with best practice, the evaluation framework is designed in tandem with the design of the intervention approach. The benefit of this approach is a clear alignment of monitoring and evaluation processes with the goals and objectives of the intervention. At the same time, the evaluation framework provides clarity to schools on the expected outcomes even as the schools plans for the implementation of LSM.

5 BUILDING TEACHER CAPACITY

There is an abundance of research that links teacher effectiveness to the quality of educational outcomes for pupils (e.g., Ellett & Teddlie, 2003; Jones et al., 2003). Hence, building teacher capacity has been and will remain a focus of efforts to improve educational outcomes for pupils in LSM.

5.1 *Teacher Selection*

Teaching in LSM is highly demanding, both in terms of the technical knowledge and skills of mathematics instruction, and the teacher's empathy, patience, and confidence. School leaders are encouraged to identify teachers who are experienced and effective classroom teachers, who enjoy teaching mathematics. The teachers should also have an interest and affinity for working with pupils with learning difficulties. As it is common to have only one LSM teacher in each school, this teacher also needs to be independent and to demonstrate initiative and creativity in his/her work.

5.2 *In-service Training*

Where teachers have the necessary attributes, training enhances their ability to provide effective intervention for their pupils. Training for LSM is targeted at two broad areas: Screening and Monitoring, and Intervention. In the former, teachers are trained in the use of the SRT (Math) and the SMART-Alt instruments. In the latter, the focus is on pedagogical principles and the application of the 4-Pronged Intervention Approach.

The design of the training for LSM is informed by the Lancaster Cycle (Burgoyne, 1992), which delineates three types of learning cycles: reflection, discovery, and reception. Table 1 presents a sample of a typical training plan that incorporates the three learning cycles.

Table 1
Sample plan for training of LSM teachers

Cycle	Activity
Reflection	Group discussion on common errors made by LSM pupils.
Reception	Trainers to share strategies to overcome these difficulties (categorised by the 4-prongs).
Reception	Group Activity
Discovery	Teachers to distil prongs and principles from sample lessons given (2-3 linked lessons). Teachers to discuss how prongs and principles: • can support the learning objectives • guide the selection of teaching approaches, strategies and resources • can be applied to address difficulties raised in previous activity.
Discovery	Teachers to simulate a lesson to a group of pupils in LSM.
Reflection Reception	For each simulated lesson, participants to discuss: • Strengths of the lesson • Possible improvements (e.g., strategies and the corresponding prongs/principles which can be applied to the lesson).
Discovery	Based on the day's discussion and guided by the 4-PIA, teachers plan a lesson.
~ Teachers return to school for a week, during which they conduct the planned lesson with one or more groups of pupils ~	
Reflection	Teachers return to the course the following week. Reflection and Sharing: What went well? What would you do differently?

In the training of LSM teachers, the predominant modes of learning are problem-based learning and experiential learning. Problem-based learning encourages teachers to engage in discussions to tackle true-to-life scenarios, to find solutions to real problems. These discussions surface participants' knowledge and rich experiences, and facilitates peer learning. According to Coles (1991), problem-based learning enables the active construction of knowledge by learners, thereby resulting in deeper knowledge. The scenarios provided in the course of the training also serve to provide teachers with a sense of the challenges they are likely to encounter as LSM teachers, so that they are mentally and emotionally better prepared for their role.

The use of experiential learning is supported by Kolb (1984), who contends that concrete experiences encourage reflection and meaningful learning. Experiential learning on the course is encouraged through assignments at the end of each day, which requires teachers to plan and conduct a lesson with a group of pupils. This allows them to consolidate their learning, and apply what they have learnt. A subsequent debriefing session provides opportunities for teachers to reflect on their own learning, and to extend their own learning and the learning of others through feedback and discussion.

A unique feature of the training of LSM teachers is that field practitioners (a Master Teacher in mathematics and experienced LSM teachers) are recruited to contribute to the design of the training programme, and to conduct the training. Their involvement is crucial to providing new LSM teachers with relevant and authentic learning experiences, as well as firsthand accounts of the challenges and rewards of being an LSM teacher.

5.3 *Mentoring and Continuous Professional Development*

In recognition of the need for the continuous support of new LSM teachers, a formal Mentoring Programme was established in 2007. Through this programme, experienced LSM teachers are selected to serve as mentors to new LSM teachers in their first year, providing professional advice on pedagogy, practical advice on operational issues such as planning the time-table and providing emotional support and encouragement.

The mentoring process consists of four key features, namely identification of learning goals, ongoing coaching, progress monitoring, and outcome evaluation. New LSM teachers on the programme are required to identify their learning goals through a needs analysis with their respective mentors. The mentor and mentee then work together to achieve the learning goals through agreed communication channels, e.g., e-mails, phone consultations, face-to-face meetings. LSM teachers on the programme are encouraged to monitor their learning progress and to evaluate their learning outcomes at the end of the year. Feedback gathered through surveys of mentors and mentees indicates that the Mentoring Programme is well received; both mentors and mentees reported that the process has been effective in supporting their professional development.

To encourage on-going professional sharing, LSM teachers come together annually to share best practices. At this platform, teachers are encouraged to share their experiences in the application of the 4-Pronged Intervention Approach. In the course of their preparations for this event, teachers usually work in groups, providing yet another opportunity for professional exchange and mutual support.

The involvement of experienced LSM teachers in the training and mentoring of new teachers is also strategic to their professional development. In teaching others, experienced LSM teachers are encouraged to reflect more deeply on their practice, clarify concepts and hone their own skills. As they serve as trainers and mentors for new LSM teachers, they continue to grow the fraternity, enabling more and more teachers to provide effective support to LSM pupils.

6 ENGAGING PARENTS

Numerous studies have demonstrated that parents can contribute significantly to pupil achievement (e.g., Epstein, 2001). As described in the 4-Pronged Intervention Approach, the involvement of the family and community is an important aspect of intervention in LSM.

In 2002, the Enhancing Parental Involvement Committee (EPIC) was formed. It comprised LSM teachers and educational psychologists, and its aim was to promote the involvement of parents in mathematics education. A secondary aim for the formation of this committee was to promote networking amongst LSM teachers and promote a culture of collaboration that would sustain them in their role.

One of the first initiatives of EPIC was the organisation of a Math Carnival. The carnival aimed to present mathematics concepts to children in a fun and engaging way, providing them with positive experiences of learning mathematics. At the same time, as every child participating in the carnival was accompanied by a parent, parents too were exposed to the range of activities that could be used to support learning. As part of the carnival, educational psychologists also conducted talks for parents, sharing ideas on how to use daily activities to extend the children's mathematics learning. The effects of the Math Carnival have been multiplied since, with LSM teachers organising similar events at their own schools, to engage the parents of their pupils.

Another the key contribution of EPIC has been the development and promotion of Math Packs, modelled after Orman's (2000) "Mathematics Backpacks." The 13 Math Packs produced contain materials and instructions for mathematics activities to be carried out at home. They are designed to encourage pupils to practise at home what they have learnt in school, in a fun way, and with their parents. Math Packs strengthen the home-school link, keep parents informed of current instructional practices, and facilitate their involvement in their child's learning (see Figure 6 for an example).

The Math Packs have been well received by teachers as they cover the key topics in the Primary 1 mathematics syllabus and are easily reproducible. They are equally well received by parents, who appreciate the opportunity to support their child's learning. Comments from parents include the following:

> *Both of us enjoyed playing the game. My child learnt through play. And I had fun quality time with him!*
>
> *My child enjoyed the game and picked up concepts like bigger/more and smaller/less quickly.*
>
> *Although my child was not as fast in getting the shapes in the picture, he was accurate in the end. He knew where the shapes should go. He had fun. Thanks!*
>
> *.... it really helps to reinforce the addition and subtraction concept...*

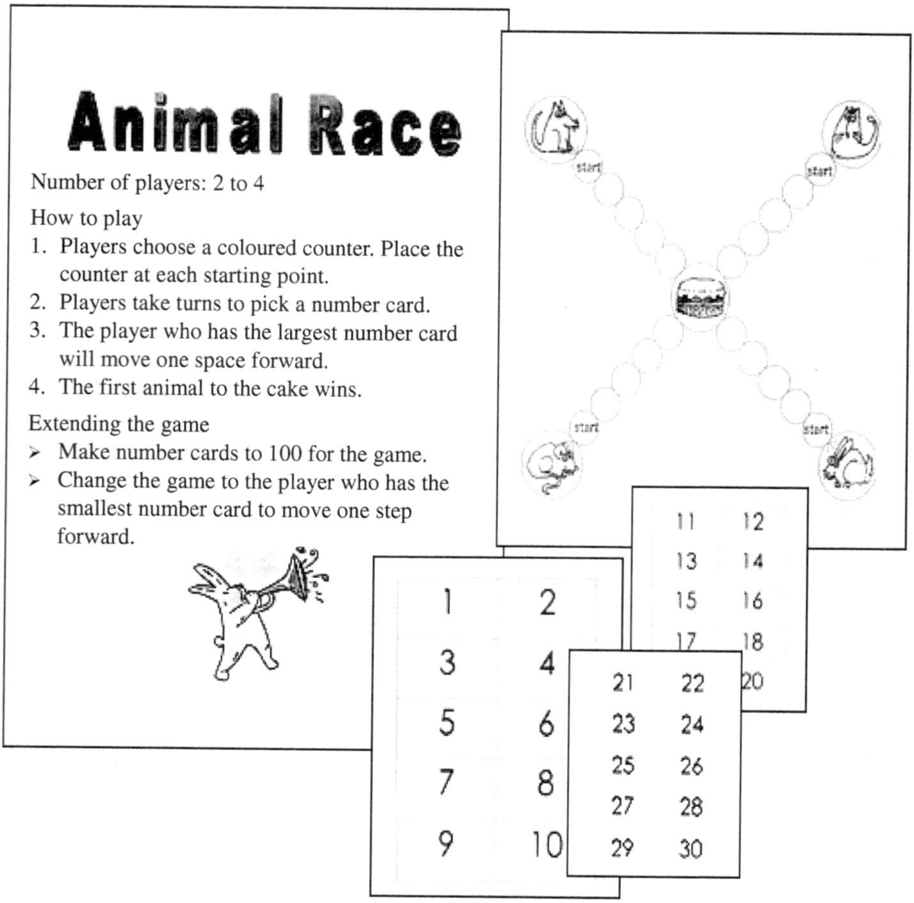

Figure 6. Math pack on comparison of numbers (Largest or Smallest).

7 CONCLUSION

The conceptualisation and implementation of LSM has been guided by a rigorous evidence-based approach. This approach ensures that practices associated with pupil identification, intervention, and teacher training are sound, with demonstrated efficacy. Accompanying this commitment to rigour is the readiness to invest in research, manpower, and teacher training, to build the capacity of our schools to provide quality education.

The driving force behind these efforts is the vision of enabling all pupils to optimise their learning and to benefit from the range of educational opportunities available for them, so that they may fulfil their potential.

References

Assor, A., Kaplan, H., & Roth, G. (2002). Choice is good, but relevance is excellent: Autonomy-enhancing and suppressing teacher behaviours predicting students' engagement in schoolwork. *British Journal of Educational Psychology, 72*(2), 261-278.

Bronfenbrenner, U. (1989). Ecological systems theory. *Annals of Child Development, 6,* 187-249.

Burgoyne, J. G. (1992). Frameworks for understanding individual and collective professional development. *Educational and Child Psychology, 9*(2), 42-52

Coles, C. (1991). Is problem-based learning the only way? In D. Boud & G. Feletti (Eds.), *The challenge of problem based learning*. London: Kogan Page.

Colin, E., Smith, P., & McCulloch, K. (1996). *British Ability Scales (BAS)* (2nd ed.). London: NFER-Nelson.

Desforges, C., & Abouchaar, A. (2003). *The impact of parental involvement, parental support and family education on pupil achievement and adjustment: A literature review.* Department for Education and Skills. Retrieved February 23, 2005, from http://www.dfes.gov.uk/research/data/uploadfiles/RR433.pdf.

Desoete, A., Roeyers, H., & De Clercq, A. (2003). Can offline metacognition enhance mathematical problem solving? *Journal of Educational Psychology, 95*(1), 188-200.

Ellett, C.D., & Teddlie, C. (2003). Teacher evaluation, teacher effectiveness and school effectiveness: Perspectives from the USA. *Journal of Personnel Evaluation in Education, 17,* 101 -128.

Epstein, J.L. (2001). *School, family and community partnerships: Preparing educators and improving schools.* Colorado: Westview Press.

Ford, J. K., Smith, E.M., Weissbein, D. A., Gully, S. M., & Salas, E. (1998). Relationships of goal orientation, metacognitive activity, and practice strategies with learning outcomes and transfer. *Journal of Applied Psychology, 83*, 218-233.

Fuchs, L. S., Compton, D. L., Fuchs, D., Paulsen, K., Bryang, J. D., & Hamlett, C. L. (2005). The prevention, identification and cognitive determinants of math difficulty. *Journal of Educational Psychology, 97*(3), 493-513.

Gersten, R, Jordan, N. C., & Flojo, J. R. (2005). Early identification and interventions for students with mathematics difficulties. *Journal of Learning Disabilities, 38*(4), 293–304.

Jones, D.L., Powers, S.W., Cox, A., Bintz, W., Christian, B., Davis, M., Greenwalk, Y., Higgins, P., & Newsome, F. (2003). An examination of early reading intervention instructional practices for diverse groups for the primary grades. *International Journal of Learning, 10,* 2337-2357.

Kaplan, P. (1990). *Educational psychology for tomorrow's teachers.* Minneapolis: West Group.

Kolb, D. (1984). *Experiential learning.* Hemel Hempstead: Prentice Hall.

Kuhn, D. (2000). Metacognitive development. *Current Directions in Psychological Science, 9*(5), 178-181.

Lepper, M. R., & Cordova, D. I. (1992). A desire to be taught: Instructional consequences for intrinsic motivation. *Motivation and Emotion, 16*(3), 187-208.

Lucangeli, D., & Cornoldi, C. (1997). Mathematics and metacognition: What is the nature of the relationship? *Mathematical Cognition, 3*(2), 121-139.

Mayer, R. E. (1998). Cognitive, metacognitive, and motivational aspects of problem solving. *Instructional Science, 26,* 49-63.

Ministry of Education. (2001). *Primary school mathematics syllabus.* Singapore: Author.

Ministry of Education. (2006). *Nurturing every child: Flexibility & diversity in Singapore schools.* Singapore: Author.

Orman, S. A. (2000). Mathematics backpacks: Making the home-school connection. In D. Edge (Ed.), *Involving families in mathematics education* (pp. 51-53). Reston, VA: National Council of Teachers of Mathematics.

Ormrod, J. E. (1998). *Educational psychology: Developing learners* (2nd ed.). New Jersey: Prentice-Hall, Inc.

Pappas, S., Ginsburg, H. P., & Jiang, M. (2003). SES differences in young children's metacognition in the context of mathematical problem solving. *Cognitive Development, 18,* 421-450.

Robertson, A., Robertson, A., Fisher, J., Henderson, A., & Gibson, M. (1997). *Quest* (2nd ed.). London: NFER-Nelson.

Schweinle, A., Meyer, D K., & Turner, J.C. (2006). Striking the right balance: Students' motivation and affect in elementary mathematics. *The Journal of Educational Research, 99*(5), 271-293.

Topping, K.J., Campbell, J., Douglas, W., & Smith, A. (2003). Cross-age peer tutoring in mathematics with seven- and 11-year old: Influence on mathematical vocabulary, strategic dialogue and self-concept. *Educational Research, 45*(3), 287-308.

Ysseldyke, J., & Christenson, S. (2002). *Functional assessment of academic behavior (FAAB).* Colorado: Sopris West.

Chapter 17

Numeracy Matters in Singapore Kindergartens

Pamela SHARPE

This chapter focuses on three local studies by the author. The first two highlight difficulties faced by kindergarten children, not exposed to a developmental approach to mathematics learning, who were trained to solve mathematical problems using "workbooks" and "worksheets." Problems discussed include the children's inability to apply their learning and the kindergarten teachers' lack of knowledge and understanding of how children think and learn about mathematics. The third study investigates the effects of a more developmental approach with a focus on interaction between teacher and child and the provision of manipulatives. Knowledge and information gained from conducting these studies provided the basis for the development of the preschool numeracy programme, as part of the Framework for a Kindergarten Curriculum in Singapore. The final section of the chapter mentions some current practices in mathematics education in Singapore kindergartens aimed to ensure quality teaching.

Key words: Kindergarten, preschool numeracy, developmental approach, teacher questioning, children strategies

1 INTRODUCTION

Becoming numerate involves thinking mathematically about problems and their solutions. It also necessitates that children use numbers to make connections and realise the relationships between them in relation to real life situations. This involves using and understanding a system of signs and symbols. How then do children develop numeracy skills and become numerate? A developmental approach stresses the need to build on children's natural ways of representing their ideas about numbers by guiding them on from perceptually dominated experiences to thinking logically about numbers and their relationships. Children need to be given opportunities to investigate, discover, and apply their own solutions to mathematical problems. Using this approach, a new framework for a developmental kindergarten curriculum (Preschool Education Unit, 2003) was constructed where active discovery learning and holistic development are key features. The effects of the new curriculum on children's learning and development, especially for numeracy development, were found to be very positive. The development of the numeracy curriculum was based, by and large, on some of the findings and recommendations of the three studies described below.

2 THE FIRST STUDY: STRATEGIES ON ROUTINE ACTIVITIES

The impetus for the study (Sharpe, 1998) stemmed from a concern to identify and account for some reported difficulties some children experienced in adjusting to the demands of the mathematics syllabus early on in primary school. In a climate where children were rarely asked to explain how they arrived at solutions to problems, it was envisaged that the results and the prescriptions, might be useful to teachers and parents in both understanding more clearly just how young children think, and, how they make sense of their experiences.

Some time ago, Elkind (1988, 1989) warned of the dangers of adults focusing their expectations on what young children think and learn whilst neglecting to observe the process of how they think and learn. He proposed that children are only able to extend their thinking and learning when given opportunities to talk about and reflect on these, and not by being constantly exposed to new skills, knowledge, information and experiences per se.

Provision of such opportunities for reflection and generalisation of children's own successful solutions to problem solving activities has been stressed by Boulton - Lewis and Tait (1994) and Kamii (1989) who proposed that children only appear to make sense of number operations when using their own strategies. These findings are supported by Carraher, Carraher, and Schliemann (1990) and Cowan and Foster (1993) who concluded that merely telling children how to solve a problem is not effective unless they are convinced of the value of the strategy in terms of the likelihood of its success. Furthermore, in his review of the research into young children's number development, Durkin (1993) highlighted the adult's role as one of several factors shown to account for the difficulties young children face. He notes that, in spite of their inevitable progress, their learning of the models and strategies for solving mathematical problems used by adults can often intimidate them as they attempt to use and understand numbers in a problem solving context. Riem (quoted in Durkin 1993) found that in asking questions such as, "What comes after…?" and "How many…?" adults assume that with the child's repertoire of number words, s/he must therefore know their cardinal value. Bryant (1995), in describing young children's difficulties in combining their understanding and use of simple mathematical relationships and the number sequence, proposed that the informal and ambiguous verbal information about numbers provided by adults may cause children to have difficulties with the basic meaning of number words, even though they naturally grasp quantitative relationships between numbers.

There are problems, however, with the view that children grasp mathematical relationships from experience alone. Experience needs to be structured, as Carraher et al. (1990) show. Nunes and Bryant (1996) suggested that there seems to be an ordered sequence to children's number understanding and a need for adult intervention. Young-Loveridge (1989) also highlighted the adult's influential role, a theme extended by Smith (1993) into an appeal for child educators to extend the developmental milestones view of learning and development in favour of a view of learning in a more social context where children's competencies are challenged and extended, with priority given to the scaffolding role of adults. However, building on the theoretical ideas of Vygotsky, Shulman (1986, 1987) argued that to provide this scaffolding adequately teachers especially must have a sound grasp of subject matter and organisation. Simply put, what teachers know will be related to the kinds of experiences they offer to children.

In this regard, Aubrey (1997) attributed a mismatch between preschool children's responses and teachers' questions in structured mathematics teaching situations to teachers' lack of subject knowledge. Kleinberg and Menmuir (1995) have noted teachers' inability to identify preschool children's incidental mathematical opportunities in informal pre-school settings. The naming of "colours" was noted as being mathematical, whilst the building of patterns was not. However, Young-Loveridge, Peter, and Carr (1998) found that once researchers had worked with teachers to increase the amount of mathematical activity in preschool classrooms, young children achieved significant gains in mathematical skills.

Building on the brief review of research above, the following section provides details of the initial investigation to investigate differences in the strategies young children use to manage the routine workbook activities at both kindergarten and primary school in Singapore. A second aim was to ascertain the extent of teacher involvement in assisting children both to understand the requirements of the tasks and to select the appropriate strategies for successfully completing the tasks. Structured observations were made of children's responses to some of the counting, addition and subtraction problems typically encountered in kindergarten and Primary 1.

The tasks involved counting and addition and subtraction tasks and were designed to be completed over the eight kindergarten terms. They comprised pictures for countables and required the following skills: count all (counting all the objects in the array), count on (adding on to the first addend by counting on to the second addend), counting up (counting up from the value of the numeral to be subtracted to the value of the numeral from which the subtraction is to be made), and counting down (beginning with the value of the numeral from which the subtraction is to be made the child counts down or back to the value of the numeral which is to be subtracted); matching sets, making sets, and comparing the cardinal value of sets using more, few, and less, more than, less than; use of symbols "+", "-", and "="; matching numerals to numeral sums, addition and subtraction with countables, additive composition, using money values to establish additive composition, and word problems.

The tasks presented to the primary school children, without pictures as countables, required knowledge of numbers in a series; comparison of sets using "less than," "greater than;" number decomposition; creation of addition and subtraction sentences with numerals and symbols, and word problems.

Individual interviews and observations were carried out with eight children. All eight children had progressed through the same kindergarten. The tasks presented to the kindergarten children were selected from the syllabus used at the time (1998) by the majority of kindergartens in Singapore. The tasks were presented individually and orally, with the children being asked to justify and explain their solutions, which were recorded verbatim. The primary school children were first given written tasks and then questioned individually about these same tasks but with different number values, the purpose being to compare their understanding of the tasks under different conditions. The resulting observations are described below.

(a) Two K1 Children

For addition, they counted all rather than by set with count on and they could not identify a number of sets of similar objects arranged in sets. For subtraction, they were able to cover a given quantity and count up, but they could not apply the cover up strategy when counting on was required. For example, the children were shown a picture of eight cakes on a plate and a box with seven more cakes inside. When asked about the total number of cakes, they counted all the cakes instead of counting on from seven (the set of seven cakes in the box). When a count up strategy was needed for comparing sets to establish which set had more items, they also counted all the items.

When required to add sets as part of a number sentence with numerals and words, they counted all, and when it was unclear where to start counting, they miscounted. Only one of the children was able to solve a word problem using countables: "*If you have 6 blocks and you hide 3, how many are left?*"

(b) Two K2 Children

One child was able to complete the majority of tasks mentally but experienced difficulties with more complex tasks described in pictures and requiring the use of additive composition involving money values: *"The bananas cost 17 cents. I give the fruit seller 20 cents. How much change will I get?"* Here the child counted the total number of coins in the pictures rather than focusing on their values to solve the subtraction sum. The other child employed strategies similar to the K1 children. For addition, he counted all rather than by set with count on. He used count all for continuous quantities when comparing more and less but he was able to compare quantities arranged in bar charts for "more," "less" and "fewer." He miscounted when he was unclear where to start counting, and he relied on counting all instead of comparing sets in one to one correspondence. He was able to match addition sums with numerals less than 5 to appropriate solutions and he used fingers rather than pictures of countables for addition sums where count all was required.

For subtraction, he used fingers and pictures of countables to cover the smallest number and counted up. But he could not apply the cover up strategy when counting on was required. He was unaware of the meaning of the subtraction symbol referring to it as dash (a word used by the teacher to indicate a missing item); but when told of the meaning he covered the smallest quantity and counted up. He was able to complete word problems with countables. This particular child was able to apply available strategies to new concepts, whilst the other child was well ahead of the syllabus.

(c) Four Primary 1 Children

All of them were able to complete numbers in series up and down, and one child orally repeated the number sequence up to its highest value in order to locate the required number for counting down : *"Fill in the missing numbers: 10 __ __ 13 __ 15__ __ 18, 19 __;* She counted up to 18 and then down to 10.

None of the children, however, were able to locate the smallest number in a series without reminders: *"Arrange these numbers in order.*

Begin with the greatest number: 11, 19, 6, 8, 14. The smallest number is __?."

All but one child were unable to use "less than" when comparing sets of dots, and were unable to identify numerals "greater than," although they were able to apply "bigger than."

All but one child was unable to complete word problems involving comparison of sets (with dots) in one to one correspondence. The successful child counted on her fingers under the table and counted on "in her brain," she said. Only one child was able to decompose numbers between 10 and 20, and another succeeded with reminders from the experimenter. Both of these children were able to identify tens and units when reminded of decomposition and count on.

Three children were able to compare sizes of letters:

X Y Z "Y is smaller than _____"

Two children were able to write and complete a number sentence with matching sets of countables:

Complete the number sentence about hearts

♥ ♥ ♥ ♥ ⟶ ♥ ♥ ♥ ♥ ♥

(The child would be expected to complete the number sentence by filling in the blanks below:

4	+	5	=	9
	O		=	

For this task, a third child made a subtraction sum and then added sets when she was checked. All but one child were able to identify sums equal to a given quantity but were unsure when asked for those giving a larger quantity.

None of the children were able to solve word problems involving equality:

Iskander buys 6 fish. He puts 1 fish in each jar. He only has 4 jars. How many more jars does he need?

(The child would be expected to complete the number sentence by filling in the blanks below:

6	-	4	=	2
	O		=	

Finally, all of the children had difficulties interpreting the written instructions for completing the question paper.

It seems clear that even though the Primary 1 children had experienced the requirements for many of these tasks whilst in kindergarten classes, they were largely unable to apply the required strategies unless reminded. Although the kindergarten children were progressing well, they needed to be reminded of their previously successful strategies and they were clearly ready and capable enough to move on to use more sophisticated ones with more challenging tasks. Hence, these observations suggested to the author that some attention should be given to the complexity of tasks kindergarten children are capable of, and to the kinds of support that teachers may be capable of providing, given their knowledge of subject matter and knowledge of the differences in children's mathematical abilities.

To explore the extent of teachers' understanding of children's mathematical abilities, questionnaires were completed by 60 kindergarten principals and 43 kindergarten teachers. When asked about the need to monitor and identify difficulties children might face, the majority were unsure why children had problems, especially with symbols and numerals. The teachers and principals proposed that although individual help was given in school, parents were often asked to help their children at home. They also reported that homework was not generally given nor was assistance extended to parents in ways to help their children. Even though all were trained in the teaching of preschool mathematics, they were reluctant to indicate where mathematics arises incidentally during the preschool day other than during the daily mathematics lesson. Identification of shapes, numerals, and matching, sorting and counting skills were frequently mentioned as taking place during block play, art and craft, and music and movement, whilst other skills such as pattern recognition, enumeration, and time which might be realised during daily routines, songs and rhymes, and story telling were less frequently noted. However, a number of teachers were able spontaneously to provide examples of mathematics skills children might use during play and how mathematics might arise during "sharing time."

Although the teachers realised the importance of monitoring children's work which was not individualised, they rarely checked the strategies the children used. Children who finished their work earlier than others were directed to read books or complete puzzles. There appeared to be no real provisions for children who need to take more time than others, or who have difficulties. In addition to the questionnaires, three teachers were observed teaching. There was found to be a total reliance merely on children repeating by rote the instructions given for completing the example they were later to complete in their activity books.

The observations of the children's mathematical skills revealed the typical strategies which emerge naturally in young children and those which have to be taught (Bryant 1995, Nunes and Bryant 1996). Particularly striking was the primary children's failure to link counting to one-to-one correspondence unless reminded, and the kindergarten children's failure to adopt the "count on" strategy for addition with numerals, since they had not been taught this strategy. Furthermore, the primary children's inability to order a random sequence of numbers and locate the smallest number, may have indicated a problem with language concepts, or an inability to grasp the differences between the cardinal values and ordinal relationships of numbers, unless reminded.

Bryant (1995) suggested that word problems tell us much about children's number awareness, or lack of it since words represent problems in different ways. Furthermore, whilst early relational understanding as in one to one correspondence is generalised from hands-on experiences with songs, rhymes and games, true understanding as in additive composition, involving symbols, and has to be taught. In this regard, Nunes and Bryant (1996) showed that it is possible to lead children to co-ordinate their strategies and solve such problems. Likewise, Maclelland (1995) suggested that teachers should not wait until children know that addition is the inverse of subtraction before they are taught to count down, even though they are successful with counting up. She argues that teachers need to be both aware of the counting systems children use and to dissuade them from relying on concrete aids when they are capable of more sophisticated strategies. Furthermore, in calling for teachers to teach children problem solving strategies which do not rely on counting, Gray (1995) urged teachers to move children on to the learning of number facts and tasks requiring the application of these.

In addition to the need to train teachers to understand how children develop mathematically and become numerate, these findings also suggest that the emphasis could be on guiding teachers to provide for more problem solving opportunities enabling children to negotiate their own meanings, discover relationships, learn rules, ask questions, and select their own strategies with teachers directing attention, providing appropriate models and questions, and challenging children further.

3 THE SECOND STUDY: STRUCTURED OBSERVATIONS WITH A LARGER SAMPLE OF CHILDREN

The first study centred largely on children's use and knowledge of counting, addition and subtraction strategies. The second study with a larger sample of children (Sharpe 1999), set out to investigate more closely how children handle these tasks in kindergarten and, whether any difficulties persist into primary school.

Some of the research conducted into children's use of counting reveals that in preschool, whilst they are often quite competent at sharing, can count meaningfully, and know "more" and "less," some have difficulties applying these features in different situations. For example, Munn (1994) found that when children were asked for a set number of objects they grabbed a handful rather than using their knowledge of counting to count out the required number. Although most children are able to count and to know that the last count word represents the quantity of the set, Bryant (1995) shows that this may not help them to understand differences in quantity when two sets are compared.

To understand the quantitative significance of number words, what is needed, explains Bryant, is an understanding of one to one correspondence between two sets. Definite strategies need to be learned. Munn (1994) found that when taught to represent a quantity e.g., four, by writing 1, 2, 3, 4, children were able to link the number sequence to the operation of counting in one to one correspondence. They were able to relate the verbal to the written number system. They were thus able to understand that the order of words in the sequence represents their magnitude, the cardinal and ordinal properties of the number sequence. As noted earlier, Nunes and Bryant (1996) have suggested that there is an ordered sequence through which children progress in the development of early numeracy. Firstly, children learn to count with numbers in increasing order of magnitude; they are next able to count in one-to-one correspondence as a measure of set size; they then realise that numbers as measures of set size are related to addition and subtraction. This knowledge enables them to handle the complexities of problems represented in words with countables and/or pictures of things to count. Finally they are able to manipulate numbers related to set size, or transformations such as when handling money. Hence, simply being able to count does not necessarily mean that the purpose is understood or can be applied.

Other studies have shown that when mathematics concepts arise from activities rather than being imposed on children, they are more successful in transferring skills and knowledge learned in one context to another (Brown, Collins & Duguid 1989; Lave, 1988). The ability of some children to use a variety of approaches to solve problems indicates their level of attainment. Gray (1991) found that once children are able to deduce or infer solutions rather than recall taught procedures, they are better able to adapt methods to cope with new problems. However, it was noted that when children can both recall and deduce number facts, they make more progress since each strategy supports the other. Likewise, Steffe (1983) showed that low attaining children relied mainly on counting strategies with objects or representations whereas high attaining children used both recall of number facts and deduction based on recall.

Clearly, knowing how children use counting, how they handle addition and subtraction tasks, and how they apply these strategies to problems represented in words, and in real life situations, is crucial to the teacher in guiding children to select and apply appropriate strategies. Given these findings, what of the children observed in this study?

Observations were made of children in kindergarten programmes immediately prior to entry to primary school, and, six months later in primary school. The purpose was to investigate more closely how children handle the counting, addition and subtraction tasks in kindergarten and whether difficulties persist into primary school.

The first set of observations from individual interviews was conducted at 4 kindergartens with 158 children aged between 5 and 6 years during their last few weeks in kindergarten. All of the children had spent 2 years at the same 4 kindergartens and followed the same syllabus. The duration of the programme was 4 hours per day, for 5 days per week. All the children experienced one mathematics lesson per day using the workbook and worksheet approach.

The first sets of tasks for the interviews were chosen to judge the extent to which the children could use counting successfully with objects to be counted. Conservation of number tasks were also included to assess whether the children had grasped the significance of comparing sets by one to one correspondence.

The first task involved counting an array of countables with no obvious starting point in order to judge whether they could count and know that the last number name represents the cardinal value of the set. The next task involved producing a specified quantity of countables, and was to judge whether they could use their knowledge of counting. The next task required them to create an equivalent set when asked: "give me as many blue ones as red ones." This tested their ability to apply their knowledge of cardinal value. Finally, the number conservation task indicated whether or not the children had grasped the significance of counting which is knowing cardinal values and being able to mentally link one to one correspondence with cardinal numbers.

The next set of tasks was selected from the mathematics workbooks and the children were asked to respond orally as the tasks were read out. The tasks comprised pictures for countables, and required the following skills:

- count all (counting all the objects in the array),
- count on (adding on to the first addend by counting on to the second addend),
- counting up (counting up from the value of the numeral to be subtracted to the value of the numeral from which the subtraction is to be made),

- counting down (beginning with the value of the numeral from which the subtraction is to be made, the child counts down or back to the value of the numeral which is to be subtracted),
- counting on from a given set,
- comparing the cardinal value of sets using "more than" and "less than,"
- shopping tasks involving subtraction sums with pictures of coin values, e.g., "The bananas cost 17 cents. I give the fruit seller 20 cents. How much change will I get?"
- word problems such as: "If you have 6 blocks and you hide 3, how many would be left?"

Six months later, 126 of the original 158 children were interviewed again in their respective primary schools. At the time, all primary schools in Singapore followed the same mathematics syllabus, with the same topics and workbooks and lesson duration. The tasks selected were the same as those used in the first study and were taken from an assessment paper all Primary 1 children would have taken as part of their first end of term mathematics assessment.

An analysis of the children's performance on the two sets of tasks at kindergarten and primary school was conducted. For the purpose of the comparisons, the performances on the conservation tasks were analysed. This was because the relationship to success in addition and subtraction. Three discernible groups were revealed. Of these, 40 children were non-conservers, 43 children were conservers who needed to count, and 43 children were able to conserve number mentally. All of the children were able to complete the initial counting tasks, and as expected, their grouped performances on all tasks revealed the superiority of the conserving children over the non-conservers at kindergarten and in Primary 1, yet there were some interesting differences between groups on some of the tasks.

A glance at Tables 1 and 2, reveals that at kindergarten the conserving children relied less on the need to "count all," and "count on by set" when given pictures of sets to count. They also used fingers more than the non-conservers, who were not certain which kind of strategy to employ. The conservers demonstrated an understanding of "more than" and "less than." Very few of the children used "counting up" for simple subtraction tasks with pictures of countables, but more conservers were able to "count down." Just over half of the true conservers were able to complete the tasks involving coins and word problems.

The tasks relating to real life situations - the shopping tasks with pictures of coins - required the children to demonstrate knowledge of part/whole relationships. These tasks were beyond the majority of children, many of whom could count on from a set for addition tasks, but were unable to see the inverse relationship between addition and subtraction in these tasks. The children were observed to look at the pictures of the coins and count each as one cent instead of recognising the differences in the values between one cent, five cents, and ten-cent coins. Furthermore, the word problems required the children to compute addition or subtraction sums with very small numbers and they were encouraged to use countables. Even so, less than half of conserving children were successful.

Clearly, at the end of kindergarten, less than half of the total group (only 56% of conservers solved coin problems and 53% word problems) were able to complete the tasks expected of them by the end of kindergarten. What then are the implications of these observations for the children's performances on primary school tasks?

Table 1
Observed Responses to Addition & Subtraction Tasks in Kindergarten, Expressed in Percentages (N =126)

Kindergarten Tasks	F	CA	CS	CO	CU	CD	M/L	C	WP
Non – conservers N = 40	41	92	3	8	36	0	18	0	8
Counting conservers N = 43	70	72	28	56	40	23	51	33	42
True conservers N = 43	72	56	56	67	37	21	53	56	53

Key: F = used fingers; CA = counted all; CS = counted by set; CO = counted on; CU = counted up; CD = counted down; M/L = more/less; C = shopping tasks with pictures of coins; WP = word problems

Table 2
Observed Responses to Addition & Subtraction Tasks in Primary School, Expressed in Percentages (N = 126)

Primary One Tasks	O	M/L	EW	ND	PV	NS	CW
Non – conservers N = 40	90	69	21	46	23	8	0
Counting conservers N = 43	100	88	35	63	42	33	19
True conservers N = 43	100	86	42	81	53	40	47

Key: O = numbers in order; M/L = more/less; EW = word problem/equalise; ND = number decomposition; PV = place value; NS = number sentences; CW = word problem/compare

When interviewed later on at primary school, most of the children could arrange numbers in a series and demonstrate an understanding of "more than" and "less than," and even use "count on" strategies for the number decomposition tasks (see Table 2). The conserving children, however, were typically more accurate with the more complex tasks of number decomposition, place values and the creation of number sentences. Whilst less than half of the children, but more of the conservers, could accurately complete the word problem involving equalisation with dots to count (Siti has 7 and Poh Cheng has 3, how many must Siti give to Poh Cheng so they both have the same?), a large percentage of the true conservers, in comparison, could complete the word problem involving a comparison with nothing to count (Iskander has 6 fish, he puts one fish in each jar, he only has 4 jars, how many more jars does he need?)

The more complex of the primary school tasks: number decomposition, creation of number sentences, and the equalisation word problem, required a working knowledge and understanding of whole–part relationships: (6 = 5 + 1; 3 + 3; 4 + 2), and, commutativity: (6 + 2 = 2 + 6). The compare problem needed additional awareness of the inverse relationship between addition and subtraction (reversibility and transitivity), necessary ingredients for conservation, which probably accounts for the superior success on this task by the true conservers compared with the rest.

The success on these more complex tasks by the majority of the conserving children compared with the non-conservers can perhaps be explained by the increasing sophistication of the strategies they use to find solutions. Although discouraged by their teachers for doing so, they are more likely to use their previously reliable tools, their fingers, and count by sets when "counting on," "up" and "down" in kindergarten. The unsuccessful children were by and large confused or uncertain as to which strategy to employ. For the compare word problem, many children were unable to recognise the need to subtract, or arrive at the correct solution verbally. Some children found the correct solution but provided an inaccurate number sentence.

In sum, a comparison of performances of the same children at kindergarten and Primary 1, indicated that all of the children made some progress, especially with tasks involving "more/than" and "less/than," and an understanding of number decomposition which their primary school teachers had taught them. However, Table 2 shows that more than half of the children failed the more complex tasks, tasks that they would have been expected to have mastered after four months in primary school.

At the time, these observations suggested the need for an intervention programme with an emphasis on guiding teachers to provide for more challenges and problem solving opportunities in real life situations where children might mentally rehearse and test strategies. The implementation of these proposals is discussed in the next section.

4 STUDY THREE: AN INTERVENTION PROGRAMME

Studies in Britain (DfEE 1998; OFSTED 1993; SCAA 1997) have shown that when young children are given opportunities to explain their mental processes verbally, this has a positive effect on achievement. Furthermore, factual and conceptual knowledge is increased, leading to improvements in confidence. Aubrey (1997), in particular, has shown that young children enter school with a wide range of strategies for learning mathematics. She proposed that teachers encourage children to use a variety of strategies and provide opportunities for children to use the problem solving skills they already have and at the same time extend their knowledge of number facts.

In order to gauge the usefulness of the intervention programme, which is outlined later in this discussion and which was based on the perspective outlined above (Sharpe 2001), interview tasks were devised for oral interviews with another group of Kindergarten 2 children before the intervention programme and six months later at the end of the intervention. The end of the intervention coincided with the end of the kindergarten programme. The intervention programme was devised to guide children through the numeracy development sequence and to provide an ordered sequence of structured activities based on talk and discussion about aspects of numeracy, which linked with everyday activities and routines.

The children in this study comprised 107 Kindergarten 2 children from one PAP Community Foundation Kindergarten. They were all following the same mathematics syllabus provided by the PAP Community Foundation for all of its kindergartens. The teachers had all attended the "basic" training programme provided by the National Institute of Education.

The interview tasks were designed to tap the children's number knowledge before and after the intervention programme. The interviews were conducted by the author who talked to the children individually about counting and addition and subtraction tasks using things to count: coins, coin cards, and some small toys.

The first set of tasks was the same as those used in the previous study. The next set of tasks assessed understanding of the number operations of addition and subtraction. Using multilink cubes, the children were asked simple word problems such as *"You have 3, I give you 3 more, how many altogether?"* A similar question was asked for subtraction: *"You have 6, you give me 2. How many are left?"* In order to judge the maturity of their responses, the children were observed to see whether they counted all, up, or down from the set, or whether they gave the answers without counting.

A further task directly assessed the extent to which children could count on from a set. Here the interviewer told the children that the doll was to be given 5 cents in its purse to buy chocolate. The children then watched as the coins were placed one by one in the purse. They were then asked: *"If we give the doll 3 more coins, how many will it have altogether?"*

Using sticks of connected multilink cubes, the next set of tasks required the children to compare three different lengths and to show which had more or less cubes in it than the stick of cubes given to the children.

The next set of tasks required the children to create sets. They were given a number of multilink cubes and were then asked: "*Give me more than you.*" After the children had responded, they were asked: "*How many more than you do I have?*" The same procedure was followed for "*less than.*"

Next the children were shown cards with different quantities of pictures of animals up to 10; next they were shown cards with the equivalent number of dots; and finally they were shown cards with equivalent numerals. They were then asked to pair up the three sets of cards. The purpose of this task was to assess their ability to recognise cardinal values represented in different ways. The next task required children to order numbers. They were shown a set of numbers and asked which number was bigger or smaller. They then were required to place the numbers in order from the smallest to the biggest.

Finally, the children were asked simple word problems involving cubes representing coins. They were asked: "*You have 8cents. This apple costs 2 cents and this orange costs 4 cents. How much will the apple and the orange cost altogether?*" They were then asked: "*How much will you have left from your 8 cents?*"

Given their success in the pre-intervention interview tasks on addition and subtraction tasks with countables, the post intervention tasks began with the doll task done in the pre-intervention interviews, requiring the children to count on from a set. This was followed by the set of tasks requiring the children to create sets and then to order numbers as before. The next task was as before but using coin cards instead of cubes.

The next set of tasks involved identifying the value of coins, using coin cards of 1cent, 5 cents, 10 cents, 20 cents, 50 cents, and 1 dollar, to purchase real objects presented to the children. The purpose of these tasks was to establish the children's competence in additive decomposition, a pre-requisite for successful performance on word problems.

The noodles cost 73 cents, which coins will you use?
Show me again with the fewest number of coins?
These toys have prices on their labels, which can you buy with just two coins?
Which 2 toys can you buy with exactly 30 cents?
Which toys can you buy with $1?
Here is $1, if you spend 70 cents, how much will you have left?

In order to maximise the effectiveness of the intervention programme, the children were assigned to groups according to the way in which they responded to the level of difficulty of the tasks. It was envisaged that activities and styles of teaching and teacher/group interactions would need to accommodate the varied levels of understanding and number competence, which was expected to differ from group to group. Hence, once the pre-intervention tasks were completed and the responses analysed, three levels were identified with two groups assigned to each level. The children were grouped for the purpose of the intervention programme according to the following criteria, beginning with the weaker groups, "blue and orange"; see Table 3.

Level 3: blue and orange groups

These children were able to complete successfully some of the tasks. The upper limits were the more complex skills required for either the coin tasks involving buying the fruit (task 16) or, the task involving the doll (task 1), where they were expected to count on from a set. They were not able to complete the tasks requiring them to say how many more or less the sets of cubes were when compared (tasks 7 and 9). Furthermore, all the children at this level used fingers and/or things to count. Within each of the four classes, two groups of children were assigned to this level, a total of 45 children.

Table 3

A Comparison of Groups' Performance in Percentages on Tasks during Interviews 1 and 2

INTERVIEW TASKS	BLUE/ORANGE N=45 (raw scores)		PINK/GREEN N=35 (raw scores)		WHITE/YELLOW N=27 (raw scores)	
	INT 1	INT 2	INT 1	INT 2	INT 1	INT 2
1. Doll has 5c in purse, I give 3x1c more, how much altogether?	9* (4)	78* (35)	80 (28)	94 (33)	93 (25)	100 (27)
2. Which row of cubes has more?	93 (42)		100 (35)		100 (27)	
3. Which row of cubes has less?	93 (42)		100 (35)		100 (27)	
4. Which row of cubes has more than yours?	87 (39)		100 (35)		100 (27)	
5. Which row of cubes has less than yours?	84 (37)		97 (34)		100 (27)	
6. Give me more cubes than you.	77* (35)	96* (43)	97 (34)	97 (34)	96 (26)	100 (27)
7. How many more do I have than you?	0**	4** (2)	9** (3)	20** (7)	22** (6)	41** (11)
8. Give me less cubes than you.	49* (22)	84* (38)	83 (29)	94 (33)	89 (24)	100 (27)
9. How many less than you do I have?	0**	7** (3)	9** (3)	17** (6)	22** (6)	37** (10)
10. How many cars on this card?	98 (44)		100 (35)		100 (27)	
11. Which card has the same number of dots as cars?	96 (43)		100 (35)		100 (27)	
12. Which number card is the same as the dots?	96 (43)		100 (35)		100 (27)	
13. Which number is bigger than this? (Larger values for Interview 2)	93 (42)	96 (43)	100 (35)	100 (35)	100 (27)	100 (27)
14. Which number is smaller than this? (Larger values for Interview 2)	91 (41)	98 (44)	100 (35)	100 (35)	100 (27)	100 (27)
15. Order the number cards from small to large. (Larger values for Interview 2)	58 * (26)	89* (40)	80 (28)	100 (35)	89 (24)	100 (27)
16. You have 8c. This costs 4c this 2c. How much will both cost altogether? (cubes - Interview 1, 1c coin cards Interview 2)	20* (9)	78* (35)	69 (24)	97 (34)	100 (27)	100 (27)
17. How much do you have left?	0*	51* (23)	17 (6)	86 (30)	56 (15)	96 (26)
18. Noodles cost 73 cents. Which coins shall I use? (combination of coin cards)		20 (9)		23 (8)		63 (17)
19.Show me again with fewest number of coin cards.		11 (5)		14 (5)		44 (12)
20. These things have prices as shown, what can you buy with 2 coins (coin cards)?		11 (5)		26 (9)		48 (13)
21. Which 2 things can you buy with exactly 30 cents? (coin cards)		13 (6)		37 (13)		56 (15)
22. Which things can you buy with $1? (coin cards)		2 (1)		9 (3)		22 (6)
23. If you spend 70 (65) cents, how much will you have left? (coin cards)		0		2 (1)		7 (2)

* Denotes improvements ** denotes room for improvement

Level 2: pink and green groups

These children were able to complete successfully most of the tasks. The exceptions were the tasks requiring them to say how many more or less the sets of cubes were when compared (tasks 7 and 9), and the task requiring a subtraction strategy in order to indicate how much would be left from 8 cents (task 17). All of these children used fingers and/or things to count, rather than solve the tasks mentally. Within each of the four classes, two groups of children were assigned to this level, a total of 35 children.

Level 1: white and yellow groups

These children were able to complete successfully most of the tasks mentally with a few exceptions resorting to fingers or things to count.

The intervention programme was designed with the aim of enabling the children to handle numerical problems more flexibly and creatively in preparation for the expectations of primary school. The two studies discussed above showed that mere mechanical training in basic skills and strategies is insufficient. Furthermore, it is now recognised that to be truly numerate in this information age, children need to perform number operations mentally with measures and money in real life situations, i.e., in context. Later on in primary school, they need to make sense of numerical information presented in a variety of ways such as charts, tables, graphs, and diagrams, and, they will need to make independent decisions about which method and how to calculate or estimate. Hence, increasing attention needs to be given to mental calculation (DfEE, 1999; OFSTED, 1993).

The intervention programme involved weekly lessons, conducted by the author from April to November 1999 (with a break during the June school holidays). Each of the four classes received 30 minutes exposure to a series of sequential lessons. The lessons aimed to build on children's knowledge of counting leading to application of this knowledge to real-life situations involving addition and subtraction strategies and strategies involving the use of coins. There was also a strong focus on discussion strategies which included activities such as:

- Telling the children you are going to ask them what they are thinking before discussing the responses to your questions.
- Providing for "wait time." That is, giving them time to think before selecting a response from individual children.
- Avoiding chorus responses, selecting children by name taking care to pitch the appropriate level of difficulty to match each child's ability.
- Letting the children know that everyone will have a turn to answer.
- Encouraging the children to think carefully before they respond.
- Encouraging the children to respond clearly using full sentences.

Questions were considered to be useful as starting points for discussion and included the following:

- How did you get that answer?
- Did anyone else get the same answer?
- How did you get your answer?
- Did anyone get the same answer using a different method?
- Did anyone get a different answer?
- Could we have a different answer?
- Why (not)?
- Which is the correct answer and why?
- Why is this answer not correct?

Cognizance was always taken of the differing needs of the three levels of children, with activities, questions, and reinforcers, pitched to their differing needs. In order to capitalise on the material presented to the children each week, follow–up activities also were planned to enable consolidation of the aim of each session. Such activities included the following:

- A "topic or question of the day" was devised and the children were continually reminded of this throughout the session.
- Discussions were frequent about how things are best remembered.
- Mistakes made were discussed, in addition to suggestions of how these could be avoided in future.

- Other discussions included how the children could help each other.
- Strategies were devised to aid memory.
- Children and their parents were encouraged to complete the activity sheets accompanying some of these lessons over the weekend.

The sequence and scope of the weekly series of lessons were adapted from Leather (1997).

Table 3 also indicates the extent of the performance of the children in each of the three levels on both the pre and post- intervention interviews.

Whilst it is by no means being assumed that any improvements or progressions are due solely to the intervention programme, there was clearly an impact on the use of specific strategies. The emphasis on oral work and hands-on activities, and the teaching of specific strategies impacted on children's handling of the tasks during the oral interviews. It was also observed that their responses to questions about the tasks at the post intervention interviews indicated that they were more confident and competent with addition and subtractions task involving counting up and counting on, and the mental strategies required for the coin tasks. However, the mental strategies required for comparing sets in the more than/less than tasks (tasks 7 and 9) still eluded the majority of children. This was also the case for the coin tasks involving mental addition of coin values and additive composition (tasks 18 to 23).

The results of the three studies outlined above then, suggested that if children were to be adequately prepared for the challenges and expectations of the Primary 1 mathematics syllabus, changes would need to be made to the experiences children are exposed to as their numeracy skills develop. Teaching approaches and strategies also need to be modified and extended. These then were the triggers for the development of the numeracy programme as part of the new curriculum for kindergartens: Nurturing Early Learners (Preschool Education Unit, 2003).

5 SOME RESULTING INITIATIVES

In his speech to mark the launch of the new kindergarten curriculum, the then Senior Minister of State for Trade and Industry and Education, Mr Shanmugaratnam mentioned some findings resulting from a pilot research study which had been conducted to gauge the initial effects of the new curriculum. Significant was the finding that the developmentally appropriate curriculum had an impact on the problem solving skills of the children (MOE press release, 20th January 2003). As a result of this and other positive findings, a number of government initiatives were put in place. Notable was the drive to raise the standards of teacher training and qualifications. As such, one programme, SEED (Strategies for Effective Engagement and Development of pupils in primary schools) was implemented in 2005. This programme aims to enable teachers to set appropriate expectations for pupils at different levels and to ensure appropriate practices and assessment modes.

In Britain, where assessment and achievement are also part of the teaching and learning process for younger children, it has been found to be difficult, however, to sustain a developmental approach when targets have to be met and parents and administrators appeased. Currently, the British debate about appropriate pedagogy lingers on with some researchers calling for less attention to goals and targets for very young children and more attention to the study of how children think and learn about mathematics in their early years at school (Munn, 2004). This inevitably gives rise to questions about how teachers of young children are trained as well as their own understanding of children's learning and of mathematics. In Singapore, this is being addressed, and unlike in Britain, kindergartens in Singapore have more autonomy since they are not required to teach to attainment targets and they are not directly linked to the formal primary school system. However, in order to adequately prepare children for the demands of the primary school, particularly in mathematics, kindergarten teachers will need to be continually supported in delivering a developmental approach and in resisting demands by parents and others to teach to the targets of workbooks and worksheets in order to be ready for the expectations of primary schools.

References

Aubrey, C. (1997). Re-assessing the role of teachers' subject knowledge in early years maths teaching. *Education 3 to13, 25,* 55-60.

Boulton-Lewis, G.M., & Tait, K. (1994). Young children's representations and strategies for addition. *British Journal of Educational Psychology, 64,* 231-242.

Brown, J.S., Collins, A., & Duguid, P. (1989). Situated cognition and the culture of learning. *Educational Researcher, 18* (1), 32-42.

Bryant, P. (1995). Children and arithmetic. *Journal of Child Psychology and Psychiatry, 36*(1), 3-32.

Carraher, T., Carraher, D., & Schliemann, A. (1990). Mathematics in streets and in schools. In V. Lee (Ed.), *Children's learning in school* (pp. 91-101). London: Hodder & Stroughton.

Cowan, R., & Foster, C.M. (1993). Encouraging children to count. *British Journal of Developmental Psychology, 11,* 411-420.

Department for Education and Skills. (DfEE). (1998). *The implementation of the National Numeracy Strategy: The final report of the numeracy task force.* London: HMSO.

Department for Education and Skills. (DfEE). (1999). *The National Numeracy Strategy framework for teaching mathematics.* London: HMSO.

Durkin, K. (1993). The representation of number in infancy and early childhood. In C. Pratt & A.F. Garton (Eds.), *Systems of representations in children: Development and use* (pp. 133-166). New York: John Wiley & Sons.

Elkind, D. (1988, October). The "Miseducation" of young children. *Education Week,* 11-14.

Elkind, D. (1989, March). Handle with care: Educating young children. *USA Today,* 66-68.

Gray, E. (1991). An analysis of diverging approaches to simple arithmetic. *Educational Studies in Mathematics, 22*(6), 551-574.

Gray, E. (1995). Children and counting: A response to Maclellan. *Education 3 to 13, 23,* 36-40.

Kamii, C. (1989). *Young children continue to reinvent arithmetic (2nd Grade): Implications of Piaget's theory.* New York: Teachers College Press.

Kleinberg, S., & Menmuir, J. (1995). Perceptions of mathematics in pre-five settings. *Education 3 to 13, 23*, 29-35.

Lave, J. (1988). *Cognition in practice: Mind, mathematics and culture in everyday life.* Cambridge: Cambridge University Press.

Leather, R. (1997). *Developing mental maths with 5-7 year olds.* Leamington Spa: Scholastic.

Maclelland, E. (1995). Counting all, counting on, counting up, counting down: The role of counting in learning to add and subtract. *Education 3 to 13, 23,* 17-21.

Ministry of Education. (2003, January). *Press release: 20th January.* Singapore: Author.

Munn, P. (1994). The early development of literacy and numeracy skills. *European Early Childhood Education Research Journal, 2*(1), 5-18.

Munn, P. (2004). The psychology of maths education in the early primary years. *The Psychology of Education Review, 28*(1), 4-7.

Nunes, T., & Bryant, P. (1996). *Children doing mathematics.* Oxford: Blackwell.

OFSTED, (1993). *The teaching and learning of number in primary schools. National curriculum mathematics AT2.* London: HMSO.

Preschool Education Unit. (2003). *Nurturing early learners: A framework for a kindergarten curriculum in Singapore.* Singapore: Ministry of Education.

School Curriculum and Assessment Authority. (SCAA). (1997). *The teaching and assessment of number from Key Stage 1 to 3.* London: Author.

Sharpe, P.J. (1998). Thinking about thinking: A study of the adult's role in providing for the development of number awareness in young children. *Early Child Development and Care, 144,* 79-89.

Sharpe, P.J. (1999). Learning about numbers in kindergarten. *The Mathematics Educator, 4*(2), 70–82.

Sharpe, P.J. (2001). Numeracy development beyond the kindergarten: Some guidelines for future numeracy practices in preschool. *The Mathematics Educator, 6*(1), 42-54.

Shulman, L.S. (1986). Those who understand: Knowledge growth in teaching. *Educational Researcher, 15* (2), 4-14.

Shulman, L.S. (1987). Knowledge and teaching: Foundations of the new reform. *Harvard Educational Review, 57,* 1-22.

Smith, A. B. (1993). Early childhood educare: Seeking a theoretical framework in Vygotsky's work. *International Journal of Early Years Education, 1*(1), 47-61.

Steffe, L.P. (1983). Children's algorithms as schemes. *Educational Studies in Mathematics, 14,* 109-125.

Young-Loveridge, J (1989). The development of children's number concepts: The first year of school. *New Zealand Journal of Educational Studies, 24*(1), 47-64.

Young-Loveridge, J. Peters, S., & Carr, M. (1998). Enhancing the mathematics of four-year-olds: An overview of the EMI - 4S Study. *Australian Research in Early Childhood Education, 1,* 82-93.

Chapter 18

Rethinking and Researching Mathematics Assessment in Singapore: The Quest for a New Paradigm

QUEK Khiok Seng **FAN Lianghuo**

The concept and method of mathematics assessment in Singapore schools have undergone some important developments, and in a sense, a paradigm shift, over the last decades. This chapter traces the events and efforts that led to a broadening of the role of student assessment in mainstream Singapore schools and that gave rise to the use of assessment methods other than the "traditional" ones. Beginning with a brief clarification of terms such as tests, examinations, and assessment, the chapter goes on to describe a new conception of assessment and its broader role in the service of student learning. It next compares the so-called alternative and traditional assessment approaches, presents local empirical studies, and highlights critical issues in the design and use of alternative assessment. The chapter ends with some lessons learned, challenges faced, and suggestions for future research in the assessment of mathematics in schools.

Key words: test, assessment, examination, alternative assessment

1 INTRODUCTION

It is more fruitful if one thinks of assessment strategies and tools in terms of their aptness to the educational purposes they are called upon to provide information, rather than in terms of the limitations of this (e.g., traditional) or that (alternative) form of assessment. It is the inappropriate use and application of the assessment approaches, methods, and results that is a matter of grave concern to many educational policy makers, practitioners, and researchers. Education for life in the 21st century requires a set of skills different from those that have traditionally served the schools well. To prepare for life in the new century, schools need the best tools or best practices available to guide the development of these skills, e.g., creative thinking skills, interpersonal skills, and organisational skills. Thus, the responsibility lies with developers of educational assessment or the users of assessment information to ensure propriety. To this end, assessment-literate users and developers are essential.

Concerning the new development of assessment, the year 2002 was a watershed moment in Singapore. That year saw the suggestions on the use of "alternative" assessment methods in the Ministry of Education's official curriculum documents (MOE, 2002a, 2002b). Until then (and to a degree even now), the notion of assessment in Singapore is almost synonymous to tests and examinations which are held at the end of a term or a year. From its tentative and exploratory beginnings in the late 1990s as part of an in-service course, alternative assessment in mathematics gradually took shape and attained a recognised status among the assessment methods used by schools. It is almost a decade now of efforts at equipping teachers with the knowledge and skills to use alternative assessment methods (Fan, 2002). A critical question is: will these relatively new-to-Singapore assessment methods continue to play a subsidiary role in the assessment of students in schools? Or, will they complement the old modes of assessment and transform the mathematics education of students in their newfound role as assessment *of*, *for* and *as* learning?

2 TESTS, EXAMINATIONS, AND ASSESSMENTS

A first step in rethinking and researching mathematics assessment in Singapore is to clarify the meanings of a plethora of terms in use in our conversations about assessment. A shared meaning of assessment terms is crucial to understanding our research efforts and utilization of research findings. The term *alternative assessment*, for example, suggests that a "new" assessment method is an "lternative" to an existing or an established method. Where the multiple-choice test has been predominant in assessing mathematics, an alternative method would be the open-ended "problem sums" (such as those in PSLE), in that they differ in the manner in which an examinee responds to the test item. An immediate task ahead for us will be to initiate discussion on the meaning in use of assessment labels such as *authentic assessment, alternative assessment, performance assessment, tests* and *examinations*. Until recently, the term *alternative assessment* is rarely mentioned in relation to the assessment of students in mathematics in Singapore schools. For a long time, and even now, terms commonly used for assessment in mathematics in schools are the ubiquitous class tests (teacher-made tests), end-of-term common tests called the CA (continuous assessment), and end-of-semester common tests called the SA (semestral assessment).

At the national level, mathematics is examined in the PSLE (Primary School Leaving Examination) for the end of primary education and the Singapore-Cambridge General Certificate of Education (GCE) Examinations for the end of secondary education, with GCE (Ordinary Level) for Special and Express courses and GCE (Normal Level) for Normal courses.

For students continuing with their mathematics education to the pre-university level, there are again the school-based continuous examinations at the end of terms or semesters, a promotional examination at the end of the first year of pre-university study, and the Singapore-Cambridge GCE (Advanced Level) Examination at the end of pre-university education.

At the secondary and post-secondary levels, there are also the mock examinations, the Prelims (preliminary examinations) for students in the national examination years.

What is common among these assessments of mathematics is that they are mainly written tests and examinations. And, as Fan (2005) pointed out, they are usually tool-limited (with paper and pencil or pen), time-limited (in a block of time), and venue-limited (in classrooms or examination halls). More importantly, however, they have in common a main role to play, namely to grade students. The controlling effect of this conception of assessment on many teachers' classroom practice was revealed by Fan's (2002) research on assessment and his experience with training in-service teachers in the area of assessment.

Common to all these forms of assessment in mathematics is their singular purpose in motivating the students to perform well in them. The fact that past papers from school-based tests and examinations and the large number and range of practice books, fondly called "assessment books," are readily available for purchase by anyone, serves as a stark reminder of the stakes involved in assessment of students in Singapore. This practice further promulgates the notion of assessment as marks and grades for the individual students' achievement on written tests.

Although the school-based CAs and SAs do provide feedback to teachers and students limited to the ends of terms or semesters, they are still perceived as "high-stakes" by many people because the CAs' and SAs' results are used in within-school decisions about students' progress, selection, or placement.

In short, traditionally, the term "assessment" in the Singapore educational context refers mainly to tests or examinations, regardless of whether they are set by the teachers or external examining agencies.

3 TENTATIVE BEGINNINGS

On May 29, 1996, the then Deputy Prime Minister Lee Hsien Loong was quoted in the local dominant English-language newspaper, *The Straits Times*, as calling for a re-look at how students are being assessed academically in Singapore. He observed that the large numbers of As earned by students in the GCE O- and A-levels ought to warrant a rethink of assessment approaches and the school curriculum that assessment served.

> When I was in school, which admittedly was a very long time ago, every year, perhaps about 50 would get four As in the A-levels, probably less. Now, every year, 1,000 pupils score four As. I know that pupils are getting progressively smarter, but I am not quite

> sure whether my generation was really so much dimmer than today's pupils. ... Our education system must encourage pupils to be creative, and reliably test their ability to innovate and solve problems.

A few days later, on June 3, *The Straits Times* drew attention to the large number students who scored distinctions or A1 in Singapore in the GCE O- and A-levels, with statistics and statistical charts to boot. Splashed across the page were headlines: *'Dime a dozen' As? Time to rethink student assessment? Too many straight As, so exams may need to test creativity, independent thinking.*

Alternatives to the assessment of students there should be, but of what shape and form? Professor Leo Tan, then Director of the National Institute of Education (NIE), in reply to the interviewer's question of what NIE can do about the problem of teachers too concerned with turning out exam-smart students, suggested open-book examinations and project work as well as changes to instructional strategies (*The Straits Times*, 1996, July 6). In 1999, the Government accepted a university admission committee's recommendation to include students' project work for their university admission process (MOE, 1999a). The main purpose of project work, according to the committee, is to "inculcate and measure qualities including curiosity, creativity and enterprise. Projects also nurture critical skills for the information age and cultivate habits of self-directed inquiry" (MOE, 1999b). Later on, a trial carried out in all junior colleges affirmed that project work is a valuable learning activity, even though both teachers and students need more time than originally scheduled to be familiar with the processes involved. An outcome of this trial was the decision to include performance on project work as a criterion for admission into local universities from 2005 (MOE, 2001; 2008).

Also from the late 1990s, new strategies for assessing students' learning in mathematics began to attract the attention of education policy makers, administrators, researchers, and practitioners (e.g., see Fan & Yeo, 2000; Fan, 2002; Fan & Yeo, 2000).

As alluded to earlier, the year 2002 saw the publication of a draft set of guidelines for assessment in primary and lower secondary schools by the Ministry of Education, Singapore (MOE, 2002a, 2002b). It should be mentioned that the policy makers produced the guidelines based, in part, on careful consultation with both practitioners and researchers (including the second author of this chapter). The guidelines point out that assessment should be "an integral component of the teaching and learning process" whose main purpose is the improvement of teaching and learning. The same documents list a generous selection of "alternative" assessment methods: journal writing, classroom observation with teacher-pupil conferencing, portfolios, self-assessment, and suggestions for using them. Within a highly centralised system of education, this move by the MOE was a crucial one to encourage the use of alternative assessment by Singapore teachers. There is now official endorsement to broaden the role of assessment and to use assessment methods other than the traditional ones, in a sense a paradigm shift from the past.

The guidelines encourage teachers to use these "new" modes of assessment to support Singapore's vision of *Thinking Schools, Learning Nation* (TSLN),

> In the spirit of TSLN, it is timely for teachers to expand the mode and scope of assessment so as to give all pupils the opportunity to show what they know and can do. The assessment strategies suggested, while still rooted in basic principles of testing, encourage teachers to try out new ways to assess pupils performance in primary mathematics, and in so doing, stimulate pupils to think and learn effectively. (p. 1)

> ... there must be a variety of assessment modes to allow pupils multiple opportunities and a range of contexts in which to demonstrate how far they have achieved the outcomes as stated in the Mathematics Syllabus. (p. 1)

In 2004, these drafts were officially published as *Assessment Guides to Primary Mathematics* (MOE, 2004a) and *Assessment Guides to Lower Secondary Mathematics* (MOE, 2004b).

Thus, to distinguish from assessment methods that are typically associated with written tests in the class, CAs, SAs and *Prelims*, the newer methods of assessment, which now have new purposes to serve in

Singapore, are qualified as "alternatives". Note that this does not preclude the use of information from the alternative methods of assessment in the grading of students.

Alongside these developments there are several small-scale research studies on alternative assessment in Singapore schools. Among these studies are Yeo's (2001) on assessment via journal writing in the mathematics classroom at the junior college level (also see Fan & Yeo, 2000), Seto's (2002) on oral presentation to assess mathematics with Primary 4 students in a neighbourhood school, and Yazilah and Fan's (2002) on journal writing with Primary 5 students in a high-performing school. The last two studies were action research in nature.

Overall, these studies obtained consistent and positive results, which suggested that alternative assessment might yet be practically viable in Singapore schools.

4 INTERNATIONAL PERSPECTIVES AND INFLUENCES

In this section we will briefly consider the impetus for change in mathematics assessment in other countries in order to set the context for the development of alternative assessment in Singapore. The concern about the appropriateness-to-the-purpose of "traditional" forms of assessment in mathematics (and other school subjects as well) was not confined to Singapore. Indeed, as Fan and Zhu (2007) pointed out that being a globalised city state and well developed country, Singapore is to a great degree open to and hence heavily influenced by the outside world in many areas. It could be also argued that the push for an assessment change in Singapore was impacted by assessment reform elsewhere.

Tests and examinations have had its fair share of support and criticism ever since they were formalised in schools. Many have questioned the purposes of assessment and condemned the ill-effects examinations have on students, but as many have spoken up for the benefits to have of tests and examinations. However, the 1980s saw mounting dissatisfaction with the so-called traditional assessment modes in mathematics in schools, and educational researchers and classroom practitioners elsewhere called for new assessment strategies in mathematics instruction that would better serve the desired educational goals and values in education (e.g., Richardson, 1988; Ruthven, 1994; Stacey, 1987; Stempien & Borasi; 1985; Stenmark, 1991). They argued for the use of a wider range of new assessment strategies in mathematics classrooms (e.g., Adam, 1998; Clarke, 1992; Gripps, 1994; Haines & Izard, 1994; Kulm, 1994; National Council of Teachers of Mathematics, 1995; Zehavi, Bruckheimer & Ben-Zvi, 1988).

Waves of reform in mathematics education, particularly in the United States of America, led to a broader conception of assessment. Assessment should go beyond the focus on the grading of student achievement in written tests to include the *how*, *what*, and *why* to bring about authentic and worthwhile student learning. The *Assessment Standards for School Mathematics* published by the National Council of Teachers of Mathematics (NTCM) offers a definition of assessment that has since been commonly cited: "[Assessment] is the process of gathering evidence about a student's knowledge of, ability to use, and disposition toward mathematics and of making inferences from that evidence for a variety of purposes" (NCTM, 1995, p. 3).

A case in point is the development of students' disposition towards mathematics. According to the *Standards*, disposition refers to the interest and appreciation students show for mathematics, the tendency with which students think, act, and react positively toward mathematics – with confidence, curiosity, perseverance, flexibility, inventiveness, and reflectivity in doing mathematics. The traditional assessment approaches, being a once-off written test taken at a specific time and place, were criticised as inadequate to the task of providing relevant and timely information required for the inculcation of desired dispositions.

Although, as Fan (2003) indicated, there were different views about alternative assessment and there is no single definition about alternative assessment, and in some extreme cases, people even argued that the labels "traditional" and "alternative" are meaningless (Romagnano & Long, 2001), there was general agreement that alternative assessment may be distinguished from the traditional paper-and-pencil tests in which students simply recall an answer from memory or choose a response from a given list as typically seen in traditional multiple-choice, true/false, or matching test items.

Alternative assessment would include any type of assessment in which students create a response to a question or task rather than select a response such as in the multiple-choice questions, or MCQs. Alternative assessment tasks are expected to be aligned with curriculum goals and embedded in realistic situations. They require teachers to employ a variety of ways or procedures for observing, collecting, and evaluating students' learning. For example, to Hatfield, Edwards, and Bitter (1997, p. 107), alternative assessment has one or more of these features:

- Students perform, create, and produce (procedures, answers, or even questions)
- Tasks require problem solving or higher-order thinking
- Problems are contextualized
- Tasks are often time-consuming and need days to complete
- Scoring rubrics or scoring guides are required.

It should be pointed out that the concept above applies to assessment in the cognitive domain rather than affective domain, and would therefore exclude the assessment of affective dimensions of the curriculum.

To conclude, the term of alternative assessment in mathematics instruction has been generally used to include the following specific techniques in student assessment: performance-based assessment, authentic assessment, portfolio assessment, journal writing, project assessment, oral presentation, interview, classroom observation, student self-assessment, student-constructed assessment, among others. These specific ways of assessment have been better defined and increasingly used in classroom practice. In a relative sense, the general label "alternative assessment" is still meaningful in reflecting the difference in purpose or in mode of response between the new ways of assessment and the traditional tests and examinations.

5 TRADITIONAL AND ALTERNATIVE COMPARED

Locally, the broader role of assessment may help attain the primary aim of the Singapore mathematics curriculum, which is to enable students to develop their ability to solve mathematics problems. The curriculum identifies five essential components, i.e., concepts, skills, processes, attitudes, and metacognition, that would support the development of mathematical problem solving ability. Whereas the traditional assessment may serve teaching for the first two components (concepts and skills) well, it is the newer assessment strategies that may be more appropriate in meeting instruction in the last three components (processes, attitudes, and metacognition). Nevertheless, how the new assessment strategies can accomplish this function depends on how they are used in the classroom.

Fan (2003, 2005) compared the main features of the traditional and newly developed concepts of student assessment, as shown in Table 1.

Table 1
Comparing the main features of the traditional and new concepts of assessment

Student assessment	Traditional concept	New concept
What (Content)	Cognitive domain (mainly knowledge and skill) and the results of learning	Both cognitive and affective domains (knowledge, skill, ability, and disposition) and both the results and process of learning
Where (Location)	Within classrooms	Within and/or outside classrooms
When (Time)	During class for a block of time (e.g., one class period)	During and/or after class. It could be days, weeks, months, or even years.
How (Method)	Conventional way (paper-and pencil or written test)	Both conventional and alternative ways
Why (Purpose)	Single (most important: grading and reporting students' learning results)	Multiple (most important: improving teaching and learning)

The comparison shown in Table 1 highlights a paradigm shift from traditional concept to new concept of student assessment.

Three inter-related reasons are offered by Fan (2003) for introducing alternative assessment in classroom practice. First, the traditional paper-and-pencil test, like all assessment methods, has its limitations and therefore is inadequate in informing teacher decisions on certain types of student learning. Second, our educational goals and values have changed in the last decade, and thereby we require the use of a wider range of assessment strategies to realise these aims. Third, it reflects the latest development in the conception of pedagogy and assessment (for more details, see Fan 2003).

6 PREPARING TEACHERS FOR ALTERNATIVE ASSESSMENT

As early as 1999, in their in-service courses for primary school teachers, Fan and colleagues began introducing ways of assessing students in mathematics that complement the traditional assessment methods. These included oral presentations and communication tasks (Fan, 2002). The uptake on these newer assessment methods then would depend largely on the interest of the teachers, who would employ them to bring about better student learning and to improve their own teaching. The MOE's initiative to encourage teachers to seek professional development led to more teachers enrolling in the in-service courses on assessment in which alternative assessment is a key component. So far, the in-service course has been offered regularly, with necessary adjustment and update.

The goals of the in-service course were to prepare teachers to apply assessment theory in the construction of mathematics tests, apply alternative methods to assess students in mathematics, apply information technology in assessment in mathematics, and evaluate mathematics programmes for low ability and high ability students.

This 30-hour in-service course was initially organised into ten once-a-week sessions. Participating teachers attended eight on-campus sessions and used the other two sessions to try out what they had learned from the course in their own classroom. The teachers were further supported by means of consultation via email with the course facilitator.

The earlier in-service courses focused on seven relatively well-accepted and widely used of the alternative assessment methods. These were:

- performance/authentic-task-based assessment
- project-based assessment
- journal-based assessment
- portfolio-based assessment
- student-presentation-based assessment

- classroom-observation-based assessment
- interview-based assessment

In reviewing the courses after each run, Fan (2002) added student self-assessment to the repertoire of methods.

For each of these alternative methods of assessment, the teachers learned about what the method is, what the advantages and disadvantages are, and how to use it in actual classroom teaching. The in-class sessions exposed the participants to practical examples of different ways of conducting alternative assessment in a variety of ways. For example, actual audio-recordings for interviews were used to allow the participants to work as authentically as possible on conducting the assessment.

One special feature of the in-service course was that "authentic assessment" (see course assignment described below) was not only taught as content of the course but also used to assess the participants' learning. The teachers were required to demonstrate the integration of one alternative method taught in the course into their daily lesson. The assignment was to be completed within three to four weeks and it was an integral part of the training. A condensed description of the assignment is shown below:

In this assignment, you are required to design one of the following four alternative assessment methods – a project-based assessment, a journal-based assessment, an interview-based assessment, or student self-assessment – and then use it to assess your students. Your submission should include the following components:

(1) *Explain what method you choose to design and implement, why you choose it, what the purposes of your using the method to assess students are.*
(2) *Describe the method itself. If it is project-based or journal-based, describe what the task is and how students should do it. If it is interview-based, describe the questions you will use for the interview. If it is student self-assessment, describe the task that students should do and what you want students to assess themselves.*
(3) *Explain how you will assess or gather information from students' work on what you designed for them. If you think*

appropriate, you should establish a rubric to serve your purpose.

(4) *Report the results of your using the assessment tool with your students and present your conclusions from your assessment.*

A survey of three intakes of teachers for the in-service course by Fan (2002) revealed that the concepts and techniques of alternative assessment were relatively new to them. The respondents also reported that the most helpful strategy in the in-service course was the authentic assessment assignment they had to carry out as part of the course assessment. Also uncovered by the survey was the need to integrate the concept and method of alternative assessment into the school curriculum, for example, through curriculum materials such as textbooks, if the alternative methods were to find a place in the scheme of assessment.

Thus, from its exploratory beginnings as part of an in-service or professional development course, alternative assessment in mathematics took shape and attained a recognised status among the assessment methods used by schools. However, despite after many years of continuous efforts at equipping teachers with the knowledge and skills to use alternative assessment methods, we think that these methods still play a subsidiary role in the assessment of students in Singapore. In a sense, this indicates the challenging nature of reform in assessment, as it is associated with many societal, cultural, and educational factors.

Against this backdrop of events in assessment in school mathematics, a major research project was launched in 2003 to investigate the use of alternative assessment strategies (Fan & Quek, 2005). The project, called simply the *Mathematics Assessment Project*, or MAP, was funded by the Ministry of Education, Singapore through the Centre for Research in Pedagogy and Practice, National Institute of Education. The 15 project team members included university researchers, school teachers, heads of mathematics department, and policy makers in the Ministry of Education.

In the next section of this chapter, we shall briefly review the MAP project, its research aims, procedures, and key findings[1] .

---

[1] Readers who are interested to know more information about the MAP project can find a list of the project publications (as of 10 June 2008) at http://www.crpp.nie.edu.sg/mod/resource/view.php?inpopup=true&id=1681 or contact the authors of this chapter.

7 MAP: SEEKING RESEARCH EVIDENCE

The MAP project aimed to provide research-based evidence, practical suggestions, and resources for the effective use of new assessment strategies in the Singapore context. The project addressed three broad research questions:

- What are the influences of "new assessment strategies" on students' learning of mathematics in their cognitive domain?
- What are the influences of "new assessment strategies" on students' learning of mathematics in their affective domain?
- How can "new assessment strategies" be effectively integrated into mathematics classrooms?

The four "new" assessment approaches investigated in MAP were:

- Project assessment
- Communication assessment
- Performance assessment
- Student self-assessment

The choice of the four types was based on the following two concerns: (1) these strategies are better defined in research community compared to others, and (2) these strategies have more relevance to Singapore educational system.

7.1 *Project Assessment*

Project tasks have been increasingly used in Singapore schools, following MOE's 1999 policy directive, as mentioned earlier. In MAP, a project is defined as a task or a series of tasks for students to carry out, which often includes some or all of the following processes: gathering data, observing, looking for references, identifying, measuring, analysing, determining patterns or (inclusive) relationships, graphing, and written and oral communication. It is believed that project assessment can better assess students in the cognitive domain, particularly students' problem solving ability and creative thinking skills.

In MAP, several types of project tasks are conceptualised: guided projects, independent projects, extended projects, mini-projects, contextualised projects, non-contextualised projects, interdisciplinary projects, mathematical projects, group projects, and individual projects. The setting of various kinds of projects is convenient for teachers to make better decisions to cater for students with different abilities.

7.2 *Communication Assessment*

This assessment strategy focuses on the communication aspect in students' learning, including written communication (through mainly journal writing tasks) and oral communication (through mainly oral presentation tasks). An example of oral communication tasks used in MAP can be found in Appendix 1 (also see Fan & Yeo, 2007).

Developing students' communication skills is highly valued in the knowledge-based society and is explicitly listed as one objective in Singapore mathematics syllabus. Through students' writing, teachers can obtain useful information about students, in the cognitive and affective domains. Through oral presentation, students can tell their teachers and fellow students (whole class or small groups) what they have learnt. This information is useful for teachers to assess students' performance and for students themselves to reflect on their learning.

7.3 *Performance Assessment*

Performance tasks are helpful in developing students' problem solving abilities and higher-order thinking. Authenticity and openness are two important features of the performance tasks used in this project.

Authenticity refers to the degree to which the mathematics problems or activities represent the context and complexity of the real-life world. Openness includes two aspects: various avenues of access, which allow students at different levels of understanding to begin working on the problems, and multiple acceptable answers to problems. An example of performance task used in MAP can be found in Appendix 2.

7.4 *Student Self-assessment*

Student self-assessment includes student self-constructed assessment, which requires students to take greater responsibility, and student self-reflection. It is believed that Singapore students need improvement in this area compared to their Western peers. In this form of assessment, students assess themselves, and their self-assessment is supported by the teachers and specially crafted prompts (see Appendix 3).

In MAP, student self-assessment was divided into three sub-types: self-evaluation, self-reflection, and self-construction (for test or problem-posing). This strategy could be used to assess students' affective domain. A sample of student self-evaluation from MAP can be found in Appendix 3.

MAP involved the participating teachers in classroom-based interventions over a period of three school semesters in 8 primary and 8 secondary schools. There were an equal number of high-performing schools and typical neighbourhood schools in the sample. Comparison classes were selected within each participating school. Within each school, the intervention and comparison classes had the same curriculum environment, including the same syllabus, the same textbooks, and the same scheme of work.

Research data were mainly collected through pre- and post-questionnaire surveys, pre- and post- tests, classroom observations, interviews with teachers and students, and school-based examination scores. Some general findings follow.

The participating teachers and students had a challenging time getting started on using the assessment methods. They required time to become familiar with the concept, value, methods, and skills about the new assessment strategies. However, with adequate help, they were capable of working on the new assessment strategies. Designing new assessment tasks and relevant rubrics remained challenging to the teachers.

Overall, both teachers and students reported quite positive views about the value and feasibility of integrating these new assessment strategies into their daily teaching and learning activities.

These new assessment strategies appeared to have either a positive or a neutral influence on students' performance in the normal school examinations. In Singapore context, this finding is important as it will assure or persuade stakeholders (e.g., parents, students, teachers, schools) that the new methods are viable. The quantitative data indicated that most of the intervention classes performed better or equally well as the comparison classes in both cognitive and affective domains. The qualitative data revealed that the new assessment tasks appeared helpful in developing students' higher-order thinking, communication skills, self-regulation, and self-reflection in learning.

MAP also found a number of curriculum-related factors which had to be overcome if the new assessment strategies were to be successfully implemented in classrooms. The biggest challenge was the lack of new assessment tasks in the normal curricular materials, including syllabus, textbooks (see Zhu & Fan, 2006), teacher manuals, and student workbooks. For the MAP study, the researchers designed and provided almost all the new assessment tasks and the assessment rubrics. Another challenge was convincing the students to engage in the tasks. Despite efforts by the researchers and teachers to explain the value of doing the new assessment tasks, the effect appeared limited because the normal curriculum environment remained unchanged. The students wanted to know why they were required to do the new assessment tasks as these tasks did not appear in their textbooks, workbooks, and other learning materials, and they would not be tested in normal school examinations. Some participating teachers expressed the same concerns as the students: they had to complete the normal scheme of work and cover the normal curriculum, and were afraid if their students would be disadvantaged in the school's CA and SA.

MAP had identified a number of implications about the use of new assessment strategies in curriculum reform and development. For curriculum policy makers and school administrators, it is important to create a school culture on curriculum environment that supports the use of these assessment strategies. These strategies must be stipulated as a curriculum requirement and integrated into curriculum guidelines and materials. In particular, the new assessment tasks should replace some other traditional and less important tasks and they should be an integral part of, but not added onto the mathematics curriculum and instruction. To help teachers implement these strategies, relevant curriculum resources must be made accessible to them.

The new assessment strategies, when integrated into the daily lesson, are appropriate as assessment *for* learning in that they can inform teaching and learning. They also function well as assessment *as* learning, in that students self-assess and self-regulate their learning. However, there is work to be done if the new assessment tasks are to be incorporated into the regular semestral examination. To this purpose, MAP recommends that schools use the new assessment tasks (e.g., open-ended and authentic tasks, project and investigative tasks) first in the formative assessment of students and with more experience, to work them into a component of the summative assessment.

8 CONCLUSION

Traditional assessment still has its own advantages and values in school mathematics. We should improve rather than abolish it. One the other hand, alternative assessment can provide new tools and ideas to realise the goals of education in the new millennium, via requiring a change in teaching practices. As Fan (2003) argued, alternative assessment helps to make assessment more consistent with educational goals and curriculum objectives, more constructive in promoting students' learning, and more integral as a part of teaching and learning. However, how much it can be helpful depends on how the teachers can implement it effectively. In this connection, further exchanges, discussions, and researches are needed.

Finally, this chapter has shown that the world of student assessment in school is constantly on the move. Another wave of assessment in the service of student learning is poised to crash on the shores of educational research and classroom practice. To the tag line of assessment *of* learning, we have moved to a calling for assessment *for* learning and now we are asking for assessment *as* learning. We conclude by reiterating that the desired conception of educational assessment should be one that would serve student learning, be it in mathematics or other subjects in schools, and not for grades alone.

References

Adam, T. L. (1998). Alternative assessment in elementary school mathematics. *Childhood Education, 74*(4), 220-224.

Clarke, D. J. (1992). Activating assessment alternatives in mathematics. *Arithmetic Teacher,* 39(6), 24-29.

Fan, L. (2002). In-service training in alternative assessment with Singapore mathematics teachers. *The Mathematics Educator, 6*(2), 77-94.

Fan, L. (2003). Making alternative assessment an integral part of instructional practice. *Mathematics Newsletter* (Online publication by MOE, Singapore), No. 6. Retrieved June 10, 2008 from http://www3.moe.edu.sg/edumall/pro_develop/doc /assessment_fan.pdf

Fan, L. (2005). Improving mathematics teaching and learning through effective classroom assessment: Experiences and perspectives from Singapore schools. In N. Y. Wong (Ed.), *Revisiting mathematics education in Hong Kong for the new millennium* (pp. 359-376). Hong Kong: Hong Kong Association for Mathematics Education.

Fan, L., & Quek, K. S. (2005, May). *Integrating new assessment strategies into mathematics classrooms: What have we learned from a CRPP mathematics assessment project?* Paper presented at the CRPP International Conference: Resigning Pedagogy: Research, Policy, and Practice, Singapore.

Fan, L., & Yeo, S. M. (2000). Journal writing as an alternative assessment. *Scientas, 34*, 16-25.

Fan, L., & Yeo, S. M. (2007). Integrating oral presentation in mathematics teaching and learning: An exploratory study with Singapore secondary students. *The Montana Mathematics Enthusiast, Monograph 3*, 81-98.

Fan, L., & Zhu, Y. (2007). From convergence to divergence: The development of mathematical problem solving in research, curriculum, and classroom practice in Singapore. *ZDM-The International Journal on Mathematics Education, 39*(5-6), 491-501.

Gripps, C. V. (1994). *Beyond testing: Towards a theory of educational assessment.* London: Falmer Press.

Haines, C., & Izard, J. (1994). Assessing mathematical communications about projects and investigations. *Educational Studies in Mathematics, 27*(4), 373–386.

Hatfield, M. M., Edwards, N. T., & Bitter, G. G. (1997). *Mathematics methods for elementary and middle school teachers*. Boston, MA: Allyn and Bacon.

Kulm, G. (1994). *Mathematics assessment: What works in the classroom*. San Francisco, CA: Jossey-Bass.

Ministry of Education, Singapore. (1999a). *Government accepts recommendations of the committee on university admission system*. Press Release (12 July). Retrieved June 10, 2008 from http://www.moe.gov.sg/media/press/1999/pr990712.htm

Ministry of Education, Singapore. (1999b). *Prelimary recommendations of the committee on university admission system*. Press Release (9 Jan). Retrieved June 10, 2008 from http://www.moe.gov.sg/media/press/1999/pr090199.htm

Ministry of Education. (2000a). *Mathematics syllabus (Primary)*. Singapore: Author.

Ministry of Education. (2000b). *Mathematics syllabus (Lower Secondary)*. Singapore: Author.

Ministry of Education, Singapore. (2001). *Project work to be included for university admission in 2005*. Press Release (20 June). Retrieved June 10 2008 from http://www.moe.edu.sg/ press/2001

Ministry of Education. (2002a). *Assessment guidelines (Primary Mathematics)*. Singapore: Author.

Ministry of Education. (2002b). *Assessment guidelines (Lower Secondary Mathematics)*. Singapore: Author.

Ministry of Education. (2004a). *Assessment Guides to Primary Mathematics*. Singapore: Author.

Ministry of Education. (2004b). *Assessment Guides to Lower Secondary Mathematics*. Singapore: Author.

Ministry of Education, Singapore. (2008). *University admission criteria*. Retrieved June 10, 2008 from http://www3.moe.edu.sg/cpdd/alevel2006/faqs.htm#035

National Council of Teachers of Mathematics. (1995). *Assessment standards for school mathematics*. Reston, VA: Author.

Richardson, K. (1988). Assessing understanding. *Arithmetic Teacher*, 35(6), 39-41.

Romagnano, L., & Long, V. (2001). The myth of objectivity in mathematics assessment. *Mathematics Teacher*, 94(1), 31-37.

Ruthven, K. (1994). Better judgment: Rethinking assessment in mathematics education. *Educational Studies in Mathematics*, 27(4), 433-450.

Seto, C. (2002). Oral presentation as an alternative assessment in mathematics. In D. Edge & B. H. Yeap (Eds.), *Mathematics education for a knowledge-based era (Vol. 2)* (pp. 33-39). Singapore: Association of Mathematics Educators.

Stacey, K. (1987). What to assess when assessing problem solving? *Australian Mathematics Teacher, 43*(3), 21-24.

Stempien, M., & Borasi, R. (1985). Students' writing in mathematics: Some ideas and experiences. *For the Learning of Mathematics, 5*(3), 14-17.

Stenmark, J. (1991). *Mathematics assessment: Myths, models, good questions, and practical suggestions.* Reston, VA: National Council of Teachers of Mathematics.

Yazilah, A., & Fan, L. (2002). Exploring how to implement journal writing effectively in primary mathematics in Singapore. In D. Edge & B. H. Yeap (Eds.), *Mathematics education for a knowledge-based era (Vol. 2)* (pp. 56-62). Singapore: Association of Mathematics Educators.

Yeo, S. M. (2001). *Using journal writing as an alternative assessment in junior college mathematics classrooms.* Unpublished Master dissertation. Singapore: National Institute of Education, Nanyang Technological University.

Zehavi, N., Bruckheimer, M., & Ben-Zvi, R. (1988). Effect of assignment projects on students' mathematical activity. *Journal for Research In Mathematics Education, 19*(5), 421-438.

Zhu, Y., & Fan, L. (2006). Focus on the representation of problem types in intended curriculum: A comparison of selected mathematics textbooks from Mainland China and the United States. *International Journal of Science and Mathematics Education, 4*(4), 609-626.

Appendix 1

An example of oral communication tasks used in MAP (Topic: Hire purchase)

1. Discuss and write down what is agreed among yourselves in the group about hire purchase.
2. Find an advertisement which has both cash and hire purchase scheme. Cut out and paste the advertisement on the worksheet.
3. Suppose you are to advise your parents about purchasing the product (that you have cut out earlier in part 2). What would you say to them and what advice would you give?

Appendix 2

An example of performance task used in MAP

Building Bricks

You have come across construction workers building a wall with bricks. In order for a wall to be very stable, the bricks must be laid in a certain arrangement as shown in the picture.

Warming Up Exercises:

1. Write down the total number of bricks shown in Fig 1 and Fig 2:

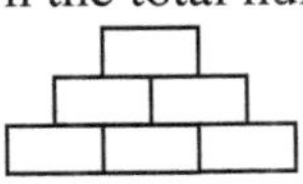

Fig 1

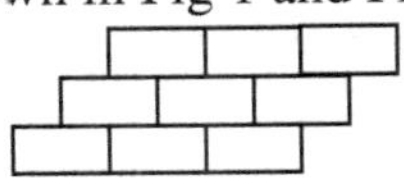

Fig 2

2. A stack of bricks is arranged in the same way as shown in Question1. It is noted that the top line contains 4 bricks and the bottom line contains 10 bricks. Find the total number of bricks in the stack.

Do you Know?

A stack of bricks is arranged in a way such that the number of bricks in each line is one less than the immediate one below. If the top line contains p bricks and the bottom line contains q bricks, the total number of bricks in the stack can be calculated using the formula:

$$\textit{Total number of bricks} = (p + q)(q - p + 1) \div 2$$

Using the formula to verify your answers in Question 1 and Question 2.

Performance Task:

There is a stack of bricks arranged in the way mentioned above. It is noted that the total number of bricks is 30 and the number of bricks at the top is even. Draw all the possible arrangements and state the corresponding numbers of bricks at the top and bottom. (Hint: You can use the formula in "*Do you Know?*")

Appendix 3

A sample of Mathematics Self-Evaluation Sheet from MAP

Dear student, We have taught Chapter 12: Arithmetical Problems last week. Please recall your experience of learning the topic and complete this self-evaluation sheet, so I can know better how you have learned about this topic and how I can help you.

A. You can just circle one of the three choices: "yes", "no", "not sure".

1. This topic is overall easy.	(Yes)	No	Not sure
2. I understand this topic well.	(Yes)	No	Not sure
3. I had difficulty in learning this topic.	(Yes)	No	Not sure
4. I feel I was quite lost in learning this topic.	Yes	(No)	Not sure
5. I can complete homework for this topic most of the time by myself.	Yes	(No)	Not sure
6. I enjoy learning this topic.	(Yes)	No	Not sure
7. I find this topic useful.	(Yes)	No	Not sure
8. I am confident about this topic.	Yes	(No)	Not sure

B. (i) What did you find most difficult in learning this topic?

Simple and Compound interest.

(ii) Do you still have problem about it? (Circle one) (Yes) No

Is there any help that you wish me to give you? If so, feel free to write it down.

Teach simple and compound Interest a few more times.

(iii) Did you encounter any other difficulties in learning the topic that you wish me to help? If so, please let me know.

I don't understand problems involving the use of tables and charts quite well.

C. What else about this topic do you wish to tell me? Please feel free to write it down. (e.g., other information, or requests, or suggestions, etc.)

Nil.

Part III

Comparative Studies in Mathematics Education

Chapter 19

Performance of Singapore Students in Trends in International Mathematics and Science Studies (TIMSS)

Berinderjeet KAUR

The performance of Singapore students in Trends in International Mathematics and Science Studies (TIMSS) of 1995, 1999, and 2003 at both Grades 4 and 8 was outstanding and captured the attention of many educators and politicians world-wide. This chapter gives the details of the participants, nature of tests, and results. Though the overall performance has been commendable, performance on several items has been of concern to some educators in Singapore. Some of these items are discussed. Three main underlying causes of difficulty that students faced in such items are stated. The possible contributing factors for Singapore's high performance are also presented.

Key words: TIMSS, Grade 4, Grade 8, benchmarks, attitude, mathematics performance, performance expectation

1 BACKGROUND

The Trends in International Mathematics and Science Study (TIMSS), originally named the Third International Mathematics and Science Study, is the most recent in the series of IEA (International Association for the Evaluation of Educational Achievement) studies to measure trends in students' mathematics and science achievement. In 1995, The Third International Mathematics and Science Study (TIMSS) was carried out (Kelly, Mullis & Martin, 2000) and 46 countries participated at the

Fourth, Eighth, and Twelfth grades. In 1999, the Third International Mathematics and Science Study (TIMSS-R) was replicated at the Eighth grade for the cohort of Grade 4 students who took the 1995 test. TIMSS-R was designed to provide trends in Eighth grade mathematics and science achievement in an international context (Mullis, Martin, Gonzalez, Gregory, Garden, O'Connor, Chrostowski & Smith, 2000). Thirty eight countries participated in TIMSS-R.

TIMSS (Trends in International Mathematics and Science Study) 2003 (Mullis, Martin, Gonzalez & Chrostowski, 2004) built on the successes of the past two TIMSS and offered participants of the earlier two studies the opportunity to study trends in Eighth grade mathematics achievement at three points over an eight-year period and trends in Fourth grade mathematics achievement at two points over the same period. A total of 49 countries participated in TIMSS 2003. Singapore participated in TIMSS 1995 at the Fourth and Eighth grades, in TIMSS 1999 at the Eighth grade, and in TIMSS 2003 at the Fourth and Eighth grades. Data were collected from students using achievement tests and questionnaires. Other relevant data were collected with the help of questionnaires from teachers and school leaders.

Although many secondary studies have been done on Singapore students' achievement data of both TIMSS 1995 and TIMSS 1999 (Ferrucci, Kaur & Carter, 2002; Kaur, 2005; Kaur, 2003a; Kaur, 2002a; Kaur, 2002b; Kaur, Ferrucci & Carter, 2003; Kaur & Pereira-Mendoza, 2000a; Kaur & Pereira-Mendoza, 2000b; Kaur & Pereira-Mendoza, 1999a, Kaur & Yap, 2002; Pereira-Mendoza & Kaur, 1999a; Pereira-Mendoza & Kaur, 1999b; Pereira-Mendoza, Kaur & Yap, 1999), and also on Singapore students' and teachers' questionnaire data (Kaur, 2003b; Kaur & Pereira-Mendoza, 1999), this chapter will focus solely on the performance of Singapore students on achievement tests of all the three TIMSS. The data and findings reported in this paper are drawn from the respective international mathematics reports of the three studies (Beaton, Mullis, Martin, Gonzalez, Kelly & Smith, 1996; Mullis, Martin, Beaton, Gonzalez, Kelly & Smith, 1997; Mullis, Martin, Gonzalez, Gregory, Garden, O'Connor, Chrostowski & Smith, 2000; Mullis, Martin, Gonzalez & Chrostowski, 2004) published by the TIMSS International Study Centre and the respective national reports (Research & Testing

Division, 1996, 1997; Research & Evaluation Branch, 2000; Research & Evaluation Section, 2004) published by the Ministry of Education (MOE) in Singapore.

2 TIMSS: 1995, 1999 & 2003

2.1 *Student Participants*

At the respective grade levels, intact classes were sampled from all the schools. Singapore adhered very strictly to the sampling design of the studies and the data collected satisfied all criteria for consideration to be apart of the international studies. Table 1 shows the number of participants by grade level at each study.

Table 1
Number of Participants in TIMSS

	TIMSS					
	1995		1999		2003	
Level	Students	Schools	Students	Schools	Students	Schools
Grade 4	7139	191	NA	NA	6668	182
Grade 8	4644	137	4966	145	6018	164

NA – not applicable
Students who participated in TIMSS 1999 were from the same cohort of students that participated at Grade 4 in TIMSS 1995.

2.2 *Achievement tests*

To ensure broad subject matter coverage without overburdening individual students, the studies used a matrix-sampling technique and distributed both the mathematics and science items among booklets prepared for the tests. For all the three studies, each student attempted one booklet containing both mathematics and science items. Thus, the same students participated in both the mathematics and science tests.

2.2.1 Grade 4 (TIMSS 1995 & TIMSS 2003)

For TIMSS 1995, there were 102 mathematics items in the test (Mullis, Martin, Beaton, Gonzalez, Kelly & Smith, 1997), distributed between eight booklets, with each student attempting only one booklet requiring 60 minutes of response time. The number of mathematics items ranged from 19 to 39 in a booklet. The items were a combination of 79 multiple choice, 15 short answer, and 8 extended response types. For TIMSS 2003 there were 161 mathematics items in the test (Mullis, Martin, Gonzalez & Chrostowski, 2004), distributed between 12 booklets, with each student attempting only one booklet requiring 72 minutes of response time. The number of mathematics items ranged from 18 to 47 in a booklet. The items were a combination of 92 multiple choice and 69 constructed response types. Constructed response items required students to generate and write their own answers. Some asked for short answers while others required extended responses with students showing their work or providing explanations for their answers. Thirty seven items in TIMSS 2003 were also used in TIMSS 1995. These items are called trend items. Table 2 shows the distribution of the items by content category and performance category.

Table 2
Content Areas and Performance Expectation of Mathematics Tests for Grade 4

Content			
TIMSS 1995		TIMSS 2003	
Whole Numbers	25%	Number	39%
Fractions & Proportionality	21%	Measurement	20%
Measurement, Estimation & Number Sense	20%	Geometry	15%
Data Rep, Analysis & Probability	12%	Data	11%
Geometry	14%	Patterns & Relationships	15%
Patterns, Relations & Functions	10%		
Performance Expectation / Cognitive Domain			
TIMSS 1995		TIMSS 2003	
Knowing	41%	Knowing Facts & Procedures	24%
Performing Routine Procedures	16%	Using Concepts	23%
Using Complex Procedures	24%	Solving Routine Problems	37%
Solving Problems#	20%	Reasoning	16%

Because results are rounded to the nearest whole number, some totals may appear inconsistent.
includes one item classified as "Justifying and proving" and 4 items classified as "Communicating".

The tests covered five main content areas, namely, number, measurement, geometry, data and patterns, relations and functions. In TIMSS 2003, there appears to be more emphasis on higher order thinking items as only a quarter of the items tested knowledge and procedures as compared to almost half in TIMSS 1995. Figures 1 and 2 show two items, one each from TIMSS 1995 and TIMSS 2003 respectively.

TIMSS 1995 – Item U5
Content Category: Whole Numbers
Performance Category: Knowing

Addition Fact
$4 + 4 + 4 + 4 + 4 = 20$

Write this addition fact as a multiplication fact.

________ x ________ = ________

Figure 1. A TIMSS 1995 Grade 4 Item.

TIMSS 2003 – Item M03-01
Content Category: Data
Performance Category: Reasoning

The graph shows the heights of four girls.

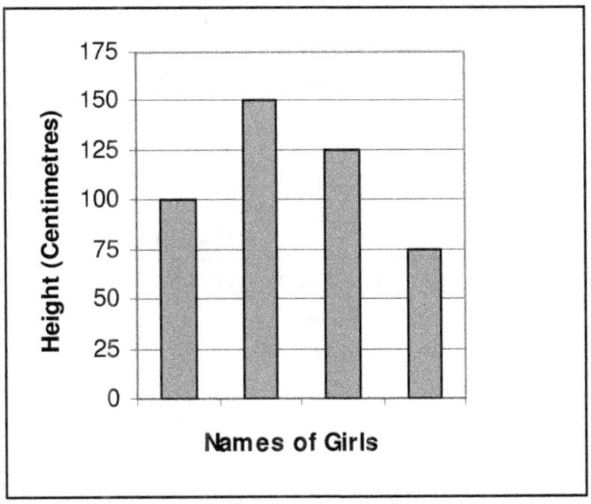

The names are missing from the graph. Debbie is the tallest. Amy is the shortest. Dawn is taller than Sarah. How tall is Sarah?

A. 75 cm B. 100 cm C. 125 cm D. 150 cm

Figure 2. A TIMSS 2003 Grade 4 Item.

2.2.2 Grade 8 (TIMSS 1995, TIMSS 1999, TIMSS 2003)

For TIMSS 1995, there were 151 mathematics items in the test (Beaton, Mullis, Martin, Gonzalez, Kelly & Smith, 1996), distributed between eight booklets, with each student attempting only one booklet requiring

90 minutes of response time. The number of mathematics items ranged from 33 to 42 in a booklet. The items were a combination of 125 multiple choice, 19 short answer and 7 extended response types. For TIMSS 1999 there were a total of 162 mathematics items in the test (Mullis, Martin, Gonzalez, Gregory, Garden, O'Connor, Chrostowski & Smith, 2000), distributed between 8 booklets, with each student attempting only one booklet requiring 90 minutes of response time. The range of mathematics items in a booklet was the same as TIMSS 1995. The items were a combination of 125 multiple choice and 37 free response types. Forty eight were trend items, i.e., the same as those in TIMSS 1995. For TIMSS 2003, there were 194 mathematics items in the test (Mullis, Martin, Gonzalez & Chrostowski, 2004), distributed between 12 booklets, with each student attempting only one booklet requiring 90 minutes of response time. The number of mathematics items ranged from 25 to 58 in a booklet. The items were a combination of 128 multiple choice and 66 constructed response types. Eighty three were trend items, i.e., 23 from TIMSS 1995 and 60 from TIMSS 1999. Table 3 shows the distribution of the items by content category and performance category.

Table 3

Content Areas and Performance Expectation of Mathematics Tests for Grade 8

Content					
TIMSS 1995		TIMSS 1999		TIMSS 2003	
Fractions & Number Sense	34%	Fractions & Number Sense	38%	Number	30%
Algebra	18%	Algebra	22%	Algebra	24%
Measurement	12%	Measurement	15%	Measurement	16%
Geometry	15%	Geometry	13%	Geometry	16%
Data Rep, Analysis & Probability	14%	Data Rep, Analysis & Probability	13%	Data	14%
Proportionality	7%				
Performance Expectation/Cognitive Domain					
TIMSS 1995		TIMSS 1999		TIMSS 2003	
Knowing	22%	Knowing	19%	Knowing Facts & Procedures	23%
Performing Routine Procedures	25%	Using Routine Procedures	23%	Using Concepts	19%
Using Complex Procedures	21%	Using Complex Procedures	24%	Solving Routine Problems	36%
Solving Problems #	32%	Investigating & Solving Problems	31%	Reasoning	22%
		Communicating & Reasoning	2%		

Because results are rounded to the nearest whole number, some totals may appear inconsistent.

includes two items classified as "Justifying and Proving" and two items classified as "Communicating."

The tests covered five main content areas, namely, number, algebra, measurement, geometry, and data. It appears that from 1995 to 2003 the tests have progressively placed more emphasis on higher order thinking. In TIMSS 1995, almost half the items tested knowledge and procedures, while in TIMSS 1999 only about 40% did so, and less than a quarter in TIMSS 2003. Figures 3, 4, and 5 show three items, one from each study.

TIMSS 1995 – Item J-11
Content Category: Geometry
Performance expectation: Knowing

A quadrilateral MUST be a parallelogram if it has

A. one pair of adjacent sides equal
B. one pair of parallel sides
C. a diagonal as axis of symmetry
D. two adjacent angles equal
E. two pairs of parallel sides

Figure 3. A TIMSS 1995 Grade 8 Item.

TIMSS 1999 – Item V04
Content category: Algebra
Performance Category: Parts A & B – Investigating and Solving Problems
Part C – Communicating and Reasoning

The figures show four sets consisting of circles.

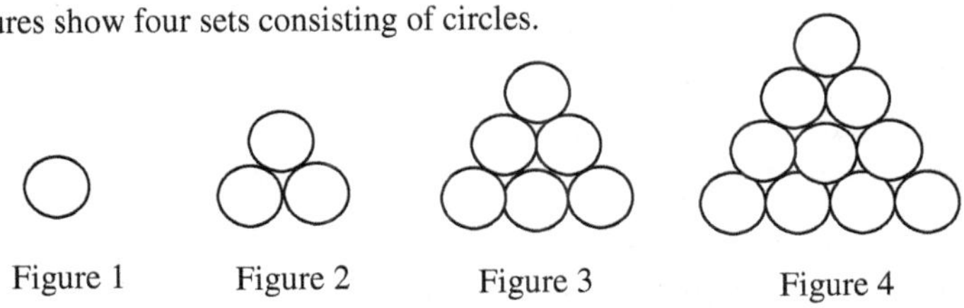

a) Complete the table below. First, fill in how many circles make up Figure 4. Then, find the number of circles that would be needed for the 5th figure if the sequence of figures is extended

Figure	Number of circles
1	1
2	3
3	6
4	
5	

b) The sequence of figures is extended to 7th figure. How many circles would be needed for Figure 7?
Answer: ________________________.

c) The 50th figure in the sequence contains 1275 circles. Determine the number of circles in the 51st circle. Without drawing the 51st figure, explain or show how you arrive at your answer.

Figure 4. A TIMSS 1999 Grade 8 Item.

TIMSS 2003 – Item M03-06
Content Category – Measurement
Performance Category: Solving Routine Problems

A thin wire 20 centimetres long is formed into a rectangle. If the width of this rectangle is 4 centimetres, what is its length?

A. 5 centimetres
B. 6 centimetres
C. 12 centimetres
D. 16 centimetres

Figure 5. A TIMSS 2003 Grade 8 Item.

2.3 *Results*

The performance of Singapore students in all the three studies was outstanding. For TIMSS 1995 and TIMSS 2003 at the Fourth grade, Singapore was ranked first. For TIMSS 1995, TIMSS 1999, and TIMSS 2003 at the Eighth grade, Singapore was also ranked first.

Table 4 shows the average scale scores of the top five countries for Grade 4. For TIMSS 1995, although Singapore was ranked first and Korea second, there was no statistically significant difference between the average scale scores of the two countries. Nevertheless, the average scale scores of both Singapore and Korea were significantly higher than all other participating countries. However, for TIMSS 2003, the average scale score of Singapore was significantly higher than all other participating countries.

Table 4
Results of TIMSS for Grade 4

TIMSS 1995			TIMSS 2003		
Country	Rank	Average Scale Score	Country	Rank	Average Scale Score
Singapore	1	625 (5.3)	Singapore	1	594 (5.6)
Korea	2	611 (2.1)	Hong Kong SAR	2	575 (3.2)
Japan	3	597 (2.1)	Japan	3	565 (1.6)
Hong Kong SAR	4	587 (4.3)	Chinese Taipei	4	564 (1.8)
Czech Republic	5	567 (3.3)	Belgium (Flemish)	5	551 (1.8)
International Average		529	International Average		495 (0.8)

( ) standard errors appear in parentheses.

Table 5 shows the average scale scores of the top five countries for Grade 8. For TIMSS 1995, Singapore was ranked first and its average scale score was significantly higher than all other participating countries. For TIMSS 1999, Singapore was again ranked first. However, its average scale score was not significantly different from Korea, Chinese Taipei, and Hong Kong SAR, but was significantly higher than all other participating countries. In TIMSS 2003, Singapore was once again ranked first and its average scale score was significantly higher than all other participating countries.

Table 5
Results of TIMSS for Grade 8

TIMSS 1995			TIMSS 1999			TIMSS 2003		
Country	Rank Avg		Country	Rank Avg		Country	Rank Avg	
Singapore	1	643 (4.9)	Singapore	1	604 (6.3)	Singapore	1	605 (3.6)
Korea	2	607 (2.4)	Korea	2	587 (2.0)	Korea	2	589 (2.2)
Japan	3	605 (1.9)	Chinese Taipei	3	585 (4.0)	Hong Kong SAR	3	586 (3.3)
Hong Kong SAR	4	588 (6.5)	Hong Kong SAR	4	582 (4.3)	Chinese Taipei	4	585 (4.6)
Belgium (Flemish)	5	565 (5.7)	Japan	5	579 (1.7)	Japan	5	570 (2.1)
International Average		513	International Average		487 (0.7)	International Average		467 (0.5)

( ) standard errors appear in parentheses.

Table 6 shows the percentages of Singapore students at each of the benchmarks for all the studies. For Grade 4 students, there was some improvement from TIMSS 1995 to TIMSS 2003 in the percent of students reaching all benchmarks, except for the Advanced benchmark at the 90th percentile. As for the Grade 8 students, the gradual improvement in the performance of the students over the three studies appears to be in the Advanced benchmark. Over the eight year period, the percent of Grade 8 Singapore students in the Advanced Benchmark, i.e., top 10%, has consistently increased.

Table 6
Percent of Students at International Benchmarks

		International Benchmarks			
Band		Advanced	High	Intermediate	Low
Scale Score		625	550	475	400
Grade 4	Singapore	38 (2.2)	70 (1.6)	89 (1.0)	96 (0.4)
TIMSS 1995	International	10 (0.3)	33 (0.4)	63 (0.4)	85 (0.3)
Grade 4	Singapore	38 (2.9)	73 (2.4)	91 (1.3)	97 (0.6)
TIMSS 2003	International	10 (0.3)	36 (0.4)	69 (0.4)	88 (0.3)
Grade 8	Singapore	40 (2.9)	84 (1.8)	98 (0.4)	100 (0.0)
TIMSS 1995	International	11 (0.3)	37 (0.4)	69 (0.4)	89 (0.3)
Grade 8	Singapore	42 (3.5)	77 (2.6)	94 (1.2)	99 (0.3)
TIMSS 1999	International	10 (0.2)	31 (0.3)	57 (0.3)	80 (0.2)
Grade 8	Singapore	44 (2.0)	77 (2.0)	93 (1.0)	99 (0.2)
TIMSS 2003	International	8 (0.2)	28 (0.3)	56 (0.3)	80 (0.3)

( ) standard errors appear in parentheses.

The following items at the respective benchmarks illustrate the performance of Singapore students in TIMSS 2003.

2.3.1 TIMSS 2003 Grade 4: Some International Benchmark Items

The items in Figures 6, 7, and 8 are representative of the respective benchmarks at Grade 4. These are items that students reaching the respective international benchmarks are likely to answer correctly. For each item, the performance of the top five countries, Singapore, and the international average is stated.

Content Area: Number **Cognitive Domain**: Knowing Facts &Procedures **Description:** Identifies the decimal representation for a fraction with a denominator of 10	Country	Percent Full Credit
Which of these means 7/10 ?	Singapore	95 (0.8)
	Hong Kong SAR	78 (1.8)
	Chinese Taipei	74 (1.8)
	United States	62 (1.8)
A. 70 B. 7 C. 0.7 D. 0.07	Japan	60 (2.2)
	International Avg	43 (0.4)

Figure 6. Example of Advanced International Benchmark (625) Grade 4 Item.

Content Area: Patterns & Relationships **Cognitive Domain**: Solving Routine Problems **Description:** Selects the expression that represents a situation involving multiplication.	Country	Percent Full Credit
□ represents the number of magazines that Lina reads each week. Which of these represents the total number of magazines that Lina reads in 6 weeks? A. 6 + □ B. 6 x □ C. □ + 6 D. (□ + □) x 6	Singapore	86 (1.4)
	Chinese Taipei	81 (1.5)
	Hong Kong SAR	76 (1.9)
	United States	72 (1.2)
	Netherlands	72 (2.7)
	International Avg	58 (0.4)

Figure 7. Example of High International Benchmark (550) Grade 4 Item.

Content Area: Number **Cognitive Domain**: Using Concepts **Description:** Recognizes a familiar fraction represented by a figure with shaded parts.	Country	Percent Full Credit
Which shows $\frac{2}{3}$ of the square shaded?	Singapore	93 (1.0)
	Hong Kong SAR	86 (1.7)
	United States	82 (1.1)
	Chinese Taipei	81 (1.5)
	Belgium-Flemish	79 (1.8)
	International Avg	57 (0.4)

Figure 8. Example of Intermediate International Benchmark (475) Grade 4 Item.

Figures 6, 7, and 8 show three items, from the Advanced, High, and Intermediate international benchmark levels respectively, that Singapore students were ranked first. The first item is on the conversion of a fraction to a tenth decimal, the second is on matching the correct mathematical expression to a given word problem situation, and the third is on the basic concept of fractions where the student had to match a representation to a given fraction. These items are based on topics Singapore students by Grade 4 would have been taught at school and similar to items they routinely practiced during their in and out of class mathematics assignments.

In contrast, Figures 9, 10 and 11 show three items, from the Advanced, High, and Low international benchmark levels respectively, that Singapore students did relatively poorly on. Figure 9 shows an item on area that requires the students to complete an irregular figure so that it has an area of 13 square centimeters. This item requires students to complete a partial irregular figure so that its area is as given. The inability of many Singapore students to complete this task successfully shows that they were not able to apply their knowledge in novel situations.

Figure 10 shows an item that requires the students to reassemble 4 triangles to form a square on a given grid. Again the inability of many Singapore students to successfully assemble the triangles to form a square demonstrates that they were not able to do tasks that are unfamiliar or require them to apply their knowledge in new situations. In addition, the wording of the item could have also contributed towards the difficulty faced by the students.

Figure 11 shows an item from the Low international benchmark that Singapore students did relatively poorly. This item requires students to draw an isosceles triangle whose base is given. By Grade 4, Singapore students are introduced to the different types of triangles and usually as part of their work they are almost always given triangles that have been drawn to work on. The poor performance on this item may be attributed once again to an unfamiliar task and the inability to apply knowledge in a novel situation.

Content Area: Measurement
Cognitive Domain: Reasoning
Description: Completes an irregular figure on a grid so that it has a given area

Country	Percent Full Credit
Japan	68 (2.1)
Chinese Taipei	66 (1.8)
Hong Kong SAR	52 (2.8)
Singapore	43 (2.2)
United States	24 (1.7)
International Avg	29 (0.4)

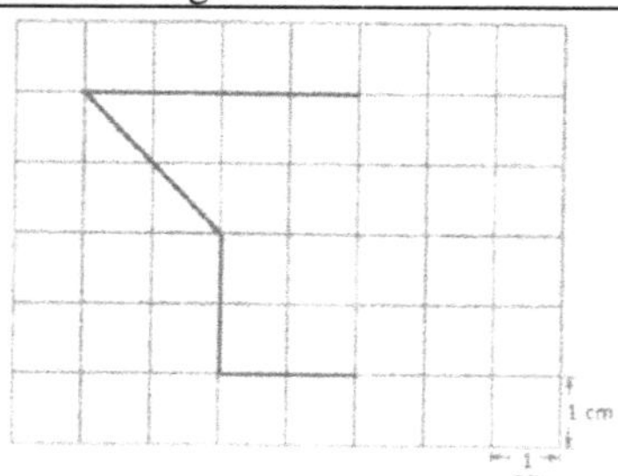

The squares in the grid above have areas of 1 square centimeter. Draw lines to complete the figure so that it has an area of 13 square centimeters.

Figure 9. Example of Advanced International Benchmark (625) Grade 4 Item.

Content Area: Geometry
Cognitive Domain: Knowing Facts and Procedures
Description: Part B – Makes and draws one square from four triangles tiles (square tiles divided diagonally into one white and one black triangle).

For this item, you have been given a piece of cardboard with 10 tiles like the ones shown below (on the left). Take the cardboard and punch out the 10 tiles.

B. Use all 4 triangle tiles to make a black square. Then show what you did with your tiles by shading in your square below (on the right).

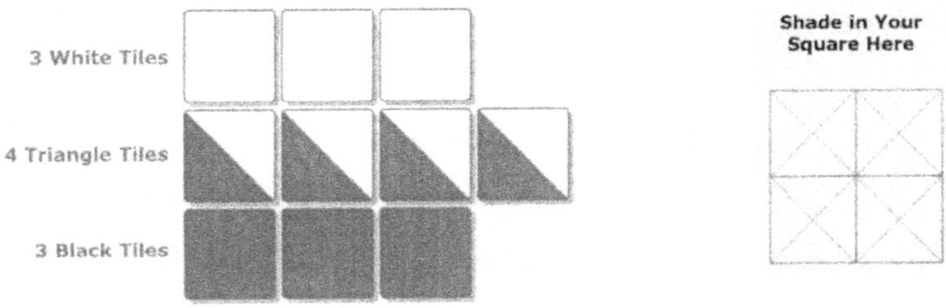

Country	Percent Full Credit
Japan	71 (2.0)
Netherlands	60 (3.2)
Russian Fed	57 (2.3)
Lithuana	57 (2.3)
Belgium-Flemish	55 (2.0)
Singapore (rank 15)	45 (2.3)
International Avg	42 (0.5)

Figure 10. Example of High International Benchmark (550) Grade 4 Item.

Content Area: Measurement
Cognitive Domain: Knowing Facts and Procedures
Description: Given the base, draw a triangle on a grid with the other two sides the same length.

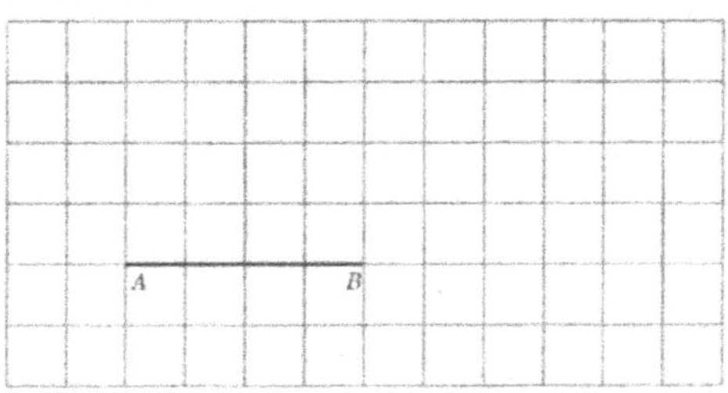

Draw a triangle in the grid so that the line AB is the base of the triangle and the two new sides are the same length as each other.

Country	Percent Full Credit
Hong Kong SAR	95 (0.9)
Latvia	84 (1.4)
Japan	80 (1.8)
New Zealand	80 (1.8)
Singapore	77 (1.8)
International Avg	67 (0.4)

Figure 11. Example of Low International Benchmark (400) Grade 4 Item.

2.3.2 TIMSS 2003 Grade 8: Some International Benchmark Items

The items in Figures 12, 13, and 14 are representative of the respective benchmarks at Grade 8. These are items that students reaching the respective international benchmarks are likely to answer correctly. For each item, the performance of the top five countries, Singapore, and the international average is stated.

Content Area: Number **Cognitive Domain**: Knowing Facts and Procedures **Description:** Solves a one-step word problem involving division of a whole number by a unit fraction.	Country	Percent Full Credit
A scoop holds $\frac{1}{5}$ kg of flour. How many scoops of flour are needed to fill a bag with 6 kg of flour? Answer:	Singapore Hong Kong SAR Chinese Taipei Netherlands Korea International Avg	79 (1.9) 76 (1.8) 75 (1.9) 74 (2.1) 68 (1.5) 38 (0.3)

Figure 12. Example of High International Benchmark (550) Grade 8 Item.

Content Area: Algebra **Cognitive Domain**: Knowing Facts and Procedures **Description:** Solves equation for missing number in a proportion	Country	Percent Full Credit
If $\frac{12}{n} = \frac{36}{21}$, then n equals A. 3 B. 7 C. 36 D. 63	Singapore Korea Hong Kong, SAR Belgium-Flemish Netherlands International Avg	93 (0.7) 89 (0.9) 88 (1.2) 86 (1.4) 85 (1.8) 65 (0.3)

Figure 13. Example of Intermediate International Benchmark (475) Grade 8 Item.

Content Area: Number **Cognitive Domain**: Using Concepts **Description:** Selects two-place decimal closest to a given whole number.	Country	Percent Full Credit
Which of these numbers is closest to 10? A. 0.10 B. 9.99 C. 10.10 D. 10.90	Netherlands Sweden Estonia Singapore Lithuania International Avg	97 (1.0) 96 (1.1) 96 (1.2) 95 (1.1) 95 (1.0) 77 (0.3)

Figure 14. Example of Low International Benchmark (400) Grade 8 Item.

Figures 12, 13, and 14 show three items, at the High, Intermediate, and Low international benchmark levels respectively, which Singapore students performed very well in. These items are very similar to the type of items Singapore students by Grade 8 are proficient in. The item in Figure 12 is on the division of a whole number by a fraction and by Grade 8 Singapore students are expected to be proficient with the algorithm to compute such items. Items similar to items in Figures 12 and 13 are part and parcel of on-going in and out of class practice, in Singapore, for both the topics equivalent fractions and solution of algebraic equations. Compared to items in Figures 12 and 13, the item in Figure 14 could not be done by a mere routine algorithm. This item requires the students to decide which number is closest to 10 by logically finding the difference between 10 and each of the given answers. Though the performance of Singapore students on this item was very high, it is somewhat lacking when compared to their peers in the Netherlands, Sweden, and Estonia.

Figures 15 and 16 each shows an example of the Advanced international benchmark items. From the figures, it is apparent that though the performance of Singapore students was relatively high compared to the international average, it is below that of their peers in Japan, Australia, and Estonia for the item in Figure 15 and below Chinese Taipei, Korea, and Hong Kong for the item in Figure 16. The item in Figure 15 requires students to make sense of the data presented and decide on the best option, while the item in Figure 16 requires students to work through it from part A to B, thereafter making a generalisation of a pattern they might observe. Compared to computational and manipulation type of items, these two items are not commonly found in both Singapore textbooks and assessment books. Although the Thinking Programme (Goh, 1997) was launched for all Singapore schools in 1997 placing much emphasis on engaging students in higher order thinking type of tasks, the results of some TIMSS 2003 items suggest that there is certainly room for improvement in this area. Perhaps, the outcome of this is manifested in the heightened emphasis on reasoning and communication in all mathematics classrooms from 2007 onwards (Ministry of Education, 2006a; 2006b).

Content Area: Data
Cognitive Domain: Reasoning
Description: Interpret data from a table, draws and justifies conclusions.

Betty, Frank and Darlene have just moved to Zedland. They each need to get phone service. They received the following information from the telephone company about the two different phone plans it offers.

They must pay a set fee each month and there are different rates for each minute they talk. These rates depend on the time of the day or night they use the phone, and on which payment plan they choose. Both plans include time for which phone calls are free. Details of the two plans are shown in the table below.

Plan	Monthly Fee	Rate per minute		Free Minutes per Month
		Day (8 am – 6 am)	Night (6 pm – 8 am)	
Plan A	20 zeds	3 zeds	1 zed	180
Plan B	15 zeds	2 zeds	2 zeds	120

Betty talks for less than 2 hours per month. Which plan would be less expensive for her?
Less expensive plan________________

Explain your answer in terms of both the monthly fee and free minutes.

Country	Percent Full Credit
Japan	49 (2.2)
Australia	44 (2.2)
Estonia	44 (2.1)
Korea	40 (1.7)
Singapore	40 (1.7)
International Avg	21 (0.3)

Figure 15. Example of Advanced International Benchmark (625) Grade 8 Item.

2.4 *Possible Contributing Factors for Singapore's High Performance*

From the national reports for Singapore on TIMSS 1995 (Research & Testing Division, MOE, 1996; 1997), TIMSS 1999 (Research & Evaluation Branch, MOE, 2000) and TIMSS 2003 (Research & Evaluation Section, MOE, 2004), the following are the possible factors that may have contributed towards the high performance of Singapore students in mathematics.

Content Area: Algebra
Cognitive Domain: Reasoning
Description: Part C – Generalising from the first several terms of a sequence growing in two dimensions, explains a way to find a specified term, e.g., the 50th.

The three figures below are divided into small congruent triangles.

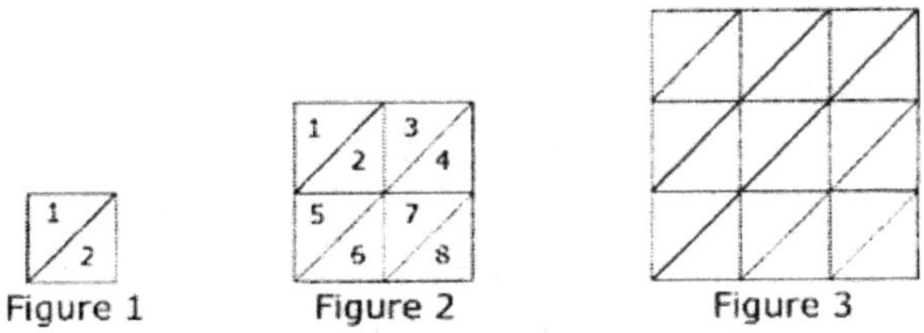

A. Complete the table below. First, fill in how many small triangles make up Figure 3. Then, find the number of small triangles that would be needed for the 4th figure if the sequence is extended.

Figure	Number of Small Triangles
1	2
2	8
3	
4	

B. The sequence of figures is extended to the 7th figure. How many small triangles would be needed for Figure 7?

C. The sequence of figures is extended to the 50th figure. Explain a way to find the number of small triangles in the 50th figure that does not involve drawing it and counting the number of triangles.

Country	Percent Full Credit
Chinese Taipei	49 (2.0)
Korea	48 (1.8)
Hong Kong SAR	45 (2.0)
Singapore	44 (2.0)
Japan	44 (2.1)
International Avg	14 (2.0)

Figure 16. Example of Advanced International Benchmark (625) Grade 8 Item.

2.4.1 The Education System

The education system in Singapore is a centralised one with a national authority providing guidelines in the areas of curriculum, textbook, and assessment. This allows for a high degree of homogeneity and coherence in curriculum coverage. As a result, teachers work in an environment which enables them to be more focused in their teaching, including the provision of suitable remedial, enrichment, and supplementary programmes to cater to the individual needs of the students. From Table 6, it is apparent that there is a substantial increase in the scores of the students between Grades 4 and 8. At the end of Grade 4, the system places students in different streams according to their academic ability to allow them to learn at their own pace. It appears that the better scores at Grade 8 bear testimony to the improved learning through differentiated instruction, catering to the individual needs of the students, provided in Singapore schools from Grade 5 onwards.

2.4.2 School Organisation and Environment

The findings on school organisation and environment indicate that Singapore schools are well organised in terms of high availability of resources (e.g., computers, space, and materials) for instruction, despite having very large school enrolments compared to other countries. Singapore was also among countries that had a relatively low proportion of students in schools reporting shortage of qualified mathematics teachers. The school atmosphere in Singapore appeared conducive to learning and teaching, as relatively fewer principals in Singapore compared to principals in other countries, reported that absenteeism, late coming and discipline were serious problems in their schools. Singapore principals also reported working long hours performing various school activities thereby indicating their high level of commitment to their role.

2.4.3 Mathematics Curriculum and Curricular Implementation

The mathematics syllabuses developed by the Ministry of Education in Singapore, which were revised in 1990, 2000, and 2006, place emphasis on the development of mathematical concepts and skills, and the ability to apply them to solve problems. The syllabuses are embedded in a framework that enables students to develop thinking processes as well as attitudes and metacognition necessary for problem solving. Heads of mathematics departments and mathematics teachers are familiarised with the framework of the mathematics syllabuses periodically through school-based, in-service and national level briefings and workshops by officials of the Curriculum Planning and Development Division of the Ministry of Education and mathematics faculty of the National Institute of Education in Singapore.

In Singapore, the gap between the intended and implemented curriculum is narrow. This is so, as the Ministry of Education closely regulates and monitors the implementation of the mathematics curriculum in all schools. It provides schools with recommended student texts, pedagogical/instructional guides for teachers, assessment guides and sets achievement standards. The Ministry also supports schools with a system of directives and notes. Teachers place major emphasis on mastering basic skills, understanding mathematics concepts, real-life applications of mathematics, communicating mathematically, solving non routine problems, and assessing student learning during mathematics lessons.

2.4.4 Qualification and Working Ethos of Teachers

From the TIMSS 1995 and TIMSS 1999 data, it appears that the qualification of teachers is related to the good performance among top performing countries. The majority (more than 80%) of the students in nearly all high performing countries, including Singapore, were taught by certified teachers with the relevant degree (i.e., BA/BSc with Mathematics at the university level). Compared to their international counterparts, mathematics teachers in Singapore put in the highest number of hours (10 hours a week) in marking and grading students' work and in lesson planning. In addition to their scheduled teaching and extra-curricular duties, they also spent another three hours per week on keeping student records and other administrative tasks. The hard work and effort on the part of Singapore teachers may have contributed in some way to the outstanding mathematics achievement of all Singapore students.

In 1997, the critical and creative thinking and use of information technology initiatives to enhance the teaching and learning in Singapore schools were launched. The TIMSS 1999 data show an increase of 15% more students claiming that their teachers felt creativity was very important to succeed in mathematics compared to TIMSS 1995 data. Similarly, in TIMSS 1999 nearly half the students reported that computers were used sometimes in their lessons compared to nearly all the students in TIMSS 1995 reporting that computers were rarely used in mathematics lessons. These suggest that Singapore teachers are responsive to new initiatives. However, as it may be noted in the discussion of items in Figures 15 and 16, the performance of Singapore students on items that demand higher order thinking is still lacking behind some of their international peers.

2.4.5 Training and Professional Development of Teachers

Professional upgrading of mathematics teachers is an on-going process in Singapore. Teachers are entitled to 100 hours of in-service training each year. The Ministry of Education continually organises workshops, in-service courses and seminars to upgrade teachers' knowledge and skills, to equip teachers with effective teaching strategies, and to keep teachers abreast of recent developments in mathematics education. Teachers are encouraged to engage in lifelong learning, and at the school and national level, many initiatives are in place to encourage the sharing of teaching ideas and good practices. The National Institute of Education is actively engaged in the constant upgrading of mathematics teachers via Diploma level, Degree level, Master Degree level, and Doctorate level courses. The Association of Mathematics Educators and the Singapore Mathematical Society also play an active role in the professional development of mathematics teachers in Singapore.

2.4.6 Home Environment and Support

In Singapore, society as a whole places a high premium on success at school. The virtue of hardwork and the need to strive for excellence are ingrained in students from an early age. Parents have high expectations of their children and are willing to invest in their children's education in terms of resources and "out of school help." The TIMSS findings showed that, compared to their counterparts in other countries, Singapore students were among the most hardworking in the world in terms of the amount of time spent studying or doing homework in mathematics. They also had basic educational resources (e.g., study table, dictionary, relevant books) for study at home. TIMSS 1999 data show that 80% of the students reported that they had a computer at home and this was 30% more than that reported in TIMSS 1995. It appears that society as a whole is well informed of educational initiatives and is prepared to support them.

2.4.7 Students' Attitude and Expectations

TIMSS 2003 data show that at least three quarters of Singapore students from both grades agreed (at least a little) that they enjoyed learning mathematics. For Grade 8 students, there was a significant upward trend from 1995 and 1999 in the percentages of students agreeing "a lot" that they enjoyed learning mathematics. For Grade 4 students, there was also a significant increase from 1995 to 2003 in the proportion of students agreeing "a lot" about enjoying mathematics (from 48% to 57%). Almost 90% of the students in TIMSS 1999 indicated that they had reasonable (i.e., medium or high level) self-concept in mathematics. This index was based on student's responses to five statements about their mathematics ability:

- I would like mathematics much more if it were not so difficult.
- Although I do my best, mathematics is more difficult for me than for many of my classmates.
- Nobody can be good in every subject, and I am just not talented in mathematics.
- Sometimes when I do not understand a new topic in mathematics initially, I know that I will never really understand it.
- Mathematics is not one of my strengths.

Students who disagreed or strongly disagreed with all five statements were assigned to the high level of the index, while students who agreed or strongly agreed with all five statements were assigned to the low level. The medium level includes all other combinations of responses.

TIMSS 2003 data also showed that Singapore students regarded doing well in their studies as important. "To finish university" was the most popular educational aspiration (56%). In addition, the TIMSS 1995 data showed that while students in Singapore felt that doing well in school is important, they also perceived their friends to place similar emphasis on academic achievement.

2.4.8 Fostering of Interest in Mathematics

Many activities are carried out by schools and other organisations to foster students' interest in mathematics. Besides the enrichment activities organised by the schools, the Singapore Mathematical Society and Association of Mathematics Educators also play an instrumental role in fostering interest in mathematics through their annual mathematics competitions and enrichment activities. Educational programmes developed by the Science Centre also complement the school's mathematics syllabuses by offering first hand observations and hands-on activities outside the classroom. The Jurong BirdPark and the Singapore Zoological Gardens also offer students an opportunity to do mathematics trails at their respective sites and engage in out-of-class mathematics learning.

3 CONCLUSION

This chapter has provided the readers with an insight into the performance of Singapore students in TIMSS over three studies spanning an eight-year period from 1995 to 2003. Though it may be claimed that the overall performance of the students was outstanding, the many secondary studies carried out specifically on the achievement data of Singapore students and also the TIMSS 2003 benchmark items and data discussed in this chapter surface three underlying causes of difficulty:

- Students do not perform well on content that is not an integral part of the local school curriculum. The school curriculum operates as an upper limit on the mathematics that students learn.
- Students have difficulty with items that require them to comprehend concepts or apply knowledge in non-familiar contexts. It must be noted that many of the TIMSS items are of a routine nature for Singapore students.
- Students have difficulty with items that contain unfamiliar language, e.g., "pace" and "fair coin" and unfamiliar formats, for example, item in Figure 10.

Soh (1999) claimed that hardworking teachers and students and a sound mathematics curriculum may be the reasons why Singapore students outshine their international peers in TIMSS. The McKinsey study (Barber & Mourshed, 2007) stated that Singapore's sound education system, which detects failing schools and students almost immediately and intervenes with the right measures, together with good teachers and right policies has resulted in high achievement among Singapore students. The author notes Soh's claim and the findings of the McKinsey study but strongly feels that the challenge for mathematics teachers in Singapore is one of striking a balance between "over practice of routine mathematical tasks to hone algorithms" and "creation and application of knowledge in novel situations."

References

Barber, M., & Mourshed, M. (2007). *How the world's best-performing school systems come out on top*. London: McKinsey & Co.

Beaton, A.E., Mullis, I.V.S., Martin, M.O., Gonzalez, E.J., Kelly, D.L., & Smith, T.A. (1996). *Mathematics achievement in the middle school years: IEA's Third International Mathematics and Science Study (TIMSS).* Boston, MA: TIMSS International Study Center, Boston College.

Ferrucci, B. J., Kaur, B., & Carter, J. (2002). An exploration of differences in mathematics achievement in Singapore and the United States. *Journal of Science and Mathematics in Southeast Asia, 25*(1), 155 – 171.

Goh, C. T. (2007). Shaping our future: "Thinking Schools" and a "Learning Nation". *Speeches, 21*(3), 12-20. Singapore: Ministry of Information and the Arts.

Kaur, B. (2002a). TIMSS-R: Mathematics achievement of eighth graders from Southeast Asian countries. *Journal of Science and Mathematics in Southeast Asia, 25*(2), 66 – 92.

Kaur, B. (2002b). Reflections on the Curriculum – TIMSS & TIMSS-R: Performance of eighth grade Singaporean students. Reflections. *Journal of the Mathematical Association of New South Wales, 27*(2), 7–12.

Kaur, B. (2003a). TIMSS-R: Performance of eighth graders from Singapore. *The Mathematics Educator, 7*(1), 62 – 79.

Kaur, B. (2003b). TIMSS–Students' and teachers' perspectives on mathematics instruction in Singapore schools. In B. Kaur, D. Edge & B.H. Yeap (Eds.), *TIMSS and comparative studies in mathematics education – An international perspective* (pp. 85 – 96). Singapore: Association of Mathematics Educators.

Kaur, B. (2005). Performance in mathematics of eighth graders from 5 Asian countries. *Hiroshima Journal of Mathematics Education, 11*, 69-92.

Kaur, B., Ferrucci, B., & Carter, J. (2003). TIMSS & TIMSS-R: Performance of eighth graders from US and Singapore in mathematics. *New England Mathematics Journal, 36*(1), 5-25.

Kaur, B., & Pereira-Mendoza, L. (1999a). Singapore primary school TIMSS data: Whole numbers, fractions and proportionality. *The Mathematics Educator, 4*(1), 52 – 69.

Kaur, B., & Pereira-Mendoza, L. (1999b). Mathematics learning: Singapore students' perspectives. In S.P. Loo (Ed.), *Proceedings of the MERA-ERA joint conference, Educational challenges in the new millennium* (pp. 1265–1269). Singapore: Educational Research Association.

Kaur, B., & Pereira-Mendoza, L. (2000a). Singapore primary school TIMSS data: Data representation, analysis and probability, and patterns, relations and functions. *The Mathematics Educator, 5*(1/2), 180–193.

Kaur, B., & Pereira-Mendoza, L. (2000b). TIMSS – Performance of Singapore secondary students part b: Proportionality, measurement, fractions and number sense. *Journal of Science and Mathematics Education in Southeast Asia, 23*(1), 54–70.

Kaur, B., & Yap, S.F. (2002). TIMSS – The strengths and weaknesses of Singapore's lower secondary pupils' performance in mathematics. In K.A. Toh & L. Pereira-Mendoza (Eds.), *The Third International Mathematics and Science Study (TIMSS) – A look at Singapore students' performance and classroom practice* (pp. 65–75). Singapore: National Institute of Education, Nanyang Technological University.

Kelly, D. L., Mullis, I. V. S., & Martin, M. O. (2000). *Profiles of student achievement in mathematics at the TIMSS international benchmarks: U.S. performance and standards in an international context.* Boston, MA: TIMSS International Study Center, Boston College.

Ministry of Education (2006a). *A guide to teaching and learning of Ordinary level mathematics*. Singapore: Author.

Ministry of Education (2006b). *A guide to teaching and learning of primary level mathematics*. Singapore: Author.

Mullis, I.V.S., Martin, M.O., Beaton, A.E., Gonzalez, E.J., Kelly, D.L., & Smith, T.A. (1997). *Mathematics achievement in the primary years: IEA's Third International Mathematics and Science Study (TIMSS).* Boston, MA: TIMSS International Study Center, Boston College.

Mullis, I.V.S., Martin, M.O., Gonzalez, E.J., Gregory, K.D., Garden, R.A., O'Connor, K.M., Chrostowski, S.J., & Smith, T.A. (2000). *TIMSS 1999 international mathematics report.* Boston, MA: International Study Center, Lynch School of Education, Boston College.

Mullis, I.V.S., Martin, M.O., Gonzalez, E.J., & Chrostowski, S.J. (2004). *TIMSS 2003 international mathematics report.* Boston, MA: International Study Center, Lynch School of Education, Boston College.

Pereira-Mendoza, L., & Kaur, B. (1999a). TIMSS – Performance of Singapore secondary students part a: Algebra, geometry and data representation, analysis and probability. *Journal of Science and Mathematics Education in Southeast Asia, 22*(2), 38 – 55.

Pereira-Mendoza, L., & Kaur, B. (1999b). Singapore primary school TIMSS data: Geometry and measurement, estimation and number sense. *The Mathematics Educator, 4*(2), 108 – 125.

Pereira-Mendoza, L., Kaur, B., & Yap, S.F. (1999). Some implications of the TIMSS data for primary mathematics teachers. In M. Waas (Ed.), *Enhancing learning: Challenge of integrating thinking and information technology into the curriculum* (pp. 452 – 457). Singapore: Educational Research of Singapore.

Research & Evaluation Branch (2000). *Third International Mathematics and Science Study (TIMSS 1999): National report for Singapore.* Singapore: Ministry of Singapore.

Research & Evaluation Section (2004). *Trends in International Mathematics and Science Study (TIMSS) 2003: National report for Singapore.* Singapore: Ministry of Singapore.

Research & Testing Division (1996). *Third International Mathematics and Science Study (TIMSS): National report for Singapore (Population 2).* Singapore: Ministry of Education.

Research & Testing Division (1997). *Third International Mathematics and Science Study (TIMSS): National report for Singapore (Population 1).* Singapore: Ministry of Education.

Soh, K.C. (1999). Three G7 and three little Asian dragons in TIMSS mathematics at the fourth grade. In M. Waas (Ed.), *Enhancing learning: Challenge of integrating thinking and information technology into the curriculum* (pp. 431 – 435). Singapore: Educational Research Association.

Chapter 20

Findings from the Background Questionnaires in TIMSS 2003

BOEY Kok Leong

In recent years, mathematics and science education in Singapore has become a focus of considerable interest among many countries. Such interest has been generated by the results of TIMSS, which assessed the performance of students in different countries at levels corresponding to Grades 4 and 8. Singapore students performed well, especially in Mathematics. Besides the assessment items, questionnaires were administered to school principals, as well as the students and their mathematics teachers. The comprehensive data provided a useful source to interpret student achievement in the participating countries. This chapter presents some information about Singapore students' learning of mathematics in TIMSS 2003 through these questionnaires.

Key words: TIMSS, home, attitude, Grade 4, Grade 8, background questionnaire, mathematics performance, school climate

1 BACKGROUND

The International Association for the Evaluation of Educational Achievement (IEA) is an independent and international cooperative of national research institutions and government research agencies. The IEA has conducted more than 20 research studies of cross-national achievement, including the learning of subjects like Mathematics and Science, as well as reading literacy, Civic Education and Information Technology in Education.

Under the auspices of the IEA, a large scale international study to measure the trends in student performance in mathematics and science was carried out in 1994-1995 with the original TIMSS (Third International Mathematics and Science Study). This study compared the performance of 41 countries at Grades 3, 4, 7, 8 and 12 (Beaton, Mullis, Martin, Gonzalez, Kelly & Smith, 1996; Mullis, Martin, Beaton, Gonzalez, Kelly & Smith, 1997; Mullis, Martin, Beaton, Gonzalez, Kelly & Smith, 1998). TIMSS was conducted again at only Grade 8 in 1998-1999 with 38 countries that participated in the first TIMSS participating again. Also known as TIMSS-R, it was designed to provide trends in the eighth-grade Mathematics and Science achievement in an international context (Mullis, Martin, Gonzalez, Gregory, Garden, O'Connor, Chrostowski & Smith, 2000). Renamed as the Trends in International Mathematics and Science Study, TIMSS was conducted for a third time at Grades 4 and 8 in 2002-2003. Students from 49 countries participated in this study (Mullis, Martin, Gonzalez & Chrostowski, 2004). Now, TIMSS 2007 which assesses Grades 4 and 8 students is underway. It extends a sequence of very successful measure of trends in student performance, providing achievement data at four time points over a 12-year-period. TIMSS provides an opportunity to examine the impact on the achievement in Mathematics and Science of each participating country's educational system. There is also awareness of practices or approaches used by others.

This chapter focuses on the background questionnaires used to gather information about students' learning of Mathematics in TIMSS 2003. While some studies on the questionnaire data have been done recently (Kaur, 2003, 2005; Kaur & Pereira-Mendoza, 2002), the data and findings reported in this chapter are drawn from the respective international mathematics reports of the three studies (TIMSS, TIMSS-R and TIMSS 2003) published by the TIMSS International Study Centre and the national report (Boey, 2004) published by the Ministry of Education (MOE) in Singapore.

2 QUESTIONNAIRES

A series of questionnaires were used to collect information about the context within which students learn Mathematics and Science. The questionnaires were administered to school principals, the students and their Mathematics and Science teachers, and curriculum specialists. The comprehensive collection of information provided a useful source to interpret student achievement in the participating countries. The data could be used to inform educational policy makers about the contexts for learning such as what school and home supports were available, as well as how the contents of the curriculum were taught. Generally, the background data could provide a good idea of what has been done in each country and how it concurred with what was thought to work the best. Strict procedures were in place to ensure the confidentiality and security of the data collected. Students, teachers, and schools were not identified in any of the final reports.

In TIMSS 2003, four types of background questionnaires were used to gather information. These questionnaires sought to identify and hence to study some of the major qualities of the educational and social context with a view to improving student learning.

The Curriculum Questionnaires. These Mathematics and Science curriculum questionnaires collected information about the intended curriculum and were completed by each country's National Research Coordinator, usually in conjunction with curriculum specialists and educators in these subjects. The questionnaires collected information about the organisation of the curriculum in each country, as well as the content intended to be covered up to the fourth and eighth grades.

The School Questionnaires. The school principals were asked about the school organisation, the instructional time, school climate for learning, and the availability of school resources.

The Teacher Questionnaires. The Mathematics and Science teachers of the students tested were asked about their preparation to teach, the classroom setting, and the school climate. Besides these, through the teacher questionnaires, large amounts of data on the coverage of the curriculum and the instructional practices in Mathematics and Science were also collected.

The Student Questionnaires. The students supplied information about their home backgrounds. They were also asked for their attitudes toward the learning of Mathematics and Science and their experiences in learning these subjects.

Table 1
Brief Description the Background Questionnaires

1.	**Curriculum Questionnaires**		
a.	Formulating the curriculum		
b.	Scope and content of the curriculum		
c.	Organisation of the curriculum		
d.	Monitoring and evaluating the implemented curriculum		
e.	Curricular materials and support		
2.	**School Questionnaires**		
a.	School organisation		
b.	School goals		
c.	Roles of the school principal		
d.	Resources to support Mathematics and Science learning		
e.	Parental involvement		
f.	Disciplined school environment		
3.	**Teacher Questionnaires**		
a.	Academic preparation and certification	i.	Time
b.	Teacher recruitment	j.	Homework
c.	Teacher assignment	k.	Assessment
d.	Teacher induction	l.	Classroom climate
e.	Teacher experience	m.	Information Technology
f.	Teaching styles	n.	Calculator use
g.	Professional development	o.	Emphasis on investigations
h.	Curriculum topics taught	p.	Class size
4.	**Student Questionnaires**		
a.	Home background		
b.	Prior experiences		
c.	Attitudes		

Source: Mullis, Martin, Smith, Garden, Gregory, Gonzalez, Chrostowski & Connor, 2003.

3 THE TIMSS BACKGROUND DATA

In view of the large volume of data collected from the background questionnaires, some information on the educational support and practices was combined to form indices. For example, indices were formed for students' attitudes towards Mathematics, and the resources available at home, teachers' emphasis on problem solving, teachers' confidence in their preparation to teach the subjects, and the availability of school resources for Mathematics instruction.

3.1 *Home Speaking Language*

The TIMSS 2003 data showed that, with some exceptions, those countries with large proportions of students from homes where the language of the test was not often spoken had lower average Mathematics achievements at Grade 8 than those who spoke it more often.

Table 2
Students Speak Language of the Test at Home

Grade	Country	Always		Almost Always		Sometimes		Never	
		%	AA	%	AA	%	AA	%	AA
8	Singapore	23	625	19	620	49	595	8	581
		(0.6)	(3.8)	(0.6)	(3.5)	(0.8)	(4.3)	(0.4)	(5.7)
	International	68	472	11	477	17	441	4	396
	Average	(0.2)	(0.7)	(0.1)	(1.0)	(0.1)	(1.4)	(0.1)	(2.0)
4	Singapore	24	610	22	625	47	580	7	551
		(1.2)	(6.2)	(1.0)	(4.9)	(1.5)	(5.7)	(0.6)	(8.6)
	International	67	499	14	501	15	471	5	435
	Average	(0.3)	(0.9)	(0.2)	(1.4)	(0.2)	(1.6)	(0.2)	(2.9)

( ): Standard errors appear in parentheses; %: % of students; AA: Average Achievement
Source: Exhibits 4.3 in Mullis, Martin, Gonzalez & Chrostowski, 2004.

About 80% of students internationally at each grade responded that they always or almost always spoke the language of the test. Contrary to the general picture, Singapore had among the lowest proportions of students (42% for Grade 8 and 46% for Grade 4) across countries, who always or almost always spoke the language of the test at home (i.e., English). This was expected as English was not the mother tongue of our students although it was the language of instruction. At both grades, students from homes where English was always or almost always spoken had higher average achievement than those who spoke it less frequently. For instance, the mean score of Singapore Grade 8 students who always spoke English at home was 44 points higher than the score of those who never spoke it (see Table 2). However, there was an increase in the percentages compared to the previous studies (by 20% from 1995 and 27% from 1999 for Grade 8, and 20% from 1995 for Grade 4) (see Figures 1 and 2).

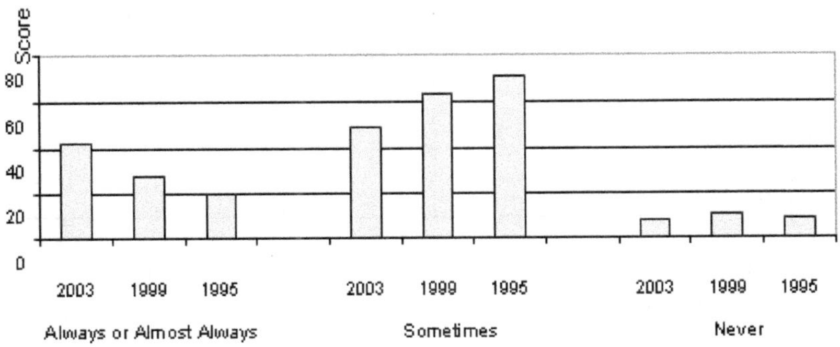

Figure 1. Trends in "Students Speak Language of Test at Home" at Grade 8.

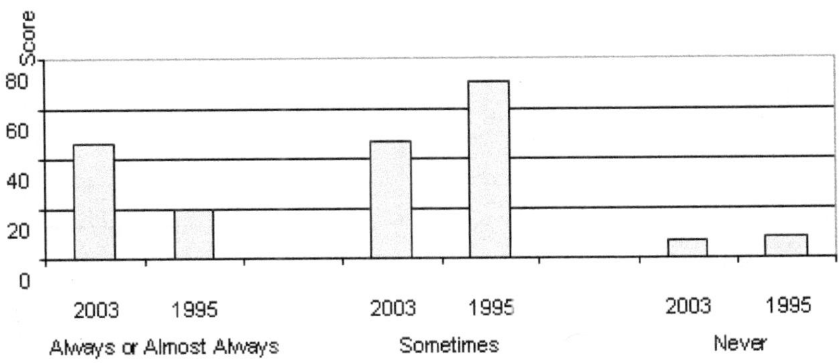

Figure 2. Trends in "Students Speak Language of Test at Home" at Grade 4.

3.2 *Attitudes toward Learning Mathematics*

At least three-quarters of the students in Singapore from both grades agreed (at least a little) that they enjoyed learning Mathematics. For the Grade 8 students, there was a significant upward trend from 1995 and 1999 in the percentages of students agreeing *a lot* that they enjoyed learning the subject (see Figure 3).

Similarly, the Singapore Grade 4 students recorded significant increases from 1995 to 2003 in the percentages of students agreeing *a lot* that they enjoyed learning the subject (from 48% to 57% for Mathematics) (see Figure 4).

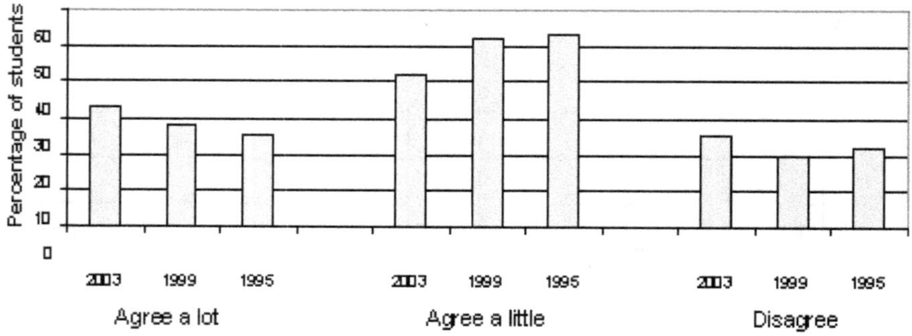

Figure 3. Trends in "I enjoy learning Mathematics" at Grade 8.

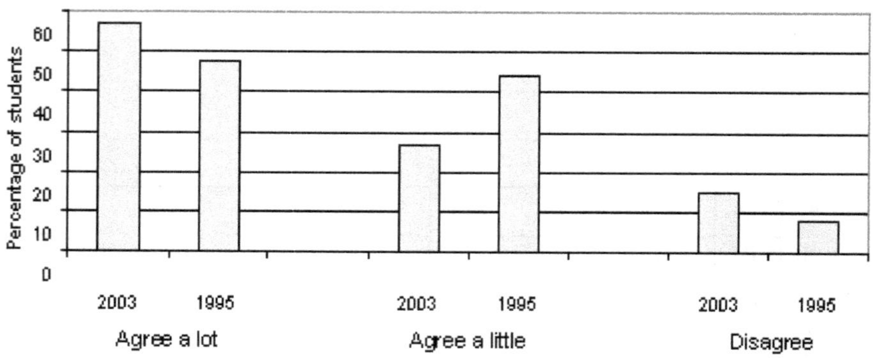

Figure 4. Trends in "I enjoy learning Mathematics" at Grade 4.
Source: Exhibits 4.11 in Mullis, Martin, Gonzalez & Chrostowski, 2004.

3.3 *Students' Educational Aspirations*

Similar to Grade 8 students internationally, students from Singapore regarded doing well in their studies as important. "To finish university" was the most popular educational aspiration for these Grade 8 students (56%) (see Figure 5).

3.4 *Home Resources*

Most of the students at both grades had access to educational resources including books, computers, and study desks for study at home. In Singapore, Grade 8 students reporting more than 200 books in their homes had an average score of 642. This score was significantly higher than the average scores of those students with 200 or fewer books at home. The difference at Grade 4 was large, ranging from 628 for students reporting more than 200 books to 528 for students reporting 10 books or less (see Table 3).

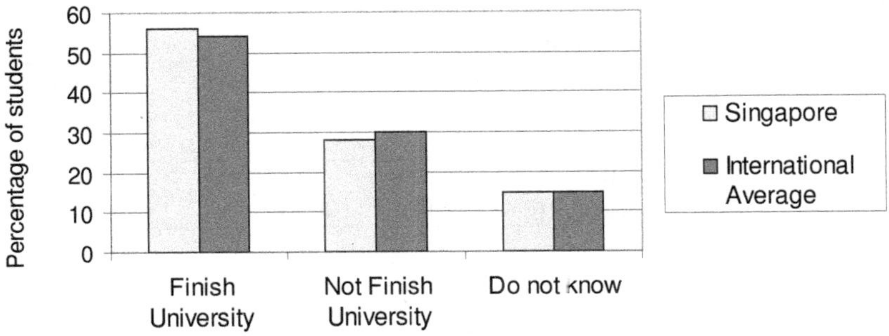

Figure 5. Grade 8 Students' Educational Aspirations.
Source: Exhibits 4.2 in Mullis, Martin, Gonzalez & Chrostowski, 2004.

Having study aids such as a computer or a study desk or table at home was associated with higher student achievement. Table 4 shows the percentage of Grades 4 and 8 students in Singapore that had a computer or study desk or table, together with their average mathematics achievement.

Table 3
Number of Books in the Home

Grade	More than 200 books		101-200 books		26-100 books		11-25 books		0-10 books	
	%	AA	%	AA	%	AA	%	AA	%	AA
8	14 (0.5)	642 (3.5)	16 (0.5)	627 (3.7)	33 (0.7)	617 (3.2)	24 (0.7)	581 (5.1)	12 (0.7)	554 (5.2)
4	10 (0.6)	628 (6.4)	17 (0.9)	622 (5.6)	40 (0.9)	609 (5.0)	22 (0.9)	569 (4.9)	11 (0.8)	528 (8.7)

( ): Standard errors appear in parentheses; %: % of students; AA: Average Achievement.
Source: Exhibits 4.4 in Mullis, Martin, Gonzalez & Chrostowski, 2004.

About 94% percent of the Grade 8 students reported having a computer at home, and a slightly lower percentage of Grade 4 students at 89%. The mathematics achievement difference between students with a computer at home and those without was substantial: 72 scale score points at Grade 8 and 55 points at Grade 4.

Similar percentages of students reported having a study desk or table at home: 91% and 90% at Grades 8 and 4 respectively. Having such a study facility was also associated with higher average mathematics achievement at each grade: a 39 point difference at Grade 8 and a 53 point difference at Grade 4.

Table 4
Computer and Study Desk/Table in the Home

Grade	Have Computer		Do Not Have Computer		Have Study Desk/Table		Do Not Have Study Desk/Table	
	%	AA	%	AA	%	AA	%	AA
8	94 (0.4)	610 (3.4)	6 (0.4)	538 (7.4)	91 (0.5)	609 (3.4)	9 (0.5)	570 (6.2)
4	89 (0.8)	601 (5.4)	11 (0.8)	546 (6.5)	90 (0.7)	600 (5.4)	10 (0.7)	547 (8.5)

( ): Standard errors appear in parentheses; %: % of students; AA: Average Achievement.
Source: Exhibits 4.5 in Mullis, Martin, Gonzalez & Chrostowski, 2004.

3.5 *Perceptions of School Climate*

To measure the extent to which schools had a positive environment, TIMSS 2003 created an *Index of Principals' Perception of School Climate (PPSC)*. This index was based on principals' responses to the following eight items:

- teachers' job satisfaction;
- teachers' understanding of the school's curricular goals;
- teachers' degree of success in implementing the school's curriculum;
- teachers' expectations for students' achievement;
- parental support for students' achievement;
- parental involvement in schools' activities;
- students' regard for school property; and
- students' desire to do well in school.

In the PPSC Index, students were assigned to the high level if their principals had high or very high reports for each aspect of school climate. On the other hand, they were assigned to the low level if their principals had low or very low reports for each aspect. See note in Table 5 for a detailed description.

Internationally, at both grades, about two thirds of the students attended schools with a medium PPSC. At Grade 8, 15% of the students were in schools with a high PPSC while at Grade 4, the figure was 23%. Compared to the international averages, more students in Singapore attended schools with a high PPSC (30% at Grade 8, 32% at Grade 4). There was a strong positive relationship between the principals' perception of school climate and the average achievement in mathematics.

Table 5
Index of Principals' Perception of School Climate (PPSC)

Grade	Country	High PPSC		Medium PPSC		Low PPSC	
		%	AA	%	AA	%	AA
8	Singapore	30 (0.0)	649 (5.1)	65 (0.0)	589 (5.5)	5 (0.0)	556 (17.7)
	International Average	15 (0.4)	495 (2.3)	67 (0.6)	466 (0.8)	18 (0.4)	446 (2.0)
4	Singapore	32 (4.1)	611 (7.5)	63 (4.1)	587 (7.3)	5 (1.6)	557 (17.3)
	International Average	23 (0.7)	515 (1.9)	66 (0.8)	492 (0.9)	11 (0.5)	468 (3.0)

( ): Standard errors appear in parentheses; %: % of students; AA: Average Achievement.
Source: Exhibits 8.4 in Mullis, Martin, Gonzalez & Chrostowski, 2004.

Note:
Based on principals' responses to 8 questions about their schools, the average was computed based on a 5-point scale: 1 = very high; 2 = high; 3 = medium; 4 = low; 5 = very low.
High level indicates average is less than or equal to 2. Medium level indicates average is greater than 2 and less than or equal to 3. Low level indicates average is greater than 3.

3.6 *Availability of School Resources*

Background data about the availability of school resources were collected during every TIMSS cycle. An *Index of Availability of School Resources for Mathematics Instruction (ASRMI)* was created for each TIMSS. The index was based on schools' average response to five questions about shortages that affected general capacity to provide instruction and five questions about shortages that affected Mathematics instruction in particular. See Table 6.

Table 6
Questions about School Resources for Mathematics Instruction

Questions about shortages that affect general capacity to provide instruction	Questions about shortages that affect mathematics instruction
• instructional materials (e.g., textbook)	• computers
• budget for supplies (e.g., paper, pencils)	• computer software
• school buildings and grounds	• calculators
• heating/cooling and lighting systems	• audio-visual resources
• instructional space (e.g., classrooms)	• library materials relevant to Mathematics instruction

Students attending schools where principals reported that both shortages (general and subject-specific) had no or little effect on instructional capacity were in the high category of the index. Students at the medium level were in schools where one type of shortage affected instruction "some" or "a lot." Students in the low category were in schools where both types of shortages affected instruction "some" or "a lot." See Table 6.

Table 7 shows changes in the percentages of Grade 8 students in the high, medium, and low categories for 1995, 1999, and 2003, and for the Grade 4 students for 1995 to 2003. At Grade 8 internationally, the trend suggested a dip in available resources in 1999. At Grade 4, there was a

Table 7
Trends in Index of Availability of School Resources for Mathematics Instruction (ASRMI)

Grade	Country	High ASRMI % of students in			Medium ASRMI % of students in			Low ASRMI % of students in		
		2003	1999	1995	2003	1999	1995	2003	1999	1995
8	Singapore	88	50Δ	55Δ	10	46▼	43▼	1	4	2
	International Average	26	19Δ	23Δ	64	64	67▼	11	19▼	10
4	Singapore	86	--	47Δ	14	--	53▼	1	--	0
	International Average	33	--	26Δ	58	--	68▼	10	--	6 Δ

Δ 2003 significantly higher; ▼ 2003 significantly lower

Source: Exhibits 8.3 in Mullis, Martin, Gonzalez & Chrostowski, 2004.

Note:

Based on principals' responses to 10 questions about shortages that affected general capacity to provide instruction as well as shortages that affected mathematics instruction, the average was computed based on a 4-point scale: 1 = none; 2 = a little; 3 = some; 4 = a lot.

High level indicates that both shortages are on average lower than 2. Low level indicates that both shortages are on average greater than or equal to 3. Medium level includes all other possible combinations of responses.

significant increase in the high category from 1995 to 2003. Among all participating countries, Singapore had the highest percentage of students attending schools with high availability of school resources for instructional use. There were significant increases between 1995 and 2003 in available resources for mathematics instruction for each grade.

3.7 *Safety in Schools*

Since school safety was central for providing an environment conducive to learning, TIMSS asked teachers to characterize their perceptions of safety in their schools. TIMSS 2003 produced an *Index of Teachers' Perception of Safety in the Schools (TPSS).* This index was based on teachers' responses to how much they agreed with the following three statements:

- This school is located in a safe neighborhood.
- I feel safe in this school.
- This school's security policies and practices are sufficient.

At both grades, a positive relationship between teachers' perception of safety in schools and achievement in mathematics was found. Internationally, at both grades, about 70% of the students were in schools with a high TPSS (see Table 8). Compared to the international averages,

Table 8
Index of Mathematics Teachers' Perception of Safety in Schools (TPSS)

Grade	Country	High TPSC		Medium TPSC		Low TPSC	
		%	AA	%	AA	%	AA
8	Singapore	91 (1.5)	609 (3.7)	8 (1.5)	582 (16.6)	1 (0.5)	--
	International Average	72 (0.5)	470 (0.8)	22 (0.5)	461 (1.3)	6 (0.3)	440 (3.1)
4	Singapore	87 (2.8)	599 (5.8)	12 (2.8)	563 (14.8)	0 (0.2)	--
	International Average.	75 (0.7)	498 (0.9)	21 (0.7)	486 (1.9)	4 (0.3)	465 (4.)

( ): Standard errors appear in parentheses; %: % of students; AA: Average achievement.
Source: Exhibits 8.7 in Mullis, Martin, Gonzalez & Chrostowski, 2004.

Note:
Index was computed based on teachers' responses to three statements about their schools. High level indicates that the teacher agrees a lot or agrees to all 3 statements. Low level indicates that teacher disagrees or disagrees a lot to all 3 statements. Medium level includes all other combinations of responses.

more Grade 4 students (87%) in Singapore attended schools with a high TPSS. Singapore had the highest percentage of Grade 8 students (91%) attending schools with a high TPSS as compared with their peers from other countries.

4 CONCLUSIONS

Singapore's participation since the first TIMSS has provided an objective measure of its mathematics education against world benchmarks. It has also enabled the measuring of the trends in our students' achievement in mathematics since 1995. The TIMSS 2003 results reaffirmed that Singapore's education system is sound and robust. Through the TIMSS 2003 background data, we saw that Singapore's education system was firmly supported by committed principals, teachers, parents, and students.

This study found that among all the participating countries, Singapore had the highest percentage of students with high availability of school resources for mathematics instructions. Despite a lack of natural resources, through our comprehensive education system and policies, we have created a competitive edge for ourselves. Singapore should build on her strengths in education so that our children will be well prepared for a future of innovation-driven growth and a changing social and economic environment.

This chapter has merely touched on some of the findings that are part of a wealth of descriptive information generated by TIMSS. It is recommended that the findings presented here are not viewed in isolation, and that readers refer to the international publications for more detail. The reports can be viewed on the international website.

References

Beaton, A. E., Mullis, I. V. S., Martin, M. O., Gonzalez, E. J., Kelly, D. L., & Smith, T. A. (1996). *Mathematics achievement in the middle school years: IEA's third International Mathematics and Science Study*. Chestnut Hill, MA: Boston College.

Boey, K. L. (2004). *Trends in International Mathematics and Science Study (TIMSS) 2003: National report for Singapore*. Singapore: Ministry of Education.

Kaur, B. (2003). TIMSS: Students' and teachers' perspectives on mathematics instruction in Singapore schools. In B. Kaur, D. Edge & B.H. Yeap (Eds.), *TIMSS and comparative studies in mathematics education: An international perspective* (pp. 85-96). Singapore: Association of Mathematics Educators.

Kaur, B. (2005). Schools in Singapore with high performance in mathematics at the eighth grade level. *The Mathematics Educator, 9* (1), 29 – 38.

Kaur, B., & Pereira-Mendoza, L. (2002). TIMSS: Singapore students' perspective on mathematics learning. In K.A. Toh & L. Pereira-Mendoza (Eds.), *The Third International Mathematics and Science Study (TIMSS): A look at Singapore students' performance and classroom practice* (pp. 76–83). Singapore: National Institute of Education, Nanyang Technological University.

Mullis, I. V. S., Martin, M. O., Beaton, A. E., Gonzalez, E. J., Kelly, D. L., & Smith, T. A. (1997). *Mathematics achievement in the primary school years: IEA's Third International Mathematics and Science Study*. Chestnut Hill, MA: Boston College.

Mullis, I. V. S., Martin, M. O., Beaton, A. E., Gonzalez, E. J., Kelly, D. L., & Smith, T. A. (1998). *Mathematics and science achievement in the final year of secondary school: IEA's Third International Mathematics and Science Study*. Chestnut Hill, MA: Boston College.

Mullis, I. V. S., Martin, M. O., Gonzalez, E. J., & Chrostowski, S. J. (2004). *TIMSS 2003 international mathematics report: Findings from IEA's trends in International Mathematics and Science Study at the fourth and eighth grades*. Chestnut Hill, MA: Boston College.

Mullis, I. V. S., Martin, M. O., Gonzalez, E. J., Gregory, K. D., Garden, R. A., O'Connor, K. M., Chrostowski, S. J., & Smith, T. A. (2000). *TIMSS 1999 international mathematics report: Findings from IEA's repeat of the Third International Mathematics and Science Study at the eighth grade.* Chestnut Hill, MA: Boston College.

Mullis, I. V. S., Martin, M. O., Smith, T. A., Garden, R. A., Gregory, K. D., Gonzalez, E. J., Chrostowski, S. J., & Connor, K. M. (2003). *TIMSS assessment frameworks and specifications 2003* (2nd ed.). Chestnut Hill, MA: Boston College.

Chapter 21

Kassel Project on the Teaching and Learning of Mathematics: Singapore's Participation

Berinderjeet KAUR **YAP Sook Fwe**

This chapter describes Singapore's participation in the international comparative study on the teaching and learning of mathematics, known as the Kassel Project. The study was a longitudinal one, and entire cohorts of Secondry 2 pupils in 1995 from seven secondary schools in Singapore participated. The study monitored the individual progress of pupils relative to groups of pupils with similar potential and initial attainment in mathematics over a period of two years. The use of the data collected was two fold. At the national level, it led to recommendations for mathematics teaching in Singapore secondary schools, and at the international level it contributed towards the international database that was used to discuss issues in mathematics curriculum, mathematics teaching and learning in the participating countries.

Key words: mathematics attainment, secondary school, Kassel project, longitudinal study, value added score, teaching approach

1 BACKGROUND

The Kassel project was initiated in September 1993 by England, Germany, and Scotland. It was originally designed to compare the mathematical progress made by secondary school pupils in these three countries and consequently to identify good practice and provide empirical evidence for incorporating these practices into mathematics teaching and learning in the United Kingdom.

Over the years, the project expanded to include Australia, Brazil, Czech Republic, Finland, Greece, Holland, Hungary, Japan, Norway, Poland, Russia, Singapore, Thailand, Ukraine, and United States. It differed from other international mathematics studies, such as the Third International Mathematics and Science Study (TIMSS), which were concerned with making inter-country comparisons on standards based on a "snapshot" approach. The project was a longitudinal study of samples of pupils in the participating countries. It was the first extensive comparative study in mathematics based on monitoring the individual progress of pupils relative to groups of pupils with similar potential and initial attainment over a three-year period for England, Germany, and Scotland and a two-year period for countries that joined in the later phase. The basic aim of the project was to carry out research into the teaching and learning of mathematics in different countries so as to identify key factors that give rise to successful progress in mathematics. Partial funding of the project was provided by The Gatsby Charitable Foundation and British Council.

2 METHODOLOGY

This section describes the methodology developed by the Central Team led by David Burghes from the Centre for Innovation in Mathematics Teaching (CIMT) in England for all the participating countries in the international study. More details may be found in Burghes, Kaur, and Thompson (2004).

2.1 *Selection of Schools and Samples*

In each country, the country coordinators were responsible for selecting schools that were broadly representative of their country. The sample size had to be sufficiently large in order to reflect the variation in schools in the country but small enough to be manageable. Although the attainment both within and between countries was compared, it was stressed that the aim of the project was not to produce a league table of countries as this was a longitudinal study seeking to explain differences.

2.2 *Testing*

At the start of the project, all pupils took a mathematics Potential Test. This was a 40-minute test with 26 questions of increasing difficulty. It was taken only once and the aim was to give a measure of the pupil's potential for mathematical thinking and reasoning. Four attainment tests in Number, Algebra, Shape and Space, and Data Handling (optional) were also taken at the start of the project and then repeated each year in the form of parallel tests in order to give a measure of progress. These 40-minute tests, each with a total of 50 marks, were designed based on the following guiding principles:

- the questions were nearly always short, requiring one answer;
- the topics were "core" topics generally found in any country's curriculum (although they were based on the intersection of the English, German, and Scottish syllabuses);
- they included questions on topics that would have been taught at an earlier age as well as questions on topics likely to be met for the first time during the project;
- where contexts were used, they were straightforward contexts with universal meaning.

The questions in each test covered a wide range of difficulty, starting with relatively easy questions and often finishing with content found in Levels 9 and 10 of the English National Curriculum. It was expected that when pupils first took the tests, even for the most able, there would be many questions on topics which they would have not yet covered.

2.3 *Questionnaires*

For each year of the project, relevant information was obtained from the participating schools and teachers through four kinds of questionnaires:

- *School Questionnaire*: to determine school details such as total enrolment, type, environment, department policies;

- *Teacher Questionnaire*: to ascertain the qualifications and experience of participating teachers;
- *Class Questionnaire*: to attempt to find out from the class teachers the style and management of the mathematics taught, including number of hours per year, class size, use of resources such as calculators and computers, and teaching approaches;
- *Pupil Questionnaire*: to seek the ideas of pupils on their enjoyment of mathematics and the way it is taught.

2.4 *Value Added Scores (Performance Indicators)*

In order that fair comparison can be made in comparing progress either within a country or between countries, a value added (VA) score was developed by comparing the yearly progress made by each pupil with the average progress made by groups of pupils with similar potential and similar initial mathematical attainment. Specifically, the formula for the value added score of a pupil is

$$\mathrm{VA} = 10\left[\frac{T_2 - T_1}{\delta T} - 1\right]$$

Here, T_1 and T_2 are the pupil's total scores on the attainment tests in Year 1 and Year 2 respectively and δT is the average increase for the group of pupils with similar potential and similar initial mathematical attainment.

2.5 *Observations and Interviews*

All coordinators were encouraged to undertake observations of the classes and record the important characteristics of the lessons. A *Lesson Observation and Review Sheet* was designed for this purpose, but coordinators had the choice to use it or adapt it to suit their method of recording.

Teachers and pupils (usually those who had made either very good or very poor progress over the year based on their VA scores) were also individually interviewed with the aim of determining the key factors that either gave rise to successful progress or hindered it.

3 SINGAPORE'S PARTICIPATION

This section reports Singapore's participation in the Kassel project. More details may be found in Kaur and Yap (1996, 1997, 1998). Singapore joined the project in January 1995. We were the coordinators of the project for Singapore.

3.1 *The Composition of the Singapore Sample*

Approximately 2300 pupils from seven secondary schools in Singapore participated in the project for a two-year period. The pupils were in Secondary 2 in 1995 and Secondary 3 in 1996.

In analysing and interpreting the Singapore data, one must note the method of sampling and the composition of the sample. The seven project schools invited to participate in this study were spread over different parts of the island and entire cohorts of Secondary 2 pupils from all the three different streams, Special, Express, and Normal (Academic), were included. We did not include pupils from the Normal (Technical) stream as the 1995 batch was the pioneer cohort of pupils for this course and it included a large number of pupils from the former Extended and Monolingual courses. Pupils from the Gifted Programme were also not included in our sample.

Table 1
Distribution of Sample by Stream

	Special	Express	Normal (Academic)	Total
Sample	415 (18.4%)	1131 (50.0%)	714 (31.6%)	2260 (100%)
National	3527 (8.0%)	25 326 (57.6%)	15 115* (34.4%)	43 968* (100%)

* There were 25 115 Normal (Academic) and Normal (Technical) stream pupils. The number in Normal (Technical) was estimated to be 10 000.

Table 1 shows the distribution of pupils from the three streams in the 1995 sample in comparison with the corresponding distribution at the national level (Ministry of Education, 1995). The percentage of Special stream pupils in the sample was much higher than that at the national level, while the percentage of Normal (Academic) stream pupils in the sample was lower than that at the national level. As such, results for the entire sample should not be taken as representative of that of the country. It would be more appropriate and meaningful to examine pupils' results and progress from the three streams separately.

3.2 *Data Collection*

The schedule for testing and administration of the questionnaires was as follows.

- *Phase I: (beginning of Secondary 2 year)*
  January 1995: First round of testing and questionnaire administration
  All Pupils took the Potential, Number, Algebra, Shape and Space tests. The head of mathematics department completed the School questionnaire.
- *Phase II: (end of Secondary 2 year)*
  October 1995: Second round of testing and questionnaire administration
  All pupils took the Number, Algebra, Shape and Space tests and completed the Pupil questionnaire.
  Mathematics teachers of the pupils in the project completed the Teacher and Class questionnaires.
- *Phase III: (end of Secondary 3 year)*
  November 1996: Third round of testing.
  All pupils took the Number, Algebra and Shape and Space tests.
  Mathematics teachers of the pupils in the project completed the Teacher and Class questionnaires.

It should be noted that the participating pupils were not in the same classes for the two years due to streaming at the end of Secondary 2 into Science, Arts, or Commerce streams. There were also some changes in the number of pupils taking the tests from each school over the two years due to pupils joining or leaving the schools.

Based on the test scores of the pupils in round 1 and round 2, the performance indicators of individual pupils were used to select pupils for interviews. During the first half of 1996, 137 pupils were interviewed. Among these pupils were those who made positive, zero, or negative progress during their first year of participation in the project. In addition, 43 mathematics lessons conducted by teachers involved in this project were also observed in 1996.

3.3 *Findings for the Topic Tests*

A summary of the topic test results by stream is shown in Table 2.

Table 2
Summary of Results of Testing by Stream

Test		Special			Express			Normal		
		No.	Mean	SD	No.	Mean	SD	No.	Mean	SD
Potential	P1	418	18.7	2.8	1139	14.9	3.7	734	10.7	3.0
Number	N1	419	41.8	3.2	1141	35.1	5.6	732	26.1	6.4
	N2	419	42.2	3.4	1148	36.5	6.4	725	27.2	7.5
	N3	419	44.1	2.9	1118	39.1	6.8	699	28.0	7.3
Algebra	A1	415	39.9	3.9	1131	23.7	6.4	714	14.8	5.5
	A2	419	42.3	3.5	1144	34.2	8.8	726	18.7	9.3
	A3	419	45.4	3.0	1118	38.6	10.1	698	19.0	9.4
Shape & Space	S1	415	21.8	3.5	1131	20.2	5.9	714	12.7	5.0
	S2	419	38.2	5.2	1150	28.4	8.9	707	17.7	7.2
	S3	419	44.1	3.2	1103	35.8	9.4	691	22.6	8.7
Total	NAS1	414	103.6	8.3	1120	79.1	15.8	682	53.8	14.6
	NAS2	418	122.7	9.6	1136	99.2	21.8	690	63.9	21.7
	NAS3	419	133.6	7.1	1082	113.8	24.2	667	69.8	22.9

The means of the topic tests for the three streams in the three rounds of testing are depicted in Figure 1 to chart the progress of the pupils in the three topics over the two-year period. Note that the vertical scale for the total of the three tests is different from those of the other topic tests.

Table 2 and Figure 1 show that pupils from all the three streams had consistently shown improvement in their scores for all the Topic Tests in the three rounds of testing. This was not unexpected since the content in the tests matched very closely with the local curriculum. For round 1, pupils had only covered approximately 88% of the content for Number, 44% for Algebra, and 32% for Shape and Space. By round 3, pupils had already been taught all the necessary content in the three tests.

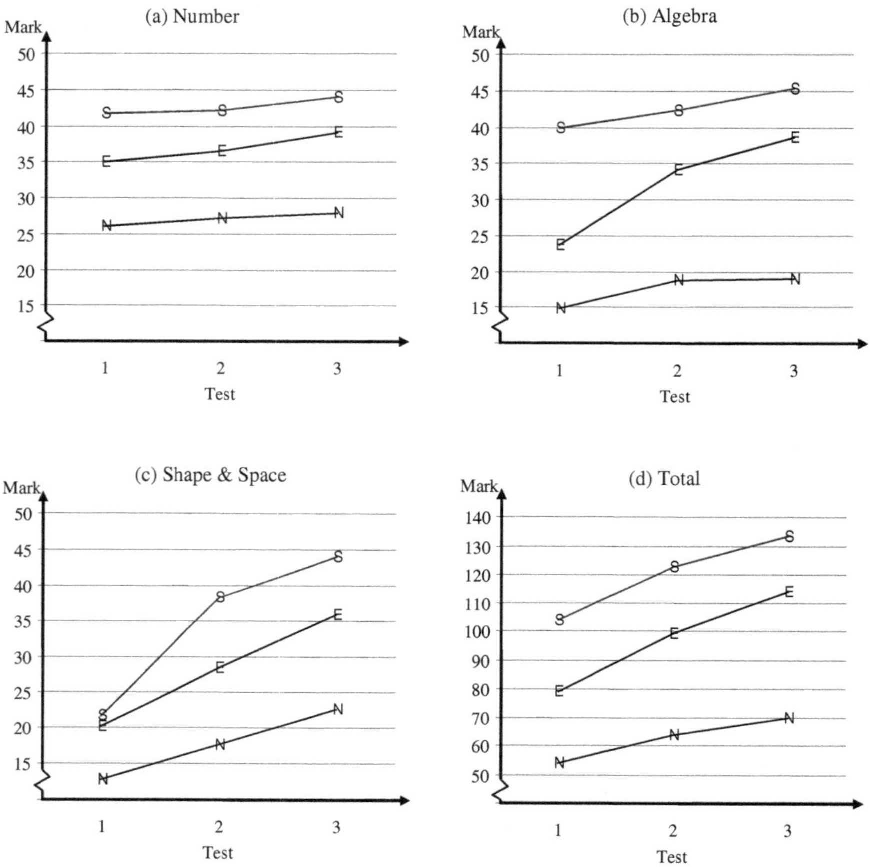

Figure 1. Topic test means by stream (S – Special, E – Express, N – Normal (Academic)).

The mean scores in the three tests for round 3 were 44 to 45 for the Special stream and 36 to 39 for the Express stream. For the Normal (Academic) stream, the mean score was 28 for Number, 19 for Algebra, and 23 for Shape and Space. The maximum possible score for each test is 50. The exceedingly low score of the Normal stream pupils for Algebra is a cause for concern since mathematical learning at the upper secondary level is heavily dependent on the mastery of Algebra skills.

For the Number Tests, the items which posed the greatest difficulty were those on approximation and estimation, followed by those on indices and surds. For the Algebra Tests, Special and Express stream pupils did not have much difficulty with most items except for items such as:

Simplify as far as possible the square root of $25x^{16}$.
Solve for x the inequality $x^2 \leq 25$.

Pupils typically gave the answer $5x^4$ to the first item and $x < 5$ or $x < \pm 5$ to the second item. Normal stream pupils had immense difficulty with various algebra items ranging from expansion and factorisation to linear equations and simultaneous linear equations. For the Shape and Space Tests, the percentage of correct answers for most items increased, as expected, in round 3 for all three streams. However, there was one geometry item (see Figure 2) that turned out to be very difficult even for Special stream pupils. Less than 8% of the Special stream pupils were able to answer this item correctly in round 3. Success in solving this non-familiar problem depends on a good understanding of the concept of similarity and good problem solving skills, which were apparently still lacking even among some of the Special stream pupils.

The diagram shows a wedge of cheese in the shape of a prism. It is cut into two equal volumes by one cut, as shown. Calculate the value of x.

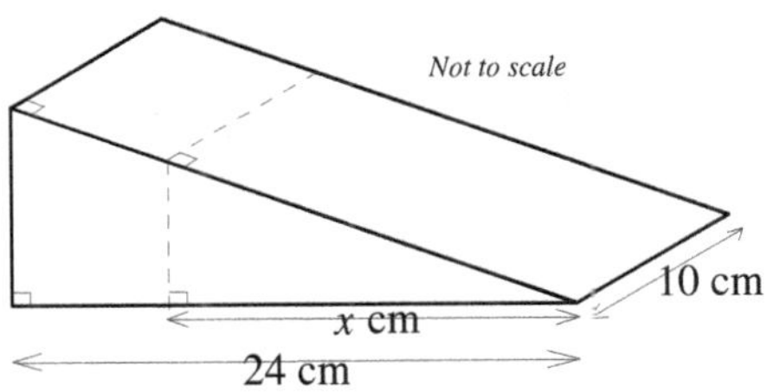

Figure 2. A difficult geometry item.

3.4 *Findings from the Qualitative Data*

The qualitative data collected in this study revealed common characteristics of the teachers and their pupils as well as their mathematics classrooms, although we did not have enough evidence to establish any direct link between these characteristics and the pupils' performance.

Whole class teaching was the main style of teaching. Most lessons were highly structured with specific objectives. Teacher exposition was the most common approach in nearly all the classes observed. The teachers presented knowledge to the pupils as a class by telling and explaining. They also often demonstrated how to solve mathematical problems step by step.

Chalkboard, textbook, and overhead projector were observed to be essential aids during the lessons. Pupils were often questioned either individually or as a class. Direct questions were often asked to check for recall of facts, algorithms, and understanding. In most of the classes, pupils did some form of class practice. Individual seatwork was preferred to pair or group work.

Teachers had high expectations of their pupils, and they felt that the main hindrances to effective teaching were pupils' attitude towards mathematics, discipline of pupils, large class size, heavy teaching load, inadequate time to prepare quality lessons, heavy load of administrative duties, too much to teach in too short a time, and examination-oriented teaching.

Most pupils claimed that they enjoyed mathematics in primary school. However, at secondary school, only less than half of the pupils said they enjoyed mathematics. Pupils from all the three streams listed *patient* and *explains clearly* as the top two qualities of the best mathematics teachers they have had.

4 FINDINGS FROM THE INTERNATIONAL DATA

The longitudinal nature of the study enabled progress in performance on the three topic tests to be examined and analysed in relation to the qualitative data obtained from questionnaire surveys and lesson observations. Thus, the project provided an international database for the participating countries to discuss issues in mathematics curriculum, mathematics teaching and learning in different countries.

A full discussion of the results from all the participating countries may be found in Burghes, Kaur, and Thompson (2004). For illustration, the test results of four of the participating countries (England, Germany, Poland, and Singapore) are shown in Table 3 and Figure 3.

Table 3

Results for England, Germany, Poland, and Singapore

Test		England (n = 1885*)		Germany (n = 497*)		Poland (n = 427*)		Singapore (n = 1943*)	
		Mean	SD	Mean	SD	Mean	SD	Mean	SD
Number	N1	19.21	8.1	24.54	6.9	25.56	8.2	34.13	7.7
	N2	22.08	8.9	27.60	7.8	30.95	8.8	35.50	7.9
	N3	24.13	9.3	31.47	9.9	35.71	7.8	37.30	8.6
Algebra	A1	12.70	7.4	12.72	6.6	19.10	10.3	24.64	10.4
	A2	16.33	9.8	18.40	9.1	27.03	9.6	32.04	11.5
	A3	19.99	11.3	25.26	11.8	34.81	8.5	34.95	13.1
Shape & Space	S1	17.17	8.6	11.67	7.0	15.14	8.7	18.45	6.3
	S2	22.55	10.6	18.38	9.0	24.54	10.1	27.89	10.3
	S3	26.11	11.4	25.67	11.9	31.78	9.5	34.14	11.1
Total	NAS1	49.07	22.7	48.94	18.3	59.80	24.9	77.22	22.3
	NAS2	60.95	27.8	64.38	23.6	82.52	25.9	95.44	28.1
	NAS3	70.23	30.7	82.41	31.3	102.30	22.5	106.39	31.3

* The sample includes only pupils who participated in all three rounds of testing.

It is obvious from Table 3 and Figure 3 that the four countries differed in their base levels and the rates of increase for the three topics. When all the participating countries were considered,

> There was great variation between countries with the high performing countries both starting at a higher base level and despite the limitations of the test, increasing at a faster rate. Whatever explanation can be found for this variation in performance, it is clear that what happens before age 13 is crucial to later progress (Burghes, Kaur & Thompson, 2004, p. 25).

For comparison between countries, the value added data were considered more valuable as the samples used were not necessarily representative of the countries. To illustrate how the value added data may provide a different view of the test results, Table 4 presents the value added mean scores and standard deviations from test round 1 to 2, and from 2 to 3 for the four countries: England, Germany, Poland, and Singapore. Note that only pupils who had taken the Potential test and the full set of Topic tests in the two rounds of testing were included.

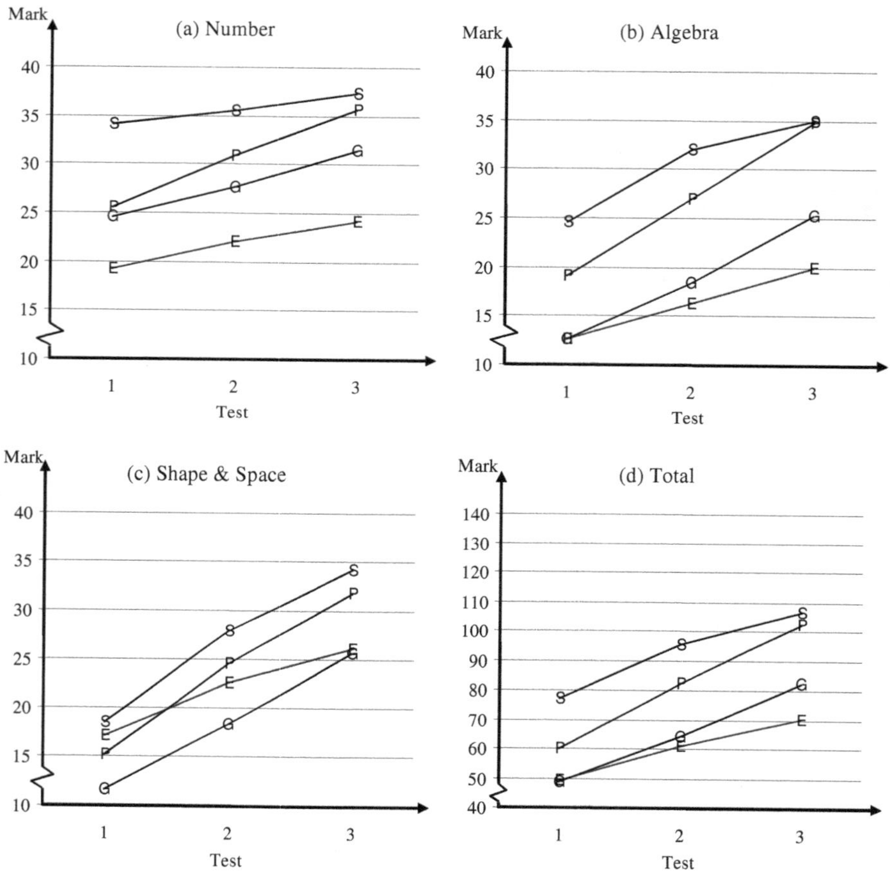

Figure 3. Topic test means for England (E), Germany (G), Poland (P), and Singapore (S).

Singapore and Germany had positive value added mean scores but they were low compared with those for Poland. England had consistently negative value added mean scores and David Burghes concluded that "England's attainment in mathematics was not impressive compared with many other countries" and that "the strategies in England are still not sufficient for enhanced progress on an international basis" (Burghes, Kaur & Thompson, 2004, p. 49 & 33).

Table 4
Value Added Scores for England, Germany, Poland, and Singapore

Country	Value Added					
	Test rounds 1 → 2			Test rounds 2 → 3		
	No.	Mean	SD	No.	Mean	SD
England	3722	-2.40	6.9	2787	-1.46	8.7
Germany	618	0.54	8.5	571	3.82	14.8
Poland	489	5.20	11.5	446	6.78	16.2
Singapore	2091	0.39	7.3	1998	1.02	12.5

By comparing the practice in the different countries based on the questionnaire survey data, interviews, and classroom observations, the Central Team at CIMT identified several characteristic features of some of the high performing countries that could possibly explain the performance of English pupils.

In countries such as Hungary, Poland, and Czech Republic, whole-class interactive teaching was observed. This style of teaching was characterised by pupils demonstrating on a regular basis at the board, pupils articulating their answers, and mistakes made by pupils being discussed with the whole class. Most lessons consisted of many varied activities, all reinforcing the theme of the lesson, and single exercises were usually set and reviewed rather than having the class tackle a range of questions at their own pace. The high achievement gained by the Singapore pupils was attributed to various factors including a structured curriculum with good resource structure (textbook, workbook, and teachers' guide), teachers' very high expectations, and pupils' diligence and hard work.

Subsequently, in making recommendations for mathematics teaching and implementing the Mathematics Enhancement Programme in England (Burghes, 2000), the Central Team at CIMT had used the teaching strategies from Hungary, Poland, and Czech Republic, together with the resource structure of Singapore.

5 CONCLUDING REMARKS

The Kassel Project was the first international mathematics project that we had participated in. The data collected from the schools were given to the Heads of Mathematics departments of the seven schools for their action and retention. Two Heads of Mathematics departments had used the data and findings to implement new curriculum initiatives in their respective schools.

The reports of the Kassel Project (Kaur & Yap, 1996, 1997, 1998) serve as a source of reference for local educators and research students to study mathematics education in Singapore. The data and findings in the reports have also been used by mathematics educators at the National Institute of Education in lectures and workshops for teachers.

We recommend that for pupils in the Normal (Academic) stream, the topic Algebra be given more emphasis or reorganised to ensure that they have a strong foundation in the basics of algebra. For all streams, more emphasis should be placed on the application of mathematical knowledge and concepts to solve problems.

The international dimension of the project provides great insight and exposure to the issues of mathematics teaching and learning in different countries. The strength and weakness of the Singapore mathematics education curriculum become more apparent when seen through lenses provided by the other participating countries.

References

Burghes, D. N. (2000). Mathematics Enhancement Programme (MEP): The first three years. *International Journal for Mathematics Teaching and Learning*. Retrieved January, 14, 2008, from http://www.cimt.plymouth.ac.uk/journal/default.htm

Burghes, D.N., Kaur, B., & Thompson, D.R. (Eds.) (2004). *Kassel project.* Budapest: Muszaki Konyvkiado, Wolterskluwer.

Kaur, B., & Yap, S.F. (1996). *Kassel project report – First phase (Jan – Oct 95).* Singapore: Division of Mathematics, National Institute of Education, Nanyang Technological University.

Kaur, B., & Yap, S.F. (1997). *Kassel project report – Second phase (Oct 95 – Jun 96).* Singapore: Division of Mathematics, National Institute of Education, Nanyang Technological University.

Kaur, B., & Yap, S.F. (1998). *Kassel project report – Third phase (Jun – Nov 96).* Singapore: Division of Mathematics, National Institute of Education, Nanyang Technological University.

Ministry of Education. (1995). *Education statistics digest 1995.* Singapore: Author.

Chapter 22

International Project on Mathematical Attainment (IPMA): Singapore's Participation

Berinderjeet KAUR **KOAY Phong Lee** **YAP Sook Fwe**

A total of 856 primary school pupils from Singapore participated in the International Project on Mathematical Attainment (IPMA) from 1999 till 2003. The performance of the pupils in a test administered at the beginning of their primary schooling showed that they generally enter Primary 1 with considerable knowledge in arithmetic. From subsequent annual tests over a period of five years, it was found that they generally had a good grasp of the knowledge and mastery of skills that they were taught at school. Low attaining Primary 1 and 2 pupils were likely to be less attentive when the teacher was teaching and were also easily distracted. They were often not on task during groupwork or individual seatwork. From lesson observations, it was apparent that nearly all the mathematics lessons were structured with specific learning objectives. The study also revealed some weaknesses in the teaching of primary mathematics in Singapore, particularly in representation, making connections, and communication of mathematical ideas.

Key words: mathematical attainment, primary school pupils, IPMA, longitudinal study, Kassel project, value added score, learning behaviours

1 BACKGROUND

The International Project on Mathematical Attainment (IPMA) was a development of the Kassel project, described in Chapter 21. The Kassel project led the participating researchers to question the gaps (both within and between countries) in mathematics performance of their secondary students. These researchers subsequently participated in the IPMA to examine mathematical attainment of pupils in the primary years of schooling. IPMA was initiated in England with some funding from Pricewaterhouse Coopers. Fifteen countries participated, namely Brazil, China, Czech Republic, England, Greece, Hungary, Ireland, Japan, Poland, Russia, Singapore, South Africa, Ukraine, the United States of America, and Vietnam. As in the Kassel project, IPMA was an extensive comparative study in mathematics. It monitored the mathematical progress of individual pupils throughout their primary years of education. Singapore participated in IPMA from 1999 till 2003. We were the coordinators of the project for Singapore. Singapore's participation was made possible with financial support from both the Academic Research Fund of the National Institute of Education and the University of Exeter in UK.

2 METHODOLOGY

The methodology of the IPMA, described in detail in Burghes, Geach, and Roddick (2004), was similar to that of the Kassel project. The country coordinators:

- selected the sample cohort of pupils according to the overall guidelines of the project;
- administered and marked the yearly tests. There were altogether six tests. The test data of the first two tests were used to provide performance indicators (i.e., value added scores) based on the progress made by pupils with same initial test scores;
- completed country questionnaires so that the general situation concerning primary education and, in particular, mathematics, could be determined;
- observed and videotaped lessons to record different teaching strategies in mathematics classes and shared the effective practices with other country coordinators;
- met yearly with other country coordinators to discuss all aspects of the project;
- made national recommendations, if any.

3 IPMA IN SINGAPORE

3.1 *The Sample*

Three schools participated in the study. School 1 is an all girls' convent school that belongs to the Good Shepherd Order. It is government-aided, which means that 95% of the funding for the school comes from the Ministry of Education. School 2 is a typical co-educational school that is fully funded by the government. The school is located in the neighbourhood of a large government-subsidised housing precinct. School 3 is a co-educational school run by Hokkien Huay Kuan, a Chinese clan. It is also a government-aided school. The clan provides 5% of the funding and the government the rest. It is also one of the best performing schools in the Primary School Leaving Examination (PSLE).

All the Primary 1 pupils (aged 6+ years) in the 1999 cohort of the three schools participated in the project. Due to various reasons (e.g., transfer in or out of the school), only 856 pupils completed all the tests administered from 1999 to 2003. Table 1 shows the distribution of the pupils by school and gender.

Table 1
Distribution of Sample by School and Gender

	n (girls)	*n* (boys)	Total
School 1	153	-	153
School 2	145	171	316
School 3	173	214	387
Total	471	385	856

3.2 *The Tests*

The tests were developed in England for the project and were modified accordingly to reflect Singapore English and Singapore's monetary system. There were six tests. Test 1 and Test 2 were administered to Primary 1 pupils at the beginning and at the end of the school year respectively. Subsequently, the same cohort of pupils sat for a test at the end of every school year. The same cohort of pupils sat for Test 6 when they were in Primary 5.

As shown in Table 2, the tests expanded over time. Test 1 was the shortest with only seven items. The number of items in each subsequent test increased with the previous test embedded in it, i.e., Test 1 items formed a subset of Test 2 items. Consequently, the total number of items in Test 6 was 81. For every test, the pupils were not told in advance about the test and hence were not expected to prepare for it. To control for language competency, in Test 1 and Test 2, the teachers who administered the tests read the items one at a time to the pupils. For Test 4, Test 5, and Test 6, the pupils were given a half-hour break between Part A and Part B. Part C of Test 6 was administered a day after Part A and Part B.

Table 2
Details of Tests Administered from 1999 – 2003

Test	Items	Level	Month / Year	Duration
1	1 – 7	Beginning of Primary 1	March 1999	30 – 40 minutes
2	1 – 17	End of Primary 1	November 1999	30 – 40 minutes
3	1 - 29	End of Primary 2	November 2000	30 – 40 minutes
4A 4B	1 – 29 30 - 42	End of Primary 3	November 2001	40 minutes for each part
5A 5B	1 – 29 30 - 56	End of Primary 4	November 2002	40 minutes for each part
6A 6B 6C	1 – 29 30 – 56 57 - 81	End of Primary 5	November 2003	40 minutes for each part

The test scripts were scored manually. A score of one was given to each correct response and zero for an incorrect response. Some items had more than one response. Consequently, the maximum test scores were as follows: Test 1: 20 marks, Test 2: 40 marks, Test 3: 60 marks, Test 4: 80 marks, Test 5: 110 marks, and Test 6: 140 marks.

3.3 *Performance Indicators (Value Added Scores)*

In addition to the test scores, a value added score for each pupil for every year of participation was computed as follows:

$$\text{PI or value added} = 10\left[\frac{S_{n+1} - S_n}{\delta S} - 1\right]$$

where

S_{n+1} and S_n are the pupil scores on Test $(n+1)$ and Test n,
δS is the average increase in score made by all pupils in the database.

Here, only prior attainment was used in the value added analysis and the progress made over one year for each pupil was compared with pupils of similar initial mathematical attainment. A pupil with an increase in his/her score that is exactly on average for the comparable group would have a value added score of zero, while a pupil with an enhanced increase would have a positive value added score.

3.4 *Questionnaires*

The performance indicators of the pupils in Test 2 and Test 3 were used to select pupils for further study. A total of 109 pupils were selected with 47 in the Low category (PI < -5), 29 in the Average category (-5 < PI < 4) and 33 in the High category (PI > 4). When these pupils were in Primary 3, their teachers were given a *Pupil questionnaire* to complete for each pupil. The questionnaire was designed to elicit information on the behaviour of the pupil during mathematics lesson and the parental and home support received by the pupil. An in-depth analysis of the questionnaire and the data is reported in Kaur, Koay, and Yap (2003).

3.5 *Lesson Observations*

As pupils were regrouped each year in the three participating schools, it was difficult to track classes that showed largest and smallest progress. In Singapore, major reassignment of pupils takes place at the end of Primary 3 and Primary 4 when pupils are selected for different educational programmes. At the end of Primary 3, high ability pupils are selected for the Gifted programme while at the end of Primary 4, low ability pupils are selected for the Foundation programme. These pupils follow mathematical programmes that are different from the main stream. In this study, an attempt was made to observe some of the classes and several mathematics lessons were videotaped.

3.6 *Results of Tests*

Table 3 shows the maximum possible scores, the means, and standard deviations of the six tests for the same cohort of pupils.

Table 3
Summary of Results of Tests (n = 856)

	Max Mark	Mean	Standard Deviation	Mean / Max Mark
Test 1	20	15.70	3.26	0.785
Test 2	40	30.41	5.58	0.760
Test 3	60	45.66	7.54	0.761
Test 4	80	59.54	9.73	0.744
Test 5	110	85.26	13.9	0.775
Test 6	140	106.12	17.23	0.772

The tests were cumulative in nature as the items in Test n form a subset of items in Test $n+1$. Table 4 shows how the pupils performed in the various subsets of test items across the years. As expected, the mean score for each subset of items increases and the standard deviation decreases.

Table 4
Mean Scores of Groups of Items for Tests 1- 6 (n=856)

Items (max mark)	Mean Scores of groups of items (Standard Deviation)					
57 – 81 (30 marks)						16.21 (5.68)
43 – 56 (30 marks)					16.10 (5.63)	18.14 (5.51)
30 – 42 (20 marks)				9.23 (3.67)	14.88 (3.50)	15.79 (3.20)
18 – 29 (20 marks)			11.61 (3.72)	14.43 (3.36)	16.95 (3.05)	17.90 (2.51)
8 – 17 (20 marks)		12.45 (3.76)	15.72 (2.87)	17.01 (2.61)	18.10 (2.32)	18.66 (1.80)
1 – 7 (20 marks)	15.70 (3.26)	17.96 (2.34)	18.32 (1.87)	18.87 (1.62)	19.24 (1.35)	19.42 (1.12)
	Test 1 Primary 1 Mar 1999	**Test 2** Primary 1 Nov 1999	**Test 3** Primary 2 Nov 2000	**Test 4** Primary 3 Nov 2001	**Test 5** Primary 4 Nov 2002	**Test 6** Primary 5 Nov 2003

The data in Table 4 also show that pupils generally entered Primary 1 with considerable knowledge in arithmetic (Kaur, Koay, Yap & Burghes, 1999) and there were some pupils who failed to master some topics that were tested repeatedly across the grade levels.

3.7 *Performance Indicators (Value Added Scores)*

Table 5 shows the value added scores across the years. The scores were computed for all the pupils who had taken all the tests.

Table 5
Value added from Test n to Test (n+1)

n	Mean Score Test *n*	Mean Score Test (*n*+1)	Mean Difference	Value Added
1	15.70	30.41	14.72	1.71
2	30.41	45.66	15.24	-0.11
3	45.66	59.54	13.88	-0.37
4	59.54	85.26	25.73	2.53
5	85.26	106.14	20.87	1.34

The trend is illustrated in the bar chart in Figure 1. The value added is positive from Test 1 to Test 2, i.e., from the beginning of Primary 1 to end of Primary 1 and again from Test 4 to Test 5, and Test 5 to Test 6.

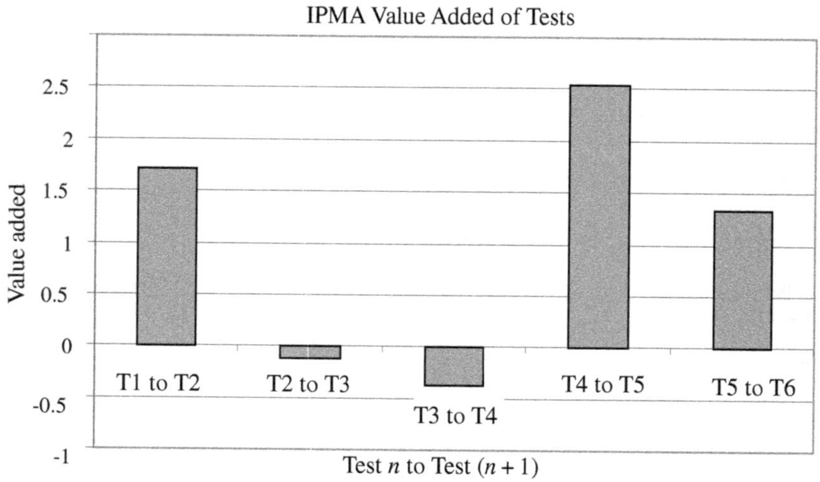

Figure 1. Value added from consecutive tests.

The greatest improvement occurs from Test 4 to Test 5. This may be due to the fact that pupils in Primary 4 were more motivated to learn in order to prepare for the high stakes streaming examination at the end of the year. The value added is marginally negative from Test 2 to Test 3, i.e., from end of Primary 1 to end of Primary 2 and again from Test 3 to Test 4, i.e., from end of Primary 2 to end of Primary 3.

In addition to the items testing topics covered in the Singapore primary mathematics syllabus, the final test (Test 6) consisted of topics not found in the syllabus.

3.8 *Items Assessing Topics not Covered in the Syllabus*

Only the analysis of the final test, Test 6, is reported here. Performance of pupils in the previous tests can be found in the reports (Kaur, Koay & Yap, 2000, 2001a, 2003; Koay, Yap & Kaur, 2003, 2004). As the test (http://www.cimt.plymouth.ac.uk/projects/ipma/test6.pdf) was designed for international study, there were items that tested concepts and skills not covered in the Singapore primary mathematics syllabus. Table 6 shows the facility indices of those items that had facility indices greater than 50%. These items involve concepts and skills in probability, algebra, number patterns, and fractions.

Although probability was not introduced in the Singapore primary mathematics syllabus, yet over 67% of the pupils in the sample were able to classify whether an event is certain, possible, or impossible (Item 46). In addition, more than 77% of the pupils were able to compute the probability of an event in the real life (Item 68). However, only about 17% of the pupils were able to compute the probability of throwing a given number with an unbiased die (Item 67, not shown). This task involves the application of specific mathematical knowledge, and pupils may not have understood what is *an unbiased die*.

Only simple algebraic expressions in one variable were formally introduced in Primary 6, and yet the data in Table 6 show that more than 91% of the Primary 5 pupils were able to use algebraic reasoning to solve simple linear equation with one as the coefficient for the unknown term. As expected, their performance deteriorated when the coefficient of the unknown term was not one, e.g., Item 70: Solve for x, $3x - 4 = 11$ (F.I. = 40.42%).

Table 6
Items not Found in the Singapore Syllabus but with Facility Indices (F.I.) Greater than 50%

Item	Topic & Objective	F.I. (%)
Probability: Classifying events as certain, possible or impossible		
46	Say whether the statements below are certain, possible or impossible	
a	It will be sunny tomorrow.	76.17
b	Next year is 1999.	67.87
c	You will be King next week.	73.60
d	England will win the next World Cup in the year 2006.	70.56
Probability: Finding probability of a real-life event		
68	The probability of it raining tomorrow is 1/4. What is the probability of it **not** raining tomorrow?	77.45
Algebra: Solving simple linear equation with one unknown		
69	Solve for x, $x + 2 = 6$	91.24
Whole numbers: Finding the terms in a number sequence with consistent second difference.		
71	Write down the next two terms of this sequence. 2, 5, 10, 17, 26, ☐ , ☐ , …	74.65 62.85
Fractions: Dividing a fraction by another fraction		
80	$\frac{1}{3} \div \frac{1}{9} =$ ☐	53.74

Item 71 involves number patterns with constant second difference and similar tasks are found in the secondary mathematics syllabus. Primary school pupils are exposed to simple number patterns only and yet more than 62% of the Primary 5 pupils in the sample were able to answer Item 71 correctly.

Fraction is a difficult concept to many people, particularly primary school pupils. The outcomes listed under the Division of Fractions for Primary 5 mathematics include only associating another fraction with division and dividing a proper fraction by a whole number. Yet more than 53% of the Primary 5 pupils were able to divide a fraction by a fraction correctly. Some teachers or tutors may have gone beyond the textbooks and taught the "invert-and-multiply" algorithm to these pupils.

The above examples illustrate clearly that some pupils were able to do mathematics beyond the prescribed syllabus by drawing upon their personal experiences and general knowledge. Hence, teachers should be aware of the informal mathematical knowledge that their pupils bring to the classroom, and help them make connections between their mathematical intuition and the formal mathematical knowledge taught.

3.9 *Items Assessing Topics Covered in the Syllabus*

Table 7 shows items that test concepts and skills found in the primary mathematics syllabus but with low facility indices (less than 50%). There are three areas of concern.

The concepts of parallel lines and symmetry are introduced in Primary 4. The poor performance on Item 45 and Item 54 revealed that the pupils lacked mastery of parallel lines and symmetry. Of concern is the communication of these two mathematical ideas. The pupils did not perform well when qualifying terms such as "exactly" and "just" are used in the questions. In school, the instructional objectives should not be just about the identification of parallel lines or symmetric figures. They must be expanded to include a deep understanding of the concepts. Teachers should provide opportunities for pupils to reflect, synthesise, and communicate what they have learnt. For example, instead of telling pupils that a rectangle has two lines of symmetry or demonstrating the lines of symmetry of the rectangle by paper folding, teachers can provide pupils with pieces of rectangular paper and get them to identify all the lines of symmetry for the rectangle. Through the discussion of examples and non-examples of lines of symmetry, teachers can lead pupils to conclude that a rectangle has more than one line of symmetry, a rectangle cannot have three lines of symmetry, and it has exactly two lines of symmetry. The pupils can then proceed to investigate the lines of symmetry for other polygons, organise and verbalise their findings using appropriate mathematical language.

Table 7

Items in the Singapore Syllabus with Facility Indices (F.I.) Less than 50%

Item	Topic, Objective and Item	F.I. (%)
Geometry: Identifying letter with one line of symmetry		
45	M A T H S	
a	Which of the letters above have just one line of symmetry?	22.43
b	Which of the letters have two lines of symmetry?	49.07
Geometry: Identifying shapes with line of symmetry and parallel sides		
54	Write the letter of each shape besides the words which describe it. Shapes can be listed more than once.	

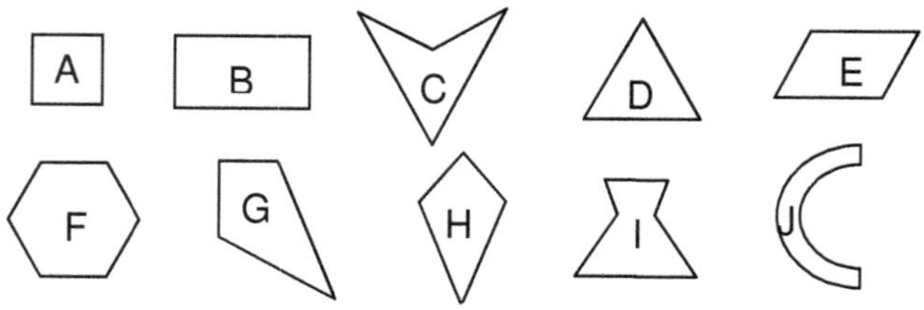

Item	Topic, Objective and Item	F.I. (%)
a	Exactly one line of symmetry	3.27
b	More than one line of symmetry	8.41
c	Exactly one pair of parallel sides	39.95
d	Exactly two pairs of parallel sides	49.30
Whole numbers: Estimating numerical expression		
50	Estimate the value of $\frac{367 \times 27}{33}$	16.71
Percentage: Finding simple interest		
81	$50 is invested in an account which pays 8% interest per year. How much interest is paid after one year?	47.78

Another concern is the lack of estimation skills among the primary school pupils. Approximation and estimation is one of the skills listed in the framework of the mathematics curriculum in Singapore. It is a skill that pupils are expected to possess for solving problems. Yet only about 17% of the Primary 5 pupils were able to answer Item 50 correctly. Examination of the answer scripts revealed that the Primary 5 pupils had the prerequisite skills of rounding numbers and operating with multiples of 10 and 100 for Item 50. However, many of them rounded up both numbers in the numerator and rounded down the denominator to get the answer 400 when the actual answer is very close to 300. They did not look at the given numerical expression as a whole and consequently overestimated the answer. With the introduction of calculators in Primary 5 and Primary 6, there is an urgent need for pupils to master the estimation skills essential for checking calculator calculations.

Another area that should not be neglected is the application of the concept of percentage. The computation of simple interest, like the Goods and Services Tax, is one way to link classroom mathematics with real life applications. Greater attempt needs to be made to teach beyond the classroom.

3.10 *Qualitative Data Analysis*

The questionnaire data of the 109 Primary 3 pupils revealed that compared to other pupils, low attaining Primary 1 and Primary 2 pupils were likely to be less attentive when the teacher was teaching and were easily distracted. They were often not on task during groupwork or individual seatwork. These pupils were also less likely to submit work promptly for checking and marking, or to give their best when doing class work or homework. In addition, they were neither neat nor organised in their work in mathematics. These pupils were not good listeners and lack proficiency in English. Generally, they were also not motivated to learn and did not have good family support compared to their peers. They were less likely to be children of diploma or degree holders. Their parents did not show great concern about their mathematical attainment and provided little help with mathematics at home.

From the lesson observations, it was apparent that nearly all the mathematics lessons were structured with specific learning objectives. Teachers attempted to use manipulatives to enhance the learning of concepts, and pupils were given opportunities to engage in hands-on activities. Generally, most pupils were receptive to the teaching and they were often on task during individual seatwork or group work. Teachers had good classroom management and appeared to review and explain mathematical concepts for considerable length of time during their lessons either through exposition or questioning. Pupils were responsive to the questions asked by their teachers but they hardly posed questions or initiated discussion.

In the lessons observed, teachers always set homework but seldom reviewed them. They seemed to place great emphasis on class practice and preferred individual seatwork to group work. Generally, pupils appeared to be active, attentive, happy and responsive. Teachers were sound in their mathematical knowledge, clear in their instructions, in control of their lessons, authoritative, enthusiastic, and warm to their pupils.

4 IMPLICATIONS FOR PRIMARY MATHEMATICS EDUCATION

The IPMA tests reveal that primary pupils in Singapore were developing proficiency in most of the concepts and skills listed in the mathematics syllabus and their achievement increased over time (see Table 4). There was significant increase in performance when a relevant topic was taught. Nevertheless, there is room for improvement.

There were pupils who had not mastered the basic concepts found in the test at the end of Primary 5. This signals the need to identify pupils at risk at an early stage so that appropriate support is provided to prevent these pupils from being left behind in their mathematics learning. Efforts must be made to help pupils cultivate good habits of learning and develop positive attitudes towards the subject. Pupils who experience repeated failure in the school mathematics assessment will not have confidence in the subject. Perhaps the schools may want to change their practice and abolish the ranking of pupils according to test performance, and review the way they assess pupils' learning. The schools may want to identify pupils who are at risk at the beginning of each year and provide customised instruction to help them attain competence for that level. In addition, the removal of school ranking may encourage more schools to assign their expert teachers to provide special assistance to the weaker pupils.

At the same time, the study also reveals some weaknesses in the teaching of primary mathematics in Singapore, particularly in representation, making connections, and communication of mathematical ideas.

4.1 *Representation*

The concrete-pictorial-abstract (C-P-A) approach in the development of concepts has long been advocated in Singapore (Ministry of Education, 1990, 2000, 2006). The current mathematics textbooks adopt this approach to introduce the topics. However, there is a need to help pupils see the links between the different representations of the same topics. For example, base 10 blocks and number discs are used in whole numbers and decimals, and fractions discs or strips are used in fractions. We would like to see more frequent use of the number line in the teaching all these topics. The number line can help pupils visualise the relationships among whole numbers, fractions, and decimals. It can be used to display operations, equivalence, and solutions to problems.

4.2 *Making Connections*

Connections refer to the abilities to "understand how mathematical ideas are interconnected" and "apply mathematics in contexts outside of mathematics" (National Council of Teachers of Mathematics, 2000, p. 402). In Singapore, mastery of procedural skills is considered essential as reflected in the high test scores on items testing them. Making connections, on the other hand, is not emphasised. For example, many Primary 5 pupils seem to have difficulty "undoing" a series of operations. This may be due to the lack of depth in their understanding of the four operations. They are not aware that:

- addition and subtraction are inverse operations;
- multiplication and division are inverse operations;
- multiplication is repeated addition; and
- division is repeated subtraction.

Teachers should make conscientious attempts to help pupils use problem solving heuristics such as working backwards and make connections among the four operations.

4.3 *Communication*

The ability to communicate through mathematics in both individual and collaborative settings is one of the essential skills for the 21st century. Teachers need to place greater emphasis on reasoning, explaining, and sense making. They must provide greater opportunities for pupils to engage in tasks that demand the use of language to communicate their mathematical ideas and to explain their solutions. Teachers often attribute the lack of communication opportunities to the lack of curriculum time. It is undeniable that pupil presentation is time consuming. However, with the recent reduction in syllabus contents, it is possible to make changes to the instructional approach to create opportunities for pupils to communicate their mathematical ideas.

4.4 *Approximation and Estimation Skills*

As mentioned before, with the introduction of calculators in the upper primary grades, there is an urgent need to improve the approximation and estimation skills of the pupils. Starting from Primary 1, pupils must develop their quantitative intuition and number sense to determine best estimate. They must be given greater opportunities to communicate their thought processes during estimation. Estimation should be included in the objectives of all the lessons on problem solving.

5 CONCLUDING REMARKS

The International Project on Mathematical Attainment (IPMA) is the second international mathematics project led by the University of Exeter that researchers at NIE participated in. The researchers were able to extend their experience from participation in the Kassel project into primary mathematics education. The Kassel project and the IPMA are longitudinal studies on secondary mathematics education and primary mathematics education respectively. Consequently, the five reports (Kaur, Koay & Yap, 2000, 2001a, 2003; Koay, Yap & Kaur, 2003, 2004), and three papers (Kaur, Koay, Yap & Burghes, 1999; Kaur, Koay & Yap,

2001b; Koay, Kaur & Yap, 2001) produced by the team of researchers at NIE provide complementary sources of reference on mathematics education in Singapore.

The teachers and pupils who participated in the study also benefited. After every round of testing, all the three schools were given detailed results on the performance of their classes and pupils. This enabled the teachers to identify pupils who merited additional help with the learning of mathematics. We also organised sharing sessions to review the findings with the teachers from the three schools.

Participation in IPMA was an invaluable experience for the local researchers. It provided opportunities for the researchers to discuss with their international counterparts the problems and issues faced by the mathematics educators. During past annual coordinators meetings, international comparisons of mathematical attainment were reviewed. Good practices in all aspects of primary mathematics teaching and learning were also shared.

Pupils in Singapore performed well in mathematics as shown in international studies such as IPMA and TIMSS. However, there is still room for improvement. We need to constantly review our instructional approaches to help our pupils acquire the abilities and skills that they would need in the 21st century.

References

Burghes, D., Geach, R., & Roddick, M. (Eds.) (2004). *Report on International Project on Mathematical Attainment*. Budapest: Muszaki Konyvkiado, Wolterskluwer.

Kaur, B., Koay, P.L., Yap, S.F., & Burghes, D. (1999). Singapore pupils' knowledge of number at the beginning of first school year. In S. P. Loo (Ed.), *Proceedings of the MERA – ERA Joint Conference: Educational challenges in the new millennium* (pp. 1562-1567). Singapore: Educational Research Association.

Kaur, B. & Koay, P.L., & Yap, S.F. (2000). *IPMA report (NIE – Exeter Joint Study) – Year one (Jan – Dec 99)*. Singapore: National Institute of Education.

Kaur, B., Koay, P.L., & Yap, S.F. (2001a). *IPMA report (NIE – Exeter Joint Study) – Year two (Jan – Dec 00)*. Singapore: National Institute of Education.

Kaur, B., Koay, P.L., & Yap, S.F. (2001b). Singapore pupils' mathematical knowledge at the end of first school year. In J. Ee, B. Kaur, N.H. Lee & B.H. Yeap (Eds.), *New literacies: Educational response to a knowledge-based society* (pp. 666 – 677). Singapore: Educational Research Association & Association of Mathematics Educators.

Kaur, B. Koay, P.L., & Yap, S.F. (2003). *IPMA report (NIE – Exeter Joint Study) – Year three (Jan – Dec 01)*. Singapore: National Institute of Education.

Koay, P.L., Kaur, B., & Yap, S.F. (2001). Numerical proficiency of Singapore students by the end of grade 2. In C. Vale, J. Horwood & J Roumeliotis (Eds.), *2001: A mathematical odyssey* (pp. 24 – 34). Australia: The Mathematical Association of Victoria.

Koay, P.L., Yap, S.F., & Kaur, B. (2003). *IPMA report (NIE – Exeter Joint Study) – Year four (Jan – Dec 02)*. Singapore: National Institute of Education.

Koay, P.L., Yap, S.F., & Kaur, B. (2004). *IPMA report (NIE – Exeter Joint Study) – Year five (Jan – Dec 03)*. Singapore: National Institute of Education.

Ministry of Education. (1990). *Mathematics syllabus – Primary*. Singapore: Author.

Ministry of Education. (2000). *Mathematics syllabus – Primary*. Singapore: Author.

Ministry of Education. (2006). *Mathematics syllabus – Primary*. Singapore: Author.

National Council of Teachers of Mathematics. (2000). *Principles and standards for school mathematics.* Reston, VA: Author.

Chapter 23

My "Best" Mathematics Teacher: Perceptions of Primary School Pupils from Singapore and Brunei Darussalam

WONG Khoon Yoong **Berinderjeet KAUR**

KOAY Phong Lee **JAMILAH Bte Hj Mohd Yusof**

This chapter describes a comparative study of the perceptions of my "best" mathematics teachers expressed by a sample of upper primary school pupils in Singapore and Brunei Darussalam. Data were collected in October 1998 as part of an inter-institutional research collaboration between the National Institute of Education, Singapore and the Sultan Hassanal Bolkiah Institute of Education, Brunei Darussalam. The pupils were asked to describe in words the qualities of their "best" mathematics teacher and to draw a picture of this teacher teaching in class. The descriptions and drawings provide interesting insights about the qualities of good mathematics teachers and teaching that can be debated by the mathematics education community.

Key words: mathematics teachers, quality of mathematics teaching, student drawings, cultural differences, Brunei Darussalam

1 QUALITY OF MATHEMATICS TEACHING

What qualities constitute good mathematics teaching has been an issue of enduring interest among policy makers, school administrators, the public, parents, teacher educators, researchers, and, of course, the teachers themselves. These qualities include generic ones (Hattie, 2003), such as being able to explain things clearly, to manage classroom effectively, and to have good rapport with the students, as well as qualities specific to mathematics, for examples, mastery of mathematical contents and processes (Hill, Schilling & Ball, 2004), use of mathematics-related technologies, and the ability to stimulate mathematical thinking. A complex combination of both generic and mathematical qualities constitutes mathematics pedagogical content knowledge, an important current research agenda among many mathematics teacher educators (e.g., Lim-Teo, Chua, Cheang & Yeo, 2006). These qualities have been codified as standards for mathematics teaching (e.g., Australian Association of Mathematics Teachers, 2006). These conclusions may be arrived at through classroom observations, interventions, and self-reports by the teachers. For example, in a previous Singapore study (Lim & Wong, 1989), 121 secondary school student teachers were asked to rate 40 teacher characteristics and these ratings were compared with those obtained from 76 qualified secondary school mathematics teachers. Both groups were very consistent in rating the top three most desirable qualities as follows:

- able to explain concepts, methods etc. clearly;
- is confident and at ease in teaching; and
- conveys an enthusiasm for mathematics to pupils.

Since teacher actions impact directly on the students, it is logical to ask the students what they perceive to be good qualities in their teachers. Unfortunately, students' perceptions in this respect are rarely reported in the literature, although most tertiary institutions routinely ask students to rate their lecturers on specific dimensions on teaching, such as clarity, knowledge of subject, and organisation of teaching. In Singapore, Lim (1991) interviewed 23 Secondary 1 students about the qualities of good teachers. She found that most students mentioned knowledge of the subject as the most important quality. Other positive attributes were: nice, warm, friendly, caring, patient, fair, and willing to spend time with the students.

The common technique to collect perceptions is the questionnaire. While questionnaire can be readily answered by a large sample of respondents, one disadvantage is that it is restricted by the items included. On the other hand, open-ended items allow the respondents to write more freely about their ideas, but this technique will not provide comprehensive data when the respondents are reluctant or unable or do not have the motivation to write extensively. A common phenomenon about using open-ended items among school students is that many returns have missing or single word answers, which do not provide useful data to the researchers. To remedy this shortcoming, some researchers have experimented with the drawing technique, with some success, to study students' perceptions of mathematicians, scientists, classrooms, and technology (Burgess, 1994; Chambers, 1983; McDonough & Pavlou, 1994; Picker & Berry, 2000; Rennie & Jarvis, 1994; Teppo, 1993; Wong, 1995, 1996).

One way to analyse student drawings is to count the frequencies of items that appear in these drawings according to certain conceptual categories. The choice of the categories depends on the research questions and theoretical framework of the study. In the Singapore study about mathematicians (Wong, 1995), 281 Secondary 1 to Secondary 4 students drew pictures about their ideas of what mathematicians are like and their work. The categories used were: *Glasses, Hair, Special dress, Equipment, Books,* and *Symbols.* Four main stereotypes about mathematicians were identified. The most predominant one was that he is a male, wears glasses, handles formulae and symbols, and has a strange and lost countenance. The second Singapore study (Wong, 1996) collected drawings of 246 "good" mathematics lessons and 200 "bad" lessons produced by secondary school students. The items that appeared in these drawings were classified under three categories: *Instructional pattern and tasks; People,* and *Tools and physical environment.* Drawings of whole class instruction appeared most frequently compared to group work (appeared in only 5 lessons). Blackboard in "good" lessons had notes and examples, whereas it was messy or blank in "bad" lessons.

The coding may be subjective, but it seems to work well. For instance, Kahle (1989) cited inter-rater reliabilities of 0.86 to 0.97 about coding drawings of scientists and commented on the close agreement between visual images and verbal images deduced from interviews. It is also fun for students who can express their feelings through the drawings. Those with language difficulties in reading questionnaire items or writing verbal responses will be able to provide visual snapshots of their perceptions. We believe that this technique is worthy of further fine-tuning.

Hence, this paper attempts to make three distinctive contributions to the study of the quality of mathematics teaching: (a) to provide insights about teaching quality from the perspective of the students, in this case, upper primary pupils; (b) to use drawing as a data collection tool in addition to open-ended descriptions; (c) to compare findings from two culturally different societies, namely secular and commercial Singapore and Islamic and agricultural Brunei Darussalam.

2 BACKGROUND OF THE STUDY

This study was part of an inter-institutional research collaboration between the National Institute of Education, Singapore and the Sultan Hassanal Bolkiah Institute of Education, Brunei Darussalam. The collaboration was called *Inter-institute Dialogue on Educational Advances* (IDEA), and it ran from 1998 to 2000. Its main purpose was to stimulate collaborative research efforts among teacher educators in the two institutes. It provides an example of bilateral collaboration using an inter-disciplinary format (Wong, Kaur, Koay & Jamilah, 2007). This study was also an extension of an earlier research by Kaur and Yap (1997), although this earlier study did not include student drawing.

3 SAMPLE AND PROCEDURE

The Singapore sample consisted of 334 pupils from four schools; 51% of them were girls. The Brunei sample of 209 pupils came from three urban schools and one rural school; 47% were girls. Pupils from both groups were about 12 years of age and in their final year of primary schooling. They took a 26-item questionnaire about their learning of mathematics, and the results of this part can be found in Wong, Kaur, Koay, and Jamilah (2007). This chapter will focus on the results from the following two items:

- Describe the qualities of the "best" mathematics teacher you have ever had.

- Draw a picture of your "best" mathematics teacher teaching in class. Each pupil drew one picture only.

The Singapore pupils took the questionnaire in English *after* their Primary School Leaving Examination (PSLE) in 1998, while the Brunei pupils did it in Bahasa Melayu (their national language) *before* their Primary Certificate in Education (PCE) examination in 1998.

4 FINDINGS AND DISCUSSION

4.1 *Written Descriptions of the Qualities of "Best" Mathematics Teachers*

The key words in these descriptions were identified and then classified into four main categories: *Instruction/Pedagogy*, *Personal Qualities, Relationship/Rapport* and *Homework policies/Expectations*. For example, the category, *Personal qualities,* includes descriptions such as *helpful, patient, hardworking, nice,* and *serious.* The same pupil may write about more than one quality. See the examples below; the Brunei descriptions were translated from Bahasa Melayu.

- *Pupil 2071* (Singapore): Patient, hardworking, caring, understanding, able to solve challenging problem sums.
- *Pupil 2081* (Singapore): Patient and explains till we understand; experienced and knows which method is the best; strict, makes sure we learn and do our homework.
- *Pupil 088* (Brunei): She is a nice person, who explains clearly, gives written and oral tests and is able to solve any mathematical problems.
- *Pupil 100* (Brunei): At first the teacher explains. Then she asks us to open our books and re-explains. Finally she asks us whether we understand or not.

Quite clearly, these pupils felt that their "best" mathematics teacher should possess several desirable qualities, and they must use a variety of teaching techniques to help them learn. The frequency for each description was obtained, and the results are given in Table 1, where the key descriptions are ranked according to the Singapore data.

Table1

Descriptions of Teacher Qualities

	Singapore		Brunei Darussalam	
Instruction/Pedagogy	Freq	%	Freq	%
Explain clearly	103	30.3	51	36.7
Ensures pupil understand	64	18.8	25	18.0
Good in math/clever	44	12.9	2	1.4
Provides individual help	31	9.1	11	7.9
Not boring/interesting/funny	25	7.4		
Teach slowly/show more examples/details	16	4.7	4	2.9
Make math meaningful/different approaches	15	4.4	1	0.7
Experienced	11	3.2	2	1.4
Group work/discussion/pupil participation	10	2.9	24	17.3
Simple and easy exposition	9	2.6	11	7.9
Good notes/worksheets	6	1.8		
Correct pacing	2	0.6	1	0.7
Show working step by step	2	0.6	3	2.2
Open-minded	1	0.3		
Systematic planning/well prepared	1	0.3	4	2.9
Total	340	100	139	100
Relationship/Rapport				
Caring/kind/supportive	94	40.5	2	6.7
Understanding	51	22.0		
Not fierce/approachable/encouraging/motivating/friendly	30	12.9	10	33.3
Strict/firm	28	12.1	2	6.7
Answer questions	7	3.0	14	46.7
Lenient	7	3.0		
Listen to pupils	5	2.2		
Never give up hope on pupils	4	1.7	1	3.3
Fair	4	1.7		
Authority	1	0.4		
Ensures class is attentive	1	0.4	1	3.3
Total	232	100	30	100

	Singapore		Brunei Darussalam	
Personal Qualities	Freq	%	Freq	%
Patient	112	39.7		
Humorous	44	15.6		
Helpful	38	13.5	23	35.4
Serious/dedicated	24	8.5	1	1.5
Courteous/pleasant/nice	15	5.3	6	9.2
Hardworking	13	4.6		
Calm/good temper	13	4.6	30	46.2
Responsible	8	2.8	2	3.1
Considerate	5	1.8		
Appearance (pretty/handsome)	5	1.8		
Cheerful	2	0.7	1	1.5
Enthusiastic	2	0.7		
Humble	1	0.4		
Neat			2	3.1
Total	282	100	65	100
Homework policies/Expectations				
Demanding/a lot of homework	12	29.3	9	23.7
Check homework/goes through homework	6	14.6	2	5.3
Constant review/reinforcement	5	12.2	6	15.8
Give difficult problems	5	12.2		
Moderate homework	5	12.2		
Encourage pupils to do more sums	4	9.8	7	18.4
Not too much work	2	4.9		
Do work in class	1	2.4	10	26.3
Regular tests/assessment	1	2.4	4	10.5
Total	41	100	38	100

On average, each Singapore pupil provided 2.7 descriptions against 1.3 descriptions by the Brunei pupils. Thus, the Singapore pupils more than the Brunei pupils might have been exposed to more different teaching techniques, had more ideas about what might constitute good mathematics teaching, or were more articulate about expressing these ideas in writing.

Both groups wrote more about instruction than personal qualities, relationships with pupils, and work expectations. This is not surprising because the instructional moves are more readily open to pupils' observation than qualities and relationships which require strong awareness and subjective interpretations.

From the descriptions about *Instruction/Pedagogy* shown in Table 1, both groups placed the strongest emphasis on the teacher's ability to explain things clearly. This was followed by being able to help pupils understand the work. These two descriptions, clarity and help pupils to understand, together account for half of the responses among the Singapore pupils and 55% of the Brunei descriptions. These qualities are often considered by educators and researchers to be important teacher competence. An interesting difference between the two groups is that about 17% of the Brunei descriptions mentioned group work and participation whereas only 3% of the Singapore descriptions did so. This could reflect the practices in the sampled schools at the time when the study was conducted, and it will be interesting to follow up with new studies, especially in Singapore, where group work has received much attention in recent years.

The most important *Personal* quality, according to the Singapore pupils, was patience, found in about 40% of the descriptions, but this quality was not mentioned at all by the Brunei pupils. We do not know the real reasons for this surprising difference, but wish to put forth a plausible hypothesis for further discussion. The Malay culture emphasises "sopan santun," which means being courteous and polite, and this quality often appears as being patient. The classroom pace in many Brunei classes is also much slower than in Singapore classes, and this difference seems to reflect the tempo of life in the communities. This suggests that the Brunei pupils may not be particularly aware of this teacher quality. Indeed, the most frequently mentioned Brunei description is about being calm, which reflects the "sopan santum" characteristic.

Being patient can be actualised in terms of a caring and supportive relationship between the teacher and her pupils, and this link is reflected in the Singapore case, where nearly 40% of the descriptions in the *Relationship/Rapport* category were about this kind of rapport. On the other hand, being strict and firm was not seen as a key relationship displayed by good mathematics teachers (12%); perhaps, more teachers should shed the image of being in control most of the times to "win" over their pupils. The number of responses from the Brunei pupils was so few (30) that it is not possible to deduce general comments from them. Nevertheless, the Brunei pupils focussed on the teacher's willingness to answer pupil's questions (modal response) and being approachable and friendly.

The number of descriptions under *Homework policies/Expectations* was low for both groups (41 and 38). Both groups mentioned giving a lot of homework, but the Brunei pupils might like to do the mathematics work in class.

4.2 *Draw A Picture of Your "Best" Mathematics Teacher Teaching in Class*

Three features were identified and counted from the drawings: *Teacher and teaching; Pupils and learning conditions;* and *Physical environment.* Some of the features could not be accurately identified or inferred, for example, the gender of the teacher depicted in the drawing, because these drawings were not well done (after all, this was not a test of drawing skill). As a consequence, some of the percentages do not add up to 100. For the Brunei data, about 22% did not do the drawing. On the other hand, a drawing may include more than one feature, and this gives rise to a sum of percentages exceeding 100.

Figure 1 shows two drawings, one by a Singapore pupil and one by a Brunei pupil. The Singapore drawing shows features such as: gender of the teacher (male), neatness, mathematical work on the board, and asking question in front of the class; the pupils were not shown. The Brunei drawing shows a female teacher in Muslim attire, interacting with one pupil, and a board filled with correct mathematical work. Both drawings show happy or encouraging expressions, indicating strong rapport with the class or pupils.

(a) (b)

Figure 1. Pupil Drawing of "Best" Mathematics Teacher: (a) Singapore (b) Brunei.

The results about the teacher and teaching are given in Table 2. For the Brunei drawings, female teachers outnumbered male teachers in the ratio 5:1, whereas the Singapore drawings had a more balanced distribution. Since the pupils were asked to draw their teachers, the numbers in Table 2 suggest that there were more female than male mathematics teachers in Brunei who had made strong impressions on their pupils.

In both countries, the predominant teaching style depicted was whole class instruction with the teacher standing and explaining in the front of the class (69.4% and 62.2%). In another part of the survey, the pupils were asked to rank five types of teaching methods. About 42% of the Singapore pupils and 71% of the Brunei pupils ranked teacher explanation as the top teaching method (see Wong, Kaur, Koay & Jamilah, 2007). Hence, at least for the Brunei group, there was concurrent validity about explanation as the main teaching method using pupils' rankings and drawings. The data in Table 2 also show that these pupils did not relate good mathematics teaching with the teacher working with groups of pupils or individual pupils. These two types of teaching strategies would certainly take place in these classes, but they did not seem to register as a significant part of the "best" teaching method.

Table 2
Features of Teacher and Teaching in Drawings

Features		Singapore	Brunei Darussalam
Teacher: Gender	Male	41.7	12.4
	Female	56.5	68.4
Dress	Neat	92.2	78.0
	Messy	3.0	1.4
Teaching Methods	Standing in front	8.4	27.8
	Explaining in front	61.0	34.4
	Asking questions	11.7	12.0
	Giving instructions	5.4	
	Working with group of pupils		0.5
	Working with individual pupils	0.3	2.9

Table 3 shows information about pupils and the learning conditions as depicted in the drawings. More than three quarters of the Singapore drawings did not show any pupil, so it is difficult to infer much from the drawings about pupils when they were taught by the "best" mathematics teacher. About 60% of the Brunei drawings show some pupils, with the majority of the pupils in neat rows, the predominant feature of whole class instruction. This part of the drawing does not provide conclusive findings and further exploration is required.

Table 3
Features of Pupils and Learning Conditions in Drawings

Features		Singapore	Brunei Darussalam
Number of pupils shown	0	77.2	40.7
	1 – 9	21.0	34.6
	10 and above	1.8	3.0
Gender of pupils	Equal	3.3	9.6
	More female	4.5	9.6
	More male	6.3	13.9
Seating	In neat rows	12.9	39.2
	In groups/pairs	4.5	2.9
	Individual	2.1	0.5
	Standing	3.3	
Learning conditions	Appear attentive	18.0	32.1
	Pupil at work at chalkboard	0.6	
	Pupil explaining in front of class	1.2	
	Pupil asking question	1.2	

Features about the classroom physical environments are reported in Table 4. About two thirds of the drawings from each group showed a white/chalk board with formulae, diagrams, or some work on it. Some of the work depicted in the Brunei drawings was mathematically incorrect. Manipulatives, teaching aids, and technological items were rarely shown in these drawings. We speculate that these physical items were hardly used by the "best" mathematics teachers or that these pupils did not link these items to the quality of good teaching.

Table 4
Features of Physical Environments in Drawings

Features		Singapore	Brunei Darussalam
Board	None	15.9	0.5
	Empty board	8.1	12.4
	With work/formulae etc.	68.5	64.6
	Messy	5.1	1.4
	Homework instructions	1.8	
	Value statement	0.6	
Mathematical work shown on board	None	8.1	11.0
	Correct mathematics	65.8	59.3
	Incorrect mathematics	2.7	10.9
Classroom conditions	Messy/untidy	0.6	1.0
	Neat/orderly	18.6	78.0
	Cannot infer	81.4	
Technology	None	97.9	100
	OHP	2.1	
Manipulatives/Aids	None	96.7	98.1
	Some	0.6	1.9

5 CONCLUSIONS

This small collaborative project has provided a better understanding of the similarities and differences in the perceptions about good mathematics teaching held by upper primary school pupils in two culturally very distinct systems. Despite differences in cultures, the predominant perception of good mathematics teaching is whole class instruction with emphasis on clarity and helping pupils to understand the work. Differences seem to relate to the personal relationships between pupils and teachers. This general conclusion should be investigated to see whether it applies to other grade levels.

The technique of student drawings is a promising research tool in capturing exciting and valid snapshots of the classroom environment not only about "good" mathematics teaching but also other instructional events. An important advantage of this technique over journal writing or open-ended survey is that it is well suited for pupils who cannot express themselves clearly in words. This technique should also be further validated against video-recording, which is gaining popularity in recent classroom research.

References

Australian Association of Mathematics Teachers. (2006). *AAMT Standards for Excellence in teaching mathematics in Australian schools.* Adelaide: Author. Available from http://www.aamt.edu.au/standards/

Burgess, Y. (1994). Drawing what you know and feel. *SET: Research information for teachers.* Hawthorn, Victoria: Australian Council for Educational Research.

Chambers, D. W. (1983). Stereotypic images of the scientist: The Draw A-Scientist test. *Science Education, 67,* 255-265.

Hattie, J. (2003, October). *Teachers make a difference: What is the research evidence?* Keynote address presented at the Australian Council for Educational Research Conference 'Building Teacher Quality: What Does the Research Tell Us?', 19-21 October 2003, Melbourne. Retrieved July 17, 2007, from http://www.acer.edu.au /documents/RC2003_Hattie_TeachersMakeADifference.pdf

Hill, H. C., Schilling, S. G., & Ball, D. L. (2004). Developing measures of teachers' mathematics knowledge for teaching. *The Elementary School Journal, 105*(1), 11-30.

Kahle, J.B. (1989). Images of scientists: Gender issues in science classrooms. *What Research Says to the Science and Mathematics Teacher, Number 4.* Perth: Curtin University of Technology. (ERIC Document Reproduction Service No. ED370785).

Kaur, B., & Yap, S.F. (1997) Qualities of my best mathematics teacher. In L. K. Chen & K. A. Toh (Eds.), *Research across the disciplines* (pp. 292-296). Singapore: Educational Research Association.

Lim, L. H. (1991). Good teachers: What are they? *ASCD Review (Singapore), 2*(1), 30-32.

Lim S. K., & Wong, K. Y. (1989). Perspectives of an effective mathematics teacher. *Singapore Journal of Education, Special Issue*, 101-105.

Lim-Teo, S.K., Chua, K. G., Cheang, W. K., & Yeo, J. K. (2006). The development of Diploma in Education student teachers' mathematics pedagogical content knowledge. *International Journal of Science and Mathematics Education, 5*(2), 237-261. Retrieved on September 20, 2006, from http://www.springerlink.com /content/l2h2268513g07110/

McDonough, A., & Pavlou, M. (1994). *Children's beliefs about learning mathematics well: Victorian and Canadian perspectives.* Mathematics Teaching and Learning Report, No. 10. Oakleigh, Victoria: Australian Catholic University.

Picker, S. H., & Berry, J. S. (2000). Investigating pupils' images of mathematicians. *Educational Studies in Mathematics, 43*, 65-94.

Rennie, L. J., & Jarvis, T. (1994). *Helping children understand technology*. Perth, Western Australia: Key Centre for School Science and Mathematics.

Teppo, A. (1993). Uses of children's drawings in classroom research. *PME News*, November, 10-12.

Wong, K.Y. (1995). Images of mathematicians. In R.P. Hunting, G.E. Fitzsimons, P.C. Clarkson & A.J. Bishop (Eds.), *Regional collaboration in mathematics education 1995* (pp. 785 - 794). Melbourne: Monash University.

Wong, K.Y. (1996). Assessing perceptions using student drawings. In M. Quigley, P.K. Veloo & K.Y. Wong (Eds.), *Assessment and evaluation in science and mathematics education: Innovative approaches* (pp. 370-379). Bandar Seri Begawan: Universiti Brunei Darussalam.

Wong, K. Y., Kaur, B., Koay, P. L., & Jamilah binti Hj Mohd Yusof. (2007). Singapore and Brunei Darussalam: Internationalisation and globalisation through practices and a bilateral mathematics study. In B. Atweh, A. Calabrese Barton, M. Borba, N. Gough, C. Keitel, C. Vistro-Yu & R. Vithal (Eds.), *Internationalisation and globalisation in mathematics and science education* (pp. 441-473). New York: Springer.

Looking Forward and Beyond

A journey of a thousand miles begins from beneath one's feet.
千里之行，始于足下。 Lao Zi (*Dao De Jing,* Chapter 64)

Sail forth – steer for the deep waters only,
Reckless O soul, exploring, I with thee, and thou with me,
For we are bound where mariner has not yet dared to go,
And we will risk the ship, ourselves and all.

Walt Whitman (*A Passage to India*)

The first epigraph highlights that a journey should begin with what one already possesses right here and now. The 23 chapters of this book thus form the foundation on which new landscapes about the Singapore mathematics education journey can be extended or created. These chapters of variable length, coverage, depth of analysis, and quality of writing, have captured a large corpus of work completed at a specific period of the journey, namely at the beginning of the 21st century. However, this coverage cannot be completely exhaustive. There are studies about Singapore mathematics education that have not been included this book but may have been referenced by these authors. Other information about research in Singapore mathematics education can be obtained from the websites of the Learning Sciences Laboratory (LSL) at http://lsl.nie.edu.sg/ and the Centre for Research in Pedagogy and Practice (CRPP) at http://www.crpp.nie.edu.sg/course/view.php?id=424

The second epitaph calls for partnership when one embarks on an uncharted, perhaps risky, journey. We invite collaborators to come on board the Singapore mathematics education journey so that we can build on mutual strengths and enjoy the journey in each other's company. Some may choose to travel along familiar paths already described in this book, perhaps en route to certain laudable goals, while others may prefer to strike new treks as detours from what have already been well trodden. Both types of journey are worthwhile and can be enlightening.

The remainder of this chapter attempts to answer the question implicit in its title: *What is the future direction of Singapore mathematics education in the coming decade*s? The current emphasis to strive for greater flexibility in the rapidly changing education system will encourage diverse approaches in praxis and research. Within this scenario, Singapore mathematics education is at a critical juncture of its journey. The next phase of the journey should move toward creating more innovative instructional practices, employing sophisticated, mixed-methods research methodologies, conducting evidence-based intervention studies that have implications for policies, strengthening professional capacity of the key players, in particular the school teachers, devising programmes that encourage active involvement of parents and the local community, and widening networks with the international mathematics education community. These areas are discussed in the following sections under the three headings of *students, teachers,* and *internationalisation of mathematics education*. We hope that these ideas will be helpful to those who embark on the Singapore mathematics education journey into an uncertain future.

1 STUDENTS ARE WHAT THE EDUCATION JOURNEY IS ALL ABOUT

The education journey should begin in kindergartens and aim to cultivate responsible and productive adults who will contribute to the welfare of their self, family, community, nation, and humanity at large. Mathematics is widely recognised as an important cognitive tool (one of the three R's) as well as a desirable outcome of this journey.

In this book there is only one chapter about early numeracy, and this suggests a huge gap in our knowledge about how best to help young children make the transition from less-structured acquisition of mathematics at kindergartens to more formal instruction at Primary 1 and beyond. The Learning Support in Mathematics (LSM) programme has given us some glimpses about the learning difficulties faced by those who need special assistance, while the Gifted Education Programme (GEP) has catered for the more talented through differentiated instruction and enriched learning. What is needed is a comprehensive description of how the majority of students who are "average" in mathematics can cope with the transition in mathematics at various periods of their schooling. We may "apply" ideas from overseas research (e.g., National Research Council, 2005) and cognitive development theories (e.g., Piaget and the van Hieles), fine-tune the Concrete → Pictorial → Abstract (CPA) instructional sequence, include more activities on visualisation, pay attention to children's ways of mathematising (e.g., Carpenter, Fennema, Franke, Levi & Empson, 1999), do more out-of-school mathematics activities, and experiment with new instructional techniques. But these are acts by the teachers, and measuring the impacts of these changes does not necessarily help us better understand mathematics learning from the students' perspective, especially in terms of differences by gender, ethnic groups, and socio-economic backgrounds. A promising path is to conduct more longitudinal studies similar to the International Project on Mathematical Attainment (IPMA). To map the mathematical development of students from kindergartens to the end of secondary schooling is an ambitious project, but it might just be feasible within the well-structured Singapore education system. An alternative approach to understand life-long development of mathematics among people is to analyse the mathematics autobiographies of adults who are successful in mathematics, in particular, the mathematicians (Fitzgerald & James, 2007), and those who have mathematics phobia beyond their school days.

Given the central focus on mathematics problem solving in the Singapore curriculum, it is not surprising that problem solving is a popular research topic among mathematics educators and graduate students. Foong has provided a comprehensive review of studies about problem solving that cover the perspective of teaching *for, about,* or *via* problem solving. The challenge is to apply the insights from this review and to gather new and richer data to depict the way a "typical" Singapore student might behave when he or she solves mathematics problems. If this "typical" problem solving behaviour is not productive in stimulating the students to want to solve more problems and in providing them the necessary cognitive tools to effectively solve a wide range of different problems, then how can the current teaching and learning methods be changed to help them become more motivated and better problem solvers? How can teachers' concerns about practical matters such as the requirement but a lack of time to complete the syllabus, competing demands on curriculum time, and lack of competence in alternative pedagogies be addressed?

In Singapore, the so-called model (drawing) method was introduced in mid 1980 to help students solve structurally complex word problems. Although drawing pictures to represent word problems is also used in many countries, this method, as it evolves in Singapore, has at least two distinctive features: (i) it is applied systematically across many topics ranging from whole numbers to fractions to ratios; (ii) its structure can make the solutions intuitively evident rather than just illustrative. After more than twenty years of teaching and researching this method, several issues have turned up, waiting to be resolved and requiring more fine-grained studies and use of technology. These issues are discussed by several authors throughout this book. The story about the model method illustrates how an initial assessment issue of trying to discriminate among the high percentages of A* scorers in the PSLE mathematics examination turns into a curriculum policy and an instructional challenge for the teachers and their students. As this method becomes more widely known outside Singapore, we hope that future studies conducted collaboratively with international researchers will provide new evidence and strong theories to determine its "future" in the Singapore curriculum.

To analyse students' written solutions is a well-established instructional strategy, though not often used systematically. It is also a fruitful research tool to understand students' mathematical thinking and problem solving processes. This technique is time-consuming. Instead, international studies such as TIMSS, the Kassel Project, and IPMA tend to use multiple-choice items for efficiency and ease of comparison. On the other hand, the public examinations in Singapore have predominantly short-answer items and constructed-response items. With the latest emphasis on real-life applications of mathematics and mathematical modelling, it becomes increasingly important to design constructed-response items with real-life contexts that can assess this competence and yet will minimise "non-construct-relevant variance in performance" as cautioned by the U.S. National Mathematics Advisory Panel (2008, p. xxv). This new test content should be assessed using the so-called alternative assessment techniques, which are gaining popularity in Singapore. Powerful ICT tools can become part of this new assessment process to widen the scope of mathematics literacy that can be assessed, for example, to auto-grade student's freely written solutions and to allow students to use relevant mathematical software such as dynamic geometry, graphing, computer algebra system, and spreadsheet to solve extended, non-routine problems under examination conditions. This will integrate the current use of ICT as a presentation and exploratory learning tool to its future use in assessment. Indeed, students who use a particular tool (ICT or otherwise) to learn mathematics should be tested with that tool as a form of performance assessment.

There is strong evidence that many present-day students who grow up in a rich technological environment have different styles and needs for learning, and we do not know much about these student characteristics in relation to mathematics. Newer forms of engaged learning with technologies must be tested, and some work about games and virtual worlds is being piloted by the Learning Sciences Lab at NIE. In mathematics, the ICT path needs to balance learning through concrete manipulatives (especially for young learners) with virtual objects in order to promote the holistic development of student's mathematical thinking (Clements & McMillen, 1999).

The holistic development of a student certainly goes beyond cognition to encompass healthy physique, stable emotions, strong motivations, engaging social relationships, and even deep spirituality. Only three chapters in this book deal marginally with the emotions of mathematics learning, and this is a rich area for future exploration. Many teachers have difficulty in designing appropriate learning activities in mathematics that encompass these non-cognitive attributes, even though the Singapore mathematics curriculum includes an *Attitude* factor and stipulates the infusion of National Education into the teaching of every school subject. To move further down this path requires mathematics teachers to think of themselves first of all as *teachers* in all aspects of this noble vocation, and only secondly as *mathematics* teachers of a specific discipline. We also need a strong research programme to study how holistic development can be achieved through the totality of experiences offered through schooling, situating mathematics learning in a cross disciplinary context. It has been claimed rightly that out-of-school factors are powerful influences in this respect, but we have not yet begun to examine the impacts of out-of-school mathematics experiences on mathematics literacy, let alone on holistic development of the students. This enormous gap awaits the intrepid explorers.

2 TEACHER PRAXIS AND EDUCATION

Teachers are the key agents to help students develop holistically. In the case of elementary schools in the United States, the teacher factor accounts for about 12% of total variability in students' mathematics achievement gains (National Mathematics Advisory Panel, 2008), but there is no comparable data for Singapore. This is a gap that requires further research.

Singapore has put in place an elaborate system of training pre-service trainee teachers, allowing teachers to choose among three career paths that are in line with their individual capabilities, inclinations, and goals, and providing the opportunities for them to grow within and beyond the chosen track. Five chapters in Part I of this book have provided extensive discussions of these issues. However, the impacts of these professional activities on sustained teacher praxis have not been subjected to rigorous analysis and adequately documented. Furthermore, the lack of understanding of the links between teacher preparation and profession development on students' mathematics learning outcomes is a most enigmatic issue not only in Singapore but also internationally.

There are many issues about mathematics teacher education that are not covered in this book. One pertinent area is why young adults choose teaching as a career and why an increasing number of adults in Singapore switch from their current career to teaching. The first question was partially answered by a large survey of about one thousand trainee teachers who first joined the various NIE pre-service programmes in July

2004 (Lim-Teo, Low, Ch'ng & Goh, 2005). The four popular reasons given by them were: interest in teaching (25.1%), job fit (20.2%), to fulfil a mission (19.7%), and love for children or young people (19.6%). The findings were not further analysed by subject areas, so there is a need to conduct follow-up studies with mathematics trainee teachers. The second question about mid-career switch might be related to the ever changing job markets in Singapore. Relevant data about the working conditions of various professions, in particular those that compare mathematics teaching with other professions that use mathematics extensively such as engineering and the sciences, should be collected and analysed because the findings will have strong policy implications.

Overseas researchers have written about the "shock" felt by some trainee teachers when they begin full-time teaching on their own after graduating from teacher training. This may arise because trainee teachers who have learned the "progressive" philosophies of teaching during their training are now confronted with the heavy demands of classroom teaching and non-teaching duties that require strong teacher-centric control and established routines to "survive," especially for large classes as in Singapore. It might be helpful if both the trainee teachers and teacher educators share similar understandings of what is called the *characteristic pedagogical flow* of mathematics lessons within the country (Schmidt, et al., 1996). Singapore mathematics teaching has been characterised as teacher-centred, convergent, reliance on textbooks, and strongly aligned with well-defined examination requirements (see Clarke, Keitel & Shimizu, 2006). In a recent conference presentation, Hogan (2008) provided results drawn from 76 observed mathematics lessons at Secondary 3 level in Singapore, to show that the classroom activities were: answer checking IRE (initiate, response, evaluate) sequences (42.0%), individual seatwork (26.2%), whole class lecture (12.9%), small group work (8.5%), and others (10.4%). These are percentages across all the observed lessons, and they do not reflect the "balance" of these activities within a single lesson. Indeed, a competent teacher should use different combinations of these activities to achieve the objectives of the particular lessons and for the target students. There is an urgent need to explore more sophisticated data collection and analysis to understand this "balance" and how it affects student learning.

The above findings about classroom teaching activities are based on observations by researchers, with reflection by the participating teachers in some cases. On the other hand, what do the students think of their mathematics lessons and teachers? The small Singapore-Brunei study is an example to answer this question about primary mathematics teaching. On the basis of personal observations over several decades, Lee (2005) believed that the teaching and learning of mathematics in Singapore schools is regarded as more structured and more officially regulated compared to many other countries. The challenge to teacher educators is to use these findings and practices to build an authentic picture of the characteristic pedagogical flow of Singapore mathematics teaching (if it exists) and to discuss this characterisation with trainee teachers to help them better handle the "shock" mentioned above. Furthermore, direct instruction in general (which can take different forms; see Stein, Kinder, Silbert & Carnine, 2006) is gaining (or re-gaining) recognition in recent years through meta-analysis using effect sizes (Petty, 2006). After a review of studies in mathematics education, the U.S. National Mathematics Advisory Panel (2008) notes that high-quality research does not support the exclusive use of either teacher-directed or student-centred strategies. Perhaps, the Asian preference for "balance" may be a fruitful guiding principle on which to strike new paths in mathematics pedagogy. Subsequent studies can gather evidence to understand how this "balanced" pedagogy might explain student mathematics learning outcomes as measured by school-based assessment, public examinations, and international comparative studies.

In recent years, a different kind of "balance" at a systemic level has been set in motion with initiatives such as SAIL, SEED, and GEP, introduced by MOE to provide differentiated instruction and more engaged learning. These school-based "reforms" are promoted through three different types of activities: (a) teachers are given training to use a fairly well-specified set of materials with prescriptive instructional methods; (b) teachers are trained in the processes and theories underpinning the initiative, giving them much flexibility to design their own materials and techniques; (c) schools are provided with additional resources and basic training on how to utilise these resources. At this stage, there is no empirical evidence to show which type of activities or which combination has the strongest effects to achieve the objectives of the respective reforms. Research into different types of professional development should also take a longitudinal approach similar to that advocated for the study of students' mathematical development as mentioned earlier, in order to capture the different trajectories that mathematics teachers undergo as they develop proficiency from the novice to the competent stage.

3 INTERNATIONALISATION OF MATHEMATICS EDUCATION

The Singapore education system has evolved from the British system, but its policymakers, curriculum specialists, and researchers have also learned from other English speaking countries, such as the United States, Australia, and New Zealand, as well as non-English speaking ones, in particular, China, Japan, and Finland. This is the first form of internationalisation of the Singapore mathematics education. In addition to this, Singapore has recently shown strong interest to participate in international comparative studies. The most prominent one is the TIMSS series of mathematics and science study. Singapore will also take part in the next PISA survey in mathematics. It is currently involved in the Teacher Education and Development Study in Mathematics (TEDS-M, http://teds.educ.msu.edu/default.asp) that investigates the preparation of primary and lower secondary mathematics teachers by looking into the interactions among teacher education programmes, perceptions of teacher educators, and the beliefs and mathematics pedagogical content knowledge of the trainee teachers. As the first international study that compares the mathematics content knowledge and pedagogical content knowledge of trainee teachers across about twenty countries, TEDS-M is expected to break new grounds in research methodologies about mathematics teacher education.

These comparative studies can provide fresh perspectives on local issues and benchmark student performance and beliefs against international standards. However, the shortcomings of these studies have not escaped criticism internationally, for examples, on the nature of the mathematics items in PISA (Prais, 2003) and methodological issues (Postlethwaite, 1999). Within Singapore, highlighting to teachers and students that Singapore students have done very well in national examinations or international comparison studies might be sending the wrong message if this leads to complacency with the current instructional practices. Much will be gained by reflecting on these assessment outcomes with a critical stance.

It seems that participation in one international project might lead to other projects when the benefits become apparent to the participating researchers. For example, the Kassel Project was initiated when data were being collected for TIMSS 1995; at the time some researchers felt that a snapshot of student performance may be better interpreted with longitudinal data made up of students' mathematics scores, their attitudes, and classroom variables. Later, this project led to a bilateral comparative study with Brunei Darussalam and the multi-national International Project on Mathematical Attainment (IPMA), which in its turn has evolved into the International Comparative Study on Mathematics Teacher Training (ICSMTT), which is at the first year of implementation at the time of writing. Similarly, the Learner's Perspective Study involving video-taping selected Grade 8 mathematics lessons in 12 countries (Clarke, Keitel & Shimizu, 2006) was motivated by the TIMSS Video Study in 1999 of seven countries. Every extension of an international project seems to include additional variables and new methodologies, and this form of internationalisation opens up new vistas that are beneficial to the participating countries.

Another form of internationalisation is the participation of Singapore educators and teachers in study tours to other countries and at conferences organised by professional bodies, such as the Southeast Asian Mathematical Society and the International Commission on Mathematics Instruction (ICMI) (Lim-Teo, 2008). The Mathematics and Mathematics Education Academic Group at NIE and local schools have played hosts to delegations of overseas educators who are keen to find out the factors that have contributed to the "success" of "Singapore Math." These visitors often raise issues and make observations that are not "obvious" to the local educators, and the ensuing discussions have provided fresh lens for the latter to understand what is happening here (e.g., Dornan, 2008). These activities have stimulated exchanges among regional mathematics educators who are working to blend the best from Eastern and Western practices. To succeed along this path, we welcome overseas partners so that our joint efforts will benefit students from different countries in this increasingly globalised world.

4 FINAL REMARKS

It is generally accepted that mathematics plays a crucial role in advancing the quality of a nation's scientific, technological, economic, and social developments. The mathematics classroom is the foundation on which these developments can take off. Many chapters in this book have provided the signposts about Singapore classrooms. The journey from the past to the present and from the present to the future requires an enormous amount of talent, creativity, and the conviction to persevere of many people, including mathematics educators, teachers, parents, ministry officials, policymakers, and commercial publishers. These people must share the strong motivation to do what are best for the students. In the foreseeable future as the Singapore education system becomes more flexible and student-centred, the students themselves will have stronger input about this journey that intimately affects their future. There will be unsuspected terrains; the discovery of treks once invisible and of praxis once impossible certainly justifies the effort made to take the first step along this fascinating odyssey.

Editorial Board
May, 2008

References

Carpenter, T. P., Fennema, E., Franke, M. L., Levi, L. W., & Empson, S. B. (1999). *Children's mathematics: Cognitively guided instruction.* Portsmouth, NH: Heinemann.

Clarke, D., Keitel, C., & Shimizu, Y. (Eds.). (2006). *Mathematics classrooms in twelve countries.* Rotterdam, The Netherlands: Sense Publishers.

Clements, D. H., & McMillen, S. (1999). Rethinking "concrete" manipulatives. In A. R. Teppo (Ed.), *Reflecting on practice in elementary school mathematics: Readings from NCTM's school-based journals and other publications* (pp. 140-149). Reston, VA: National Council of Teachers of Mathematics.

Dornan, J. (2008). *Learning from Singapore: The findings of a delegation of North Carolinians that examined education and the economy.* Raleigh, NC: Center for International Understanding, The University of North Carolina.

Fitzgerald, M., & James, I. (2007). *The mind of the mathematician.* Baltimore, MD: John Hopkins University Press.

Hogan, D. (2008, June). *Pedagogical practice and pedagogical research in South East Asia: The case of Singapore.* Paper presented at the International Symposium: Teaching, Learning and Assessment. Hong Kong.

Lee, P. Y. (2005, November). *60 years (1945 – 2005) of mathematics syllabus and textbooks in Singapore.* Paper presented at the First International Mathematics Curriculum Conference. Chicago.

Lim-Teo, S. K. (2008, March). *ICMI activities in East and Southeast Asia: Thirty years of academic discourse and deliberations.* Paper presented at the ICMI Rome Symposium. Retrieved June 12, 2008 from http://www.unige.ch/math/EnsMath/Rome2008/partST.html

Lim-Teo, S. K., Low, E. L., Ch'ng, A., & Goh, K. C. (2005, May). *Student teachers' reasons for choosing teaching as a career.* Paper presented at the Redesigning Pedagogy Conference: Research, Policy, Practice. Singapore.

National Mathematics Advisory Panel. (2008). *Foundations for success: The final report of the National Mathematics Advisory Panel.* Washington, DC: Author.

National Research Council. (2005). *How students learn: Mathematics in the classroom. Committee on How People Learn.* Washington, DC: National Academy Press.

Petty, G. (2006). *Evidence based teaching: A practical approach.* Cheltenham: Nelson Thornes Ltd.

Postlethwaite, T. N. (1999). *International studies of educational achievement: Methodological issues.* Hong Kong: Comparative Education Research Centre, The University of Hong Kong.

Prais, S. J. (2003). Cautions on OECD'S recent educational survey. *Oxford Review of Education, 29*(2), 139 – 163.

Schmidt, W. H., Jorde, D., Cogan, L. S., Barrier, E., Gonzalo, I., Moser, U., Shimizu, K., Sawada, T., Valverde, G. A., McKnight, C., Prawat, R. S., Wiley, D. E., Raizen, S. A., Britton, E. D., & Wolfe, R. G. (1996). *Characterizing pedagogical flow: An investigation of mathematics and science teaching in six countries.* Dordrecht: Kluwer Academic Publishers.

Stein, M., Kinder, D., Silbert, J., & Carnine, D. W. (2006). *Designing effective mathematics instruction: A direct instruction approach* (4th ed.). Upper Saddle River, NJ: Merrill.

Contributing Authors

BOEY Kok Leong began his career as a mathematics teacher at secondary and Junior College level. He later joined the Research and Evaluation Section at the Ministry of Education (MOE), Singapore. He was the National Research Coordinator/Principal Investigator for Singapore in TIMSS 2003 and TIMSS 2007, and served in the TIMSS 2007 Questionnaire Item Review Committee. He was also involved in other collaborative studies, including the National Foundation for Educational Research (NFER) Baseline Study on English Language proficiency at key levels, Preschool Education, and the Bilingual Approach to the teaching of Chinese Language. Currently, he is seconded from MOE to the National Institute of Education (NIE), Nanyang Technological University, Singapore and teaches mathematics and mathematics education courses at NIE. He is also the Data Manager for the Teacher Education and Development Study in Mathematics (TEDS-M) for Singapore.

Fiona CHEAM joined the Ministry of Education (MOE), Singapore as a primary school teacher in 1993. She then trained as an educational psychologist at University College London and is currently a Senior Educational Psychologist in MOE. Her research interests lie in the identification and support of children with dyscalculia.

CHUA Puay Huat is currently a Teaching Fellow at the National Institute of Education (NIE), Nanyang Technological University, Singapore. He is studying for his PhD in Mathematics Education. His research interest is in mathematical problem posing.

CHUA Wan Loo Jocelyn joined the teaching profession in 1999 and started out as a secondary school teacher in Art and English. She subsequently went on to pursue further training in Educational Psychology at University College London. She is currently an educational psychologist with the Ministry of Education (MOE).

CHUNG Siu Ping holds a Bachelor Degree in Social Science with Second Class Upper Honours in Economics and Statistics from the National University of Singapore (1984). She worked in the private sector for a while before joining a statutory board. The practice of job rotation has given her exposure to many work areas like administration, operations, quality service management, and human resource management. She made a career change in 2003 to spend time on the development of the young through education. After graduating from the Nanyang Technological University in 2004 with a Graduate Diploma in Education (with Merit), she teaches Mathematics and Computer Applications and is now the Head of Mathematics Department in a secondary school.

FAN Lianghuo obtained his PhD from the University of Chicago and is currently an Associate Professor with the Mathematics and Mathematics Education Academic Group, National Institute of Education (NIE), Nanyang Technological University, Singapore. He has taught at both school and university levels. His research interest includes mathematics assessment, teacher professional development, curriculum studies and textbook development, algorithm of polynomial algebra, and educational policy and sociology. He has a wide range of publications in these areas, including as leading editor an internationally acclaimed book, *How Chinese Learn Mathematics: Perspectives from Insiders* (2004, World Scientific), and research articles in leading international journals including *Educational Studies in Mathematics*. He is editor-in-chief of a latest series of China's secondary mathematics textbooks, as well as editor-in-chief of a latest series of Singapore's secondary mathematics textbooks. He recently completed as Principal Investigator a mathematics assessment project funded by the Ministry of Education (MOE) through the Centre for Research in Pedagogy and Practice, NIE. Since 2006, he has also served as chief editor of *The Mathematics Educator*, an official publication of the Association of Mathematics Educators, Singapore.

FANG Yanping is an Assistant Professor at the Centre for Research in Pedagogy and Practice and the Curriculum, Teaching and Learning Academic Group at the National Institute of Education (NIE), Nanyang Technological University, Singapore. She received her PhD from Michigan State University (MSU). Before her studies at MSU, she had worked as a researcher at Shanghai Academy of Educational Sciences for eight years. Her research has focused on teacher learning and teacher development (in mathematics) from international comparative perspectives. She has published book chapters and peer reviewed journal articles on education reform and teacher development in China as well as on East Asian pedagogies. Currently, in addition to teaching and supervising advanced degree students, she is principal investigator of several intervention and research projects. Both projects, Lesson Study and Mathematical Problem Solving in Singapore Classrooms, involve developing web-based video cases for teacher professional development.

FOO Kum Fong is currently a Master Teacher (mathematics) in Singapore. Her professional experiences cover 12 years of teaching mathematics in a secondary school (including 6 years as Head of Mathematics Department), 3 years as a Mathematics Curriculum Planning Officer in the Ministry of Education, and 4 years as a Teaching Fellow in the National Institute of Education (NIE), Nanyang Technological University, Singapore. She has produced an online handbook on "Managing a Secondary School

Mathematics Programme" and is currently the chief editor of Math Buzz, a newsletter of the Association of Mathematics Educators.

FOONG Pui Yee holds a BSc in Mathematics and Dip Ed from the University of Singapore, a MEd and PhD in Education from Monash University. She is currently an Associate Professor in Mathematics Education at the Mathematics and Mathematics Education Academic Group, National Institute of Education (NIE), Nanyang Technological University, Singapore. Her main teaching areas include mathematics pedagogy and mathematics education research at pre-service, in-service, and Masters level. Her research interests cover problem solving, metacognition, and affect in mathematics learning.

HO Siew Yin is a lecturer at the National Institute of Education (NIE), Nanyang Technological University, Singapore. She was the project manager of the Think-Things-Through Project that investigated sense-making among pupils during problem solving. Her research interests include visualisation in problem solving and problem posing, the teaching and learning of geometry, and the teaching and learning of mathematics using children's literature.

JAMILAH binti Hj Mohd Yusof holds a PhD from Curtin University of Technology, Australia. She is the Deputy Dean of the Sultan Hassanal Bolkiah Institute of Education, Universiti Brunei Darussalam. She was a mathematics teacher in secondary and primary schools for several years before joining Universiti Brunei Darussalam as a mathematics educator. Her research interests include creativity in mathematics word problems, mathematical errors in the learning of fractions, and primary mathematics teacher education.

JIANG Chunlian is currently an Associate Professor at the Department of Mathematics and Statistics, Central China Normal University, Wuhan, China, and an Assistant Professor at the Faculty of Education, University of Macau, China. She holds a PhD in mathematics education from Nanyang Technological University, Singapore. Her research interest is in mathematical problem solving.

Berinderjeet KAUR is an Associate Professor of Mathematics Education at the National Institute of Education (NIE), Nanyang Technological University, Singapore. She began her career as a secondary school mathematics teacher. She taught in secondary schools for eight years before joining NIE in 1988. Since then, she has been actively involved in the education of teachers and heads of departments. Her primary research interests are in pedagogy, mathematical problem solving, and comparative studies. She has been involved in numerous international studies of mathematics education. As the president of the Association of Mathematics Educators (2002, 2004 - present), she has been active in the professional development of mathematics teachers in Singapore and is the founding chairperson of the Mathematics Teachers' Conference series that started in 2005.

KHONG Beng Choo is the Mathematics Department Head, Mathematics Committee Head, and Research and Evaluation Head in the Gifted Education Branch, Ministry of Education, Singapore. In her current position, she provides direction for planning the mathematics curriculum for the gifted, conducts training workshops for teachers, conducts research and evaluation studies on gifted students, and supervises research and evaluation projects of the Research and Evaluation committee.

KOAY Phong Lee is an Associate Professor at the Mathematics and Mathematics Education Academic Group, National Institute of Education (NIE), Nanyang Technological University, Singapore. She lectures on primary mathematics education in the pre-service and in-service programmes. Her research includes problem solving and the teaching and learning of mathematics topics at the primary and lower secondary levels. She is also one of the contributing authors of *Shaping Maths*, an interactive textbook for primary school pupils in Singapore.

Christine Kim-Eng LEE, an Associate Professor, is currently Head of the Curriculum, Teaching and Learning (CTL) Academic Group at the National Institute of Education (NIE), Nanyang Technological University, Singapore. She has held several other key posts prior to the inception of CTL, including Vice-Dean at the School of Arts from 1997 to 2000 and Head of Humanities and Social Studies Education from 2000 to 2006. She graduated with an EdD in 1992 from Teachers College, Columbia University and received the Milestone Award for academic excellence in her doctoral progeamme. Her current research interests include Lesson Study and Teacher Learning, Communities of Practice in Cooperative Learning, and School-based Curriculum Development and Implementation.

Kerry LEE has concurrent appointments in the Psychological Studies Academic Group and the Centre for Research in Pedagogy and Practice, National Institute of Education (NIE), Nanyang Technological University, Singapore. Before coming to Singapore, he was in the Department of Psychology at Bond University, Australia. His primary research interest is applied cognitive-developmental psychology. In the last few years he has focused on the relationships between working memory, executive functions, and mathematical performance. His articles have been published in a number of international journals including *Brain Research*, *The Journal of Experimental Child Psychology*, *Applied Cognitive Psychology*, and *The Journal of Educational Psychology*. He is a member of several professional organisations including the Society for Research in Child Development.

LEE Ngan Hoe is a Lecturer with the Mathematics and Mathematics Education Academic Group, National Institute of Education, Nanyang Technological University, Singapore. His research covers primary and secondary mathematics teaching, mathematics teacher training, the teaching of thinking, and the role of the family in mathematics learning.

LEONG Yew Hoong is a Lecturer with the Mathematics and Mathematics Education Academic Group, National Institute of Education (NIE), Nanyang Technological University, Singapore. His mathematics teaching experience includes stints at different levels and contexts: in secondary schools, Junior Colleges, and polytechnics. He continues to be interested in the work of school teaching through research in the complexities involved in the classroom teaching of mathematics.

LIM-TEO Suat Khoh is an Associate Professor and the Dean/Academic at the National Institute of Education (NIE), Nanyang Technological University, Singapore. She was formerly the Head of the Mathematics and Mathematics Education Academic Group (2000 to 2004). She has been a mathematics educator for nearly 30 years and a mathematics teacher educator for 20 years. Her earlier research interests are in general mathematics education, especially the learning of geometry and algebra at the secondary school level. Her current research and development interests are in mathematics teacher education, particularly in the development of mathematics pedagogical content knowledge of primary school teachers.

LUI Hah Wah Elena is an Associate Professor with the Psychological Studies Academic Group at the National Institute of Education (NIE), Nanyang Technological University, Singapore. She has three postgraduate degrees from Michigan State University (PhD, Educational Specialist and MA) and a B.Soc. Sc. (Social Work) from Chinese University of Hong Kong. She coordinates the NIE MEd Guidance & Counselling programme (2005 – present). She is also the Principal Investigator of OSCAR (a web-based career guidance system). She supervises teachers and counselors in postgraduate and in-service training programmes. Her research interests are in adolescents' self-esteem and well-being, career guidance, and lifelong learning. She is the co-editor of two recent local publications: "Reflections on Counseling: Developing practice in schools" and "Youth Guidance: Issues, interventions & reflections."

Juliana Donna NG Chye Huat has been an elementary school mathematics teacher for 35 years in Singapore, including 9 years as a Head of Mathematics Department. She was one of three pioneer Master teachers in Singapore and is now serving her seventh year as a Master Teacher. She was the sole recipient of the President Award for Teachers (PAT) for the Primary Section in 1999 and was awarded in 2004 the Public Administration Medal (Bronze) by the President of Singapore. She publishes a handbook on the teaching of primary mathematics.

NG Swee Fong is an Associate Professor at the Mathematics and Mathematics Education Academic Group, National Institute of Education (NIE), Nanyang Technological University, Singapore. She is the co-Principal Investigator of the Applied Cognitive Development Laboratory at the Centre for Research in Pedagogy and Practice at NIE. Prior to joining NIE, she spent 20 years in Malaysia teaching mathematics at the upper secondary level. She now works extensively with pre-service primary mathematics teachers and in-service mathematics teachers. Her other responsibilities include teaching and supervising at the Masters and Doctoral level. Although specialises in the teaching and learning of algebra, she has always been interested in improving ways to impact and acquire mathematics knowledge across the curriculum.

NG Wee Leng is an Assistant Professor at the Mathematics and Mathematics Education Academic Group, National Institute of Education (NIE), Nanyang Technological University, Singapore. Prior to this, he was an Assistant Director of the Ministry of Education, Singapore. He holds a MSc degree, a MEd degree and a PhD in mathematics. His research interests include the use of technology in teaching and learning mathematics and the role of ancient Chinese mathematics in the mathematics curriculum and he supervises graduate students in these research areas. He is the Principal Investigator of a funded research project Mastering Mathematics with the TI-Nspire.

QUEK Khiok Seng is an Assistant Professor at the Psychological Studies Academic Group, National Institute of Education (NIE), Nanyang Technological University, Singapore. His areas of study focus on assessment and teacher education. He is an active collaborator on projects that investigate the integration, influence, and use of new assessment strategies in mathematics classrooms, including the recent Mathematics Assessment Project funded by the Centre for Research in Pedagogy and Practice, NIE.

SHARIFAH THALHA Bte Syed Haron has recently graduated from the National University of Singapore (NUS) in Sociology and History. Currently she is working as a Research Assistant at the Centre for Research in Pedagogy and Practice, National Institute of Education (NIE), Nanyang Technological University, Singapore.

Pamela SHARPE was formally an Associate Professor at the Early Childhood and Special Education Academic Group, National Institute of Education (NIE), Nanyang Technological University, Singapore. She now teaches at NIE as a part time lecturer in the Diploma in Special Education and the Masters in Early Childhood Programmes. She is also a consultant for the Early Childhood and Special Needs Programmes at Ngee Ann Polytechnic and the Institute of Technical Education, Singapore. She has published work in early childhood.

TOH Tin Lam is an Assistant Professor at the Mathematics and Mathematics Education Academic Group, National Institute of Education (NIE), Nanyang Technological University, Singapore. He obtained his PhD in Mathematics (Henstock-stochastic integral) from the National University of Singapore. He continues to do research in mathematics as well as in mathematics education, and has published papers in international journals in both areas. He has taught in Junior Colleges in Singapore and was Head of Mathematics Department before he joined NIE.

WONG Khoon Yoong is an Associate Professor at the Mathematics and Mathematics Education Academic Group, National Institute of Education (NIE), Nanyang Technological University, Singapore. He has worked as a secondary school mathematics teacher in Malaysia and a mathematics educator at Curtin University of Technology, Murdoch University, and Universiti Brunei Darussalam. His teaching areas include mathematics pedagogy and mathematics education research at pre-service, in-service, and Masters level. During the past twenty years, he has served on national committees to review the mathematics curriculum in Malaysia, Brunei Darussalam, and Singapore. His research interests cover general education such as career and classroom disciplines and mathematics education, in particular, comparative mathematics curriculum, teacher education, learning strategies in mathematics, and use of technology. He is currently the National Research Coordinator for Singapore of the Teacher Education and Development Study in Mathematics (TEDS-M), conducted under the International Association for the Evaluation of Educational Achievement (IEA).

WU Yingkang holds a PhD in Mathematics Education from Nanyang Technological University, Singapore. Her research focuses on statistical education and the use of ICT in mathematics instruction. She is a Lecturer at the Department of Mathematics, East China Normal University at Shanghai and her teaching is in secondary mathematics education and use of English in mathematical sciences.

YAP Sook Fwe is an Associate Professor at the Mathematics and Mathematics Education Academic Group, National Institute of Education (NIE), Nanyang Technological University, Singapore. Her research interest includes statistics and statistics education.

YEAP Ban Har is an Assistant Professor at the Mathematics and Mathematics Education Academic Group, National Institute of Education (NIE), Nanyang Technological University, Singapore. He was the Principal Investigator of the Think-Things-Through Project. His research interest includes problem solving and problem posing, curriculum and curriculum materials, assessment, early mathematical literacy and professional development of teachers through Lesson Study. He teaches and supervises graduate students in mathematical problem solving and problem posing. He holds a PhD in mathematics education from Nanyang Technological University.

YEO Kai Kow Joseph is an Assistant Professor at the Mathematics and Mathematics Education Academic Group, National Institute of Education (NIE), Nanyang Technological University, Singapore. He is involved in training pre-service and in-service mathematics teachers at the primary and secondary levels. Before joining NIE, he was Vice-Principal and Head of Department in secondary schools and served in the Research and Evaluation Branch, Ministry of Education (MOE). His research interests include mathematical pedagogical content knowledge of teachers and mathematics anxiety.

Subject Index

www.ingramcontent.com/pod-product-compliance
Lightning Source LLC
Chambersburg PA
CBHW060632180825
31190CB00004B/32
* 9 7 8 9 8 1 2 8 3 3 7 5 4 *